MATLAB®

An Introduction with Applications

MATLAB®

An Introduction with Applications

Fourth Edition

Amos Gilat

Department of Mechanical Engineering
The Ohio State University

WILEY
JOHN WILEY & SONS, INC.

VP & EXECUTIVE PUBLISHER	Don Fowley
PUBLISHER	Dan Sayre
MARKETING MANAGER	Christopher Ruel
EDITORIAL ASSISTANT	Katie Singleton
DESIGNER	Wendy Lai
MEDIA EDITOR	Thomas Kulesa
PRODUCTION MANAGER	Micheline Frederick
PRODUCTION EDITOR	Amy Weintraub

Cover images: Amos Gilat

This book was printed and bound by Malloy Lithographers. The cover was printed by Malloy Lithographers.

This book is printed on acid free paper. ∞

Library of Congress Cataloging in Publication Data:

ISBN-13 978-0-470-76785-6

Printed in the United States of America

10 9 8 7 6 5 4 3 2 1

Preface

MATLAB® is a very popular language for technical computing used by students, engineers, and scientists in universities, research institutes, and industries all over the world. The software is popular because it is powerful and easy to use. For university freshmen in it can be thought of as the next tool to use after the graphic calculator in high school.

This book was written following several years of teaching the software to freshmen in an introductory engineering course. The objective was to write a book that teaches the software in a friendly, non-intimidating fashion. Therefore, the book is written in simple and direct language. In many places bullets, rather than lengthy text, are used to list facts and details that are related to a specific topic. The book includes numerous sample problems in mathematics, science, and engineering that are similar to problems encountered by new users of MATLAB.

This fourth edition of the book is updated to MATLAB 7.11 (Release 2010b). Other modifications/changes to this edition are: programming (now Chapter 6) is introduced before user-defined functions (now Chapter 7), applications in numerical analysis (now Chapter 9) follows polynomials, curve fitting and interpolation that is covered in Chapter 8. The last two chapters are 3D plotting (now Chapter 10) and symbolic math (Chapter 11). In addition, the end of chapter problems have been revised. There are many more problems in every chapter, and close to 80% are new of different than in previous editions. In addition, the problems cover a wider range of topics.

I would like to thank several of my colleagues at The Ohio State University. Professors Richard Freuler, Mark Walter, and Walter Lampert, and Dr. Mike Parke read sections of the book and suggested modifications. I also appreciate the involvement and support of Professors Robert Gustafson and John Demel and Dr. John Merrill from the First-Year Engineering Program at The Ohio State University. Special thanks go to Professor Mike Lichtensteiger (OSU), and my daughter Tal Gilat (Marquette University), who carefully reviewed the first edition of the book and provided valuable comments and criticisms. Professor Brian Harper (OSU) has made a significant contribution to the new end of chapter problems in the present edition.

I would like to express my appreciation to all those who have reviewed the first edition of the text at its various stages of development, including Betty Barr, University of Houston; Andrei G. Chakhovskoi, University of California, Davis; Roger King, University of Toledo; Richard Kwor, University of Colorado at Colorado Springs; Larry Lagerstrom, University of California, Davis; Yueh-Jaw Lin, University of Akron; H. David Sheets, Canisius College; Geb Thomas, University of Iowa; Brian Vick, Virginia Polytechnic Institute and State University; Jay Weitzen, University of Massachusetts, Lowell; and Jane Patterson Fife, The Ohio State University. In addition, I would like to acknowledge Daniel Sayre, Ken Santor, and Katie Singleton, all from John Wiley & Sons, who supported the production of the Fourth edition.

I hope that the book will be useful and will help the users of MATLAB to enjoy the software.

Amos Gilat
Columbus, Ohio
November, 2010
gilat.1@osu.edu

To my parents Schoschana and Haim Gelbwacks

Contents

Introduction

MATLAB is a powerful language for technical computing. The name MATLAB stands for MATrix LABoratory, because its basic data element is a matrix (array). MATLAB can be used for math computations, modeling and simulations, data analysis and processing, visualization and graphics, and algorithm development.

MATLAB is widely used in universities and colleges in introductory and advanced courses in mathematics, science, and especially engineering. In industry the software is used in research, development, and design. The standard MATLAB program has tools (functions) that can be used to solve common problems. In addition, MATLAB has optional toolboxes that are collections of specialized programs designed to solve specific types of problems. Examples include toolboxes for signal processing, symbolic calculations, and control systems.

Until recently, most of the users of MATLAB have been people with previous knowledge of programming languages such as FORTRAN and C who switched to MATLAB as the software became popular. Consequently, the majority of the literature that has been written about MATLAB assumes that the reader has knowledge of computer programming. Books about MATLAB often address advanced topics or applications that are specialized to a particular field. Today, however, MATLAB is being introduced to college students as the first (and often the only) computer program they will learn. For these students there is a need for a book that teaches MATLAB assuming no prior experience in computer programming.

The Purpose of This Book

MATLAB: An Introduction with Applications is intended for students who are using MATLAB for the first time and have little or no experience in computer programming. It can be used as a textbook in freshmen engineering courses or in workshops where MATLAB is being taught. The book can also serve as a reference in more advanced science and engineering courses where MATLAB is used as a tool for solving problems. It also can be used for self-study of MATLAB by students and practicing engineers. In addition, the book can be a supplement or a secondary book in courses where MATLAB is used but the instructor does not have the time to cover it extensively.

Topics Covered

MATLAB is a huge program, and therefore it is impossible to cover all of it in one book. This book focuses primarily on the foundations of MATLAB. The assumption is that once these foundations are well understood, the student will be able to learn advanced topics easily by using the information in the Help menu.

The order in which the topics are presented in this book was chosen carefully, based on several years of experience in teaching MATLAB in an introductory engineering course. The topics are presented in an order that allows the student to follow the book chapter after chapter. Every topic is presented completely in one place and then used in the following chapters.

The first chapter describes the basic structure and features of MATLAB and how to use the program for simple arithmetic operations with scalars as with a calculator. Script files are introduced at the end of the chapter. They allow the student to write, save, and execute simple MATLAB programs. The next two chapters are devoted to the topic of arrays. MATLAB's basic data element is an array that does not require dimensioning. This concept, which makes MATLAB a very powerful program, can be a little difficult to grasp for students who have only limited knowledge of and experience with linear algebra and vector analysis. The concept of arrays is introduced gradually and then explained in extensive detail. Chapter 2 describes how to create arrays, and Chapter 3 covers mathematical operations with arrays.

Following the basics, more advanced topics that are related to script files and input and output of data are presented in Chapter 4. This is followed by coverage of two-dimensional plotting in Chapter 5. Programming with MATLAB is introduced in Chapter 6. This includes flow control with conditional statements and loops. User-defined functions, anonymous functions, and function functions are covered next in Chapter 7. The coverage of function files (user-defined functions) is intentionally separated from the subject of script files. This has proven to be easier to understand by students who are not familiar with similar concepts from other computer programs.

The next three chapters cover more advanced topics. Chapter 8 describes how MATLAB can be used for carrying out calculations with polynomials, and how to use MATLAB for curve fitting and interpolation. Chapter 9 covers applications of MATLAB in numerical analysis. It includes solving nonlinear equations, finding minimum or a maximum of a function, numerical integration, and solution of first-order ordinary differential equations. Chapter 10 describes how to produce three-dimensional plots, an extension of the chapter on two-dimensional plots. Chapter 11 covers in great detail how to use MATLAB in symbolic operations.

The Framework of a Typical Chapter

In every chapter the topics are introduced gradually in an order that makes the concepts easy to understand. The use of MATLAB is demonstrated extensively within the text and by examples. Some of the longer examples in Chapters 1–3 are titled as tutorials. Every use of MATLAB is printed with a different font and with a gray background. Additional explanations appear in boxed text with a white background. The idea is that the reader will execute these demonstrations and tutorials in order to gain experience in using MATLAB. In addition, every chapter includes formal sample problems that are examples of applications of MATLAB for solving problems in math, science, and engineering. Each example includes a problem statement and a detailed solution. Some sample problems are presented in the middle of the chapter. All of the chapters (except Chapter 2) have a section at the end with several sample problems of applications. It should be pointed out that problems with MATLAB can be solved in many different ways. The solutions of the sample problems are written such that they are easy to follow. This means that in many cases the problem can be solved by writing a shorter, or sometimes "trickier," program. The students are encouraged to try to write their own solutions and compare the end results. At the end of each chapter there is a set of homework problems. They include general problems from math and science and problems from different disciplines of engineering.

Symbolic Calculations

MATLAB is essentially a software for numerical calculations. Symbolic math operations, however, can be executed if the Symbolic Math toolbox is installed. The Symbolic Math toolbox is included in the student version of the software and can be added to the standard program.

Software and Hardware

The MATLAB program, like most other software, is continually being developed and new versions are released frequently. This book covers MATLAB Version 7.11, Release 2010b. It should be emphasized, however, that the book covers the basics of MATLAB, which do not change much from version to version. The book covers the use of MATLAB on computers that use the Windows operating system. Everything is essentially the same when MATLAB is used on other machines. The user is referred to the documentation of MATLAB for details on using MATLAB on other operating systems. It is assumed that the software is installed on the computer, and the user has basic knowledge of operating the computer.

The Order of Topics in the Book

It is probably impossible to write a textbook where all the subjects are presented in an order that is suitable for everyone. The order of topics in this book is such that the fundamentals of MATLAB are covered first (arrays and array operations), and, as mentioned before, every topic is covered completely in one location, which makes the book easy to use as a reference. The order of the topics in this fourth edition of the book is a little bit different than in previous editions. Programming is introduced before user-defined functions. This allows using programming in user-defined functions. Also, applications of MATLAB in numerical analysis (now Chapter 9, previously 10) follow Chapter 8 which covers polynomials, curve fitting, and interpolation.

Chapter 1
Starting with MATLAB

This chapter begins by describing the characteristics and purposes of the different windows in MATLAB. Next, the Command Window is introduced in detail. This chapter shows how to use MATLAB for arithmetic operations with scalars in a fashion similar to the way that a calculator is used. This includes the use of elementary math functions with scalars. The chapter then shows how to define scalar variables (the assignment operator) and how to use these variables in arithmetic calculations. The last section in the chapter introduces script files. It shows how to write, save, and execute simple MATLAB programs.

1.1 Starting MATLAB, MATLAB Windows

It is assumed that the software is installed on the computer, and that the user can start the program. Once the program starts, the MATLAB desktop window opens (Figure 1-1). The window contains four smaller windows: the Command Window, the Current Folder Window, the Workspace Window, and the Command History Window. This is the default view that shows four of the various windows of MATLAB. A list of several windows and their purpose is given in Table 1-1. The **Start** button on the lower left side can be used to access MATLAB tools and features.

Four of the windows—the Command Window, the Figure Window, the Editor Window, and the Help Window—are used extensively throughout the book and are briefly described on the following pages. More detailed descriptions are included in the chapters where they are used. The Command History Window, Current Folder Window, and the Workspace Window are described in Sections 1.2, 1.8.4, and 4.1, respectively.

Command Window: The Command Window is MATLAB's main window and opens when MATLAB is started. It is convenient to have the Command Window as the only visible window, and this can be done by either closing all the other windows (click on the **x** at the top right-hand side of the window you want to close) or by first selecting the **Desktop Layout** in the **Desktop** menu, and then selecting **Command Window Only** from the submenu that opens. Working in the Command Window is described in detail in Section 1.2.

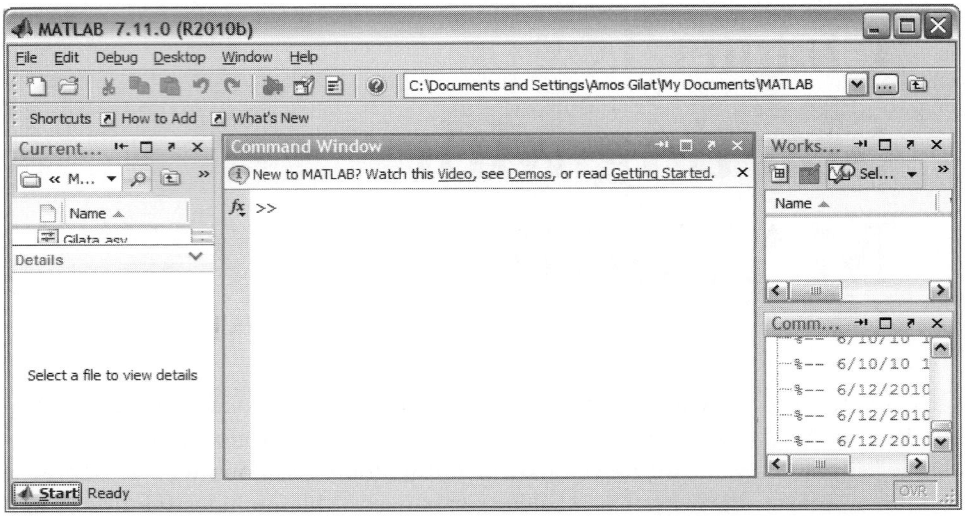

Figure 1-1: The default view of MATLAB desktop.

Table 1-1: MATLAB windows

Window	Purpose
Command Window	Main window, enters variables, runs programs.
Figure Window	Contains output from graphic commands.
Editor Window	Creates and debugs script and function files.
Help Window	Provides help information.
Command History Window	Logs commands entered in the Command Window.
Workspace Window	Provides information about the variables that are used.
Current Folder Window	Shows the files in the current folder.

Figure Window: The Figure Window opens automatically when graphics commands are executed, and contains graphs created by these commands. An example of a Figure Window is shown in Figure 1-2. A more detailed description of this window is given in Chapter 5.

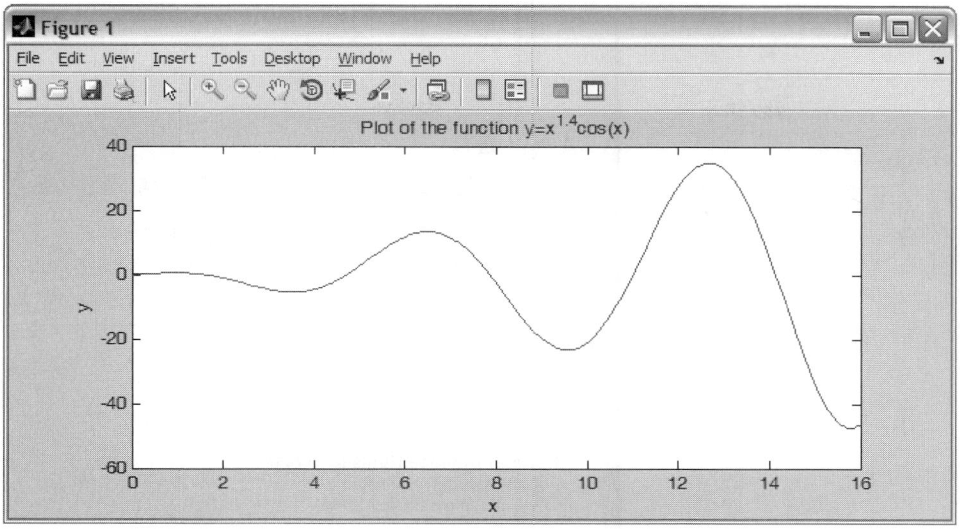

Figure 1-2: Example of a Figure Window.

Editor Window: The Editor Window is used for writing and editing programs. This window is opened from the **File** menu. An example of an Editor Window is shown in Figure 1-3. More details on the Editor Window are given in Section 1.8.2, where it is used for writing script files, and in Chapter 7, where it is used to write function files.

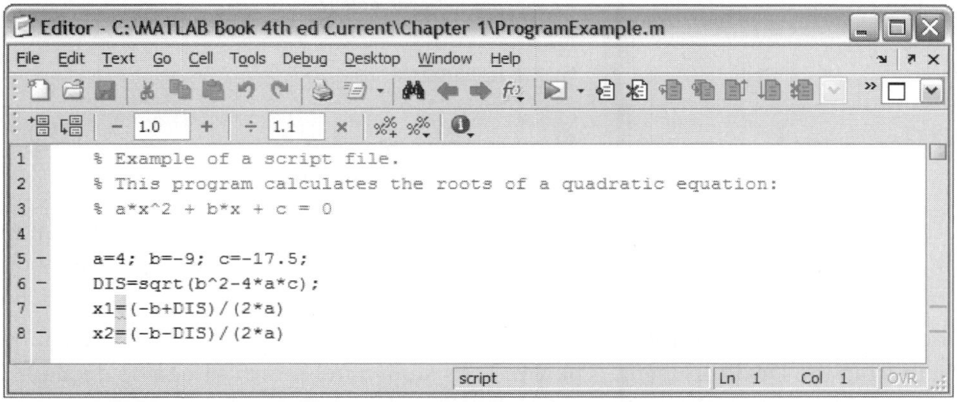

Figure 1-3: Example of an Editor Window.

Help Window: The Help Window contains help information. This window can be opened from the **Help** menu in the toolbar of any MATLAB window. The Help Window is interactive and can be used to obtain information on any feature of MATLAB. Figure 1-4 shows an open Help Window.

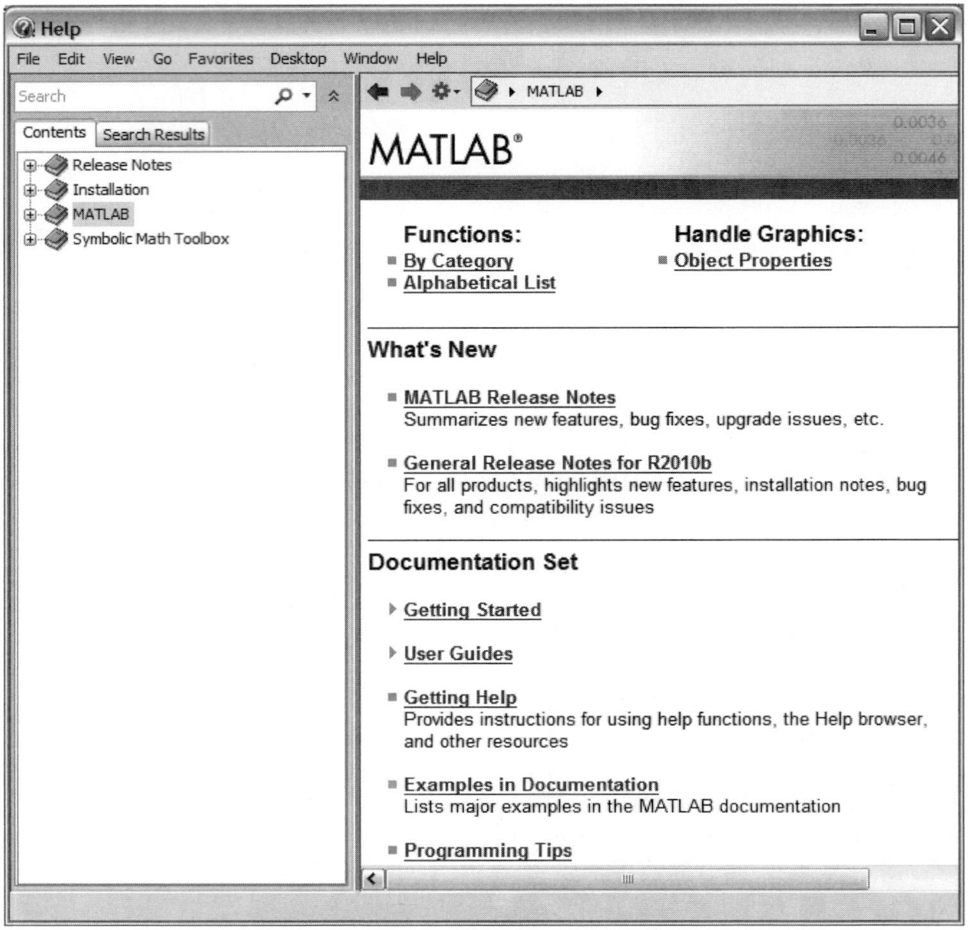

Figure 1-4: The Help Window.

When MATLAB is started for the first time the screen looks like that shown in Figure 1-1. For most beginners it is probably more convenient to close all the windows except the Command Window. (Each of the windows can be closed by clicking on the × button.) The closed windows can be reopened by selecting them from the **Desktop** menu. The windows shown in Figure 1-1 can be displayed by selecting first **Desktop Layout** in the **Desktop** menu and then **Default** from the submenu. The various windows in Figure 1-1 are docked to the desktop. A window can be undocked (become a separate, independent window) by clicking on the button on the upper right-hand corner. An independent window can be redocked by clicking on the button.

1.2 *WORKING IN THE COMMAND WINDOW*

The Command Window is MATLAB's main window and can be used for executing commands, opening other windows, running programs written by the user, and managing the software. An example of the Command Window, with several simple commands that will be explained later in this chapter, is shown in Figure 1-5.

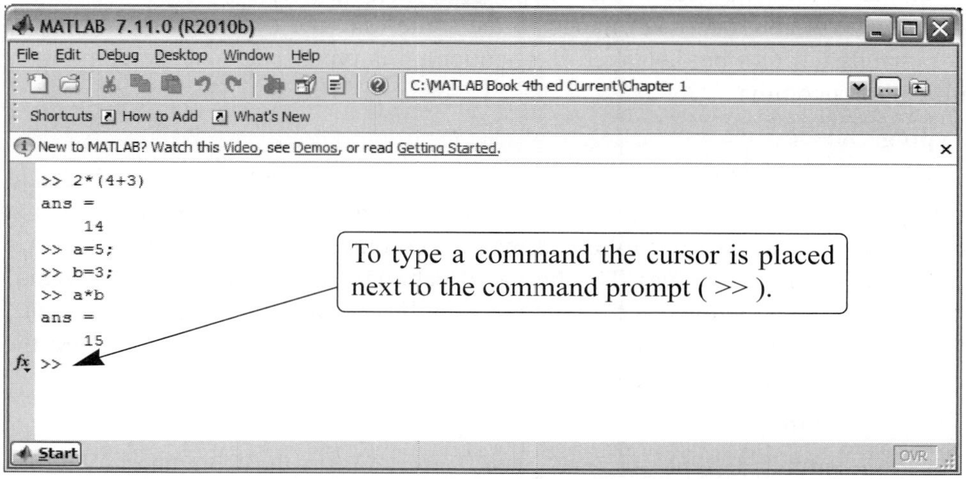

Figure 1-5: The Command Window.

Notes for working in the Command Window:

- To type a command the cursor must be placed next to the command prompt (>>).
- Once a command is typed and the **Enter** key is pressed, the command is executed. However, only the last command is executed. Everything executed previously (that might be still displayed) is unchanged.
- Several commands can be typed in the same line. This is done by typing a comma between the commands. When the **Enter** key is pressed the commands are executed in order from left to right.
- It is not possible to go back to a previous line that is displayed in the Command Window, make a correction, and then re-execute the command.
- A previously typed command can be recalled to the command prompt with the up-arrow key (↑). When the command is displayed at the command prompt, it can be modified if needed and then executed. The down-arrow key (↓) can be used to move down the list of previously typed commands.
- If a command is too long to fit in one line, it can be continued to the next line by typing three periods … (called an ellipsis) and pressing the **Enter** key. The continuation of the command is then typed in the new line. The command can continue line after line up to a total of 4,096 characters.

The semicolon (;):

When a command is typed in the Command Window and the **Enter** key is pressed, the command is executed. Any output that the command generates is displayed in the Command Window. If a semicolon (`;`) is typed at the end of a command the output of the command is not displayed. Typing a semicolon is useful when the result is obvious or known, or when the output is very large.

If several commands are typed in the same line, the output from any of the commands will not be displayed if a semicolon is typed between the commands instead of a comma.

Typing %:

When the symbol % (percent) is typed at the beginning of a line, the line is designated as a comment. This means that when the **Enter** key is pressed the line is not executed. The % character followed by text (comment) can also be typed after a command (in the same line). This has no effect on the execution of the command.

Usually there is no need for comments in the Command Window. Comments, however, are frequently used in a program to add descriptions or to explain the program (see Chapters 4 and 6).

The `clc` command:

The `clc` command (type `clc` and press **Enter**) clears the Command Window. After working in the Command Window for a while, the display may become very long. Once the `clc` command is executed a clear window is displayed. The command does not change anything that was done before. For example, if some variables were defined previously (see Section 1.6), they still exist and can be used. The up-arrow key can also be used to recall commands that were typed before.

The Command History Window:

The Command History Window lists the commands that have been entered in the Command Window. This includes commands from previous sessions. A command in the Command History Window can be used again in the Command Window. By double-clicking on the command, the command is reentered in the Command Window and executed. It is also possible to drag the command to the Command Window, make changes if needed, and then execute it. The list in the Command History Window can be cleared by selecting the lines to be deleted and then selecting **Delete Selection** from the **Edit** menu (or right-click the mouse when the lines are selected and then choose **Delete Selection** in the menu that opens).

1.3 Arithmetic Operations with Scalars

In this chapter we discuss only arithmetic operations with scalars, which are numbers. As will be explained later in the chapter, numbers can be used in arithmetic calculations directly (as with a calculator) or they can be assigned to variables, which can subsequently be used in calculations. The symbols of arithmetic operations are:

Operation	**Symbol**	**Example**
Addition	+	5 + 3
Subtraction	–	5 – 3
Multiplication	*	5 * 3
Right division	/	5 / 3
Left division	\	5 \ 3 = 3 / 5
Exponentiation	^	5 ^ 3 (means $5^3 = 125$)

It should be pointed out here that all the symbols except the left division are the same as in most calculators. For scalars, the left division is the inverse of the right division. The left division, however, is mostly used for operations with arrays, which are discussed in Chapter 3.

1.3.1 Order of Precedence

MATLAB executes the calculations according to the order of precedence displayed below. This order is the same as used in most calculators.

Precedence	**Mathematical Operation**
First	Parentheses. For nested parentheses, the innermost are executed first.
Second	Exponentiation.
Third	Multiplication, division (equal precedence).
Fourth	Addition and subtraction.

In an expression that has several operations, higher-precedence operations are executed before lower-precedence operations. If two or more operations have the same precedence, the expression is executed from left to right. As illustrated in the next section, parentheses can be used to change the order of calculations.

1.3.2 Using MATLAB as a Calculator

The simplest way to use MATLAB is as a calculator. This is done in the Command Window by typing a mathematical expression and pressing the **Enter** key. MATLAB calculates the expression and responds by displaying `ans =` and the numerical result of the expression in the next line. This is demonstrated in Tutorial 1-1.

Tutorial 1-1: Using MATLAB as a calculator.

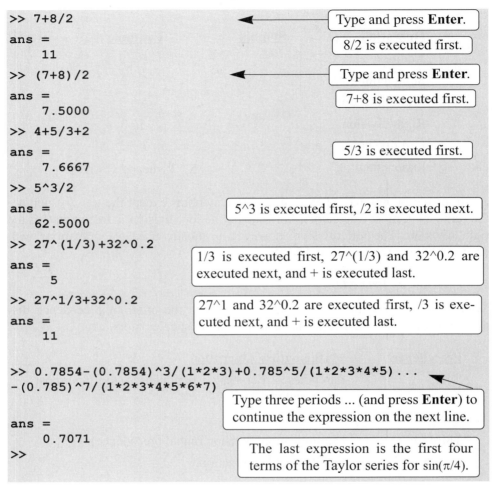

```
>> 7+8/2
ans =
    11
>> (7+8)/2
ans =
    7.5000
>> 4+5/3+2
ans =
    7.6667
>> 5^3/2
ans =
   62.5000
>> 27^(1/3)+32^0.2
ans =
     5
>> 27^1/3+32^0.2
ans =
    11

>> 0.7854-(0.7854)^3/(1*2*3)+0.785^5/(1*2*3*4*5)...
-(0.785)^7/(1*2*3*4*5*6*7)

ans =
    0.7071
>>
```

Type and press **Enter**.

8/2 is executed first.

Type and press **Enter**.

7+8 is executed first.

5/3 is executed first.

5^3 is executed first, /2 is executed next.

1/3 is executed first, 27^(1/3) and 32^0.2 are executed next, and + is executed last.

27^1 and 32^0.2 are executed first, /3 is executed next, and + is executed last.

Type three periods ... (and press **Enter**) to continue the expression on the next line.

The last expression is the first four terms of the Taylor series for sin(π/4).

1.4 Display Formats

The user can control the format in which MATLAB displays output on the screen. In Tutorial 1-1, the output format is fixed-point with four decimal digits (called `short`), which is the default format for numerical values. The format can be changed with the `format` command. Once the `format` command is entered, all the output that follows is displayed in the specified format. Several of the available formats are listed and described in Table 1-2.

MATLAB has several other formats for displaying numbers. Details of these formats can be obtained by typing `help format` in the Command Window. The format in which numbers are displayed does not affect how MATLAB computes and saves numbers.

Table 1-2: Display formats

Command	Description	Example
`format short`	Fixed-point with 4 decimal digits for: $0.001 \le number \le 1000$ Otherwise display format `short e`.	`>> 290/7` `ans =` `41.4286`
`format long`	Fixed-point with 15 decimal digits for: $0.001 \le number \le 100$ Otherwise display format `long e`.	`>> 290/7` `ans =` `41.428571428571431`
`format short e`	Scientific notation with 4 decimal digits.	`>> 290/7` `ans =` `4.1429e+001`
`format long e`	Scientific notation with 15 decimal digits.	`>> 290/7` `ans =` `4.142857142857143e+001`
`format short g`	Best of 5-digit fixed or floating point.	`>> 290/7` `ans =` `41.429`
`format long g`	Best of 15-digit fixed or floating point.	`>> 290/7` `ans =` `41.4285714285714`
`format bank`	Two decimal digits.	`>> 290/7` `ans =` `41.43`
`format compact`	Eliminates empty lines to allow more lines with information displayed on the screen.	
`format loose`	Adds empty lines (opposite of `compact`).	

1.5 ELEMENTARY MATH BUILT-IN FUNCTIONS

In addition to basic arithmetic operations, expressions in MATLAB can include functions. MATLAB has a very large library of built-in functions. A function has a name and an argument in parentheses. For example, the function that calculates the square root of a number is `sqrt(x)`. Its name is `sqrt`, and the argument is `x`. When the function is used, the argument can be a number, a variable that has been assigned a numerical value (explained in Section 1.6), or a computable expression that can be made up of numbers and/or variables. Functions can also be included in arguments, as well as in expressions. Tutorial 1-2 shows examples of using the function `sqrt(x)` when MATLAB is used as a calculator with scalars.

Tutorial 1-2: Using the `sqrt` built-in function.

```
>> sqrt(64)                          Argument is a number.
ans =
     8
>> sqrt(50+14*3)                     Argument is an expression.
ans =
    9.5917
>> sqrt(54+9*sqrt(100))              Argument includes a function.
ans =
    12
>> (15+600/4)/sqrt(121)              Function is included in an expression.
ans =
    15
>>
```

Some commonly used elementary MATLAB mathematical built-in functions are given in Tables 1-3 through 1-5. A complete list of functions organized by category can be found in the Help Window.

Table 1-3: Elementary math functions

Function	Description	Example

`sqrt(x)`	Square root.	`>> sqrt(81)`
		`ans =`
		`9`

`nthroot(x,n)`	Real *n*th root of a real number *x*. (If *x* is negative *n* must be an odd integer.)	`>> nthroot(80,5)`
		`ans =`
		`2.4022`

`exp(x)`	Exponential (e^x).	`>> exp(5)`
		`ans =`
		`148.4132`

`abs(x)`	Absolute value.	`>> abs(-24)`
		`ans =`
		`24`

`log(x)`	Natural logarithm.
	Base *e* logarithm (ln).
		`>> log(1000)`
		`ans =`
		`6.9078`

`log10(x)`	Base 10 logarithm.	`>> log10(1000)`
		`ans =`
		`3.0000`

Table 1-3: Elementary math functions (Continued)

Function	Description	Example

`factorial(x)`	The factorial function $x!$
	(x must be a positive integer.)
		`>> factorial(5)`
		`ans =`
		`120`

Table 1-4: Trigonometric math functions

Function	Description	Example

`sin(x)`
`sind(x)`
	Sine of angle x (x in radians).
	Sine of angle x (x in degrees).
		`>> sin(pi/6)`
		`ans =`
		`0.5000`

`cos(x)`
`cosd(x)`
	Cosine of angle x (x in radians).
	Cosine of angle x (x in degrees).
		`>> cosd(30)`
		`ans =`
		`0.8660`

`tan(x)`
`tand(x)`
	Tangent of angle x (x in radians).
	Tangent of angle x (x in degrees).
		`>> tan(pi/6)`
		`ans =`
		`0.5774`

`cot(x)`
`cotd(x)`
	Cotangent of angle x (x in radians).
	Cotangent of angle x (x in degrees).
		`>> cotd(30)`
		`ans =`
		`1.7321`

The inverse trigonometric functions are `asin(x)`, `acos(x)`, `atan(x)`, `acot(x)` for the angle in radians; and `asind(x)`, `acosd(x)`, `atand(x)`, `acotd(x)` for the angle in degrees. The hyperbolic trigonometric functions are `sinh(x)`, `cosh(x)`, `tanh(x)`, and `coth(x)`. Table 1-4 uses `pi`, which is equal to π (see Section 1.6.3).

Table 1-5: Rounding functions

Function	Description	Example

`round(x)`	Round to the nearest integer.	`>> round(17/5)`
		`ans =`
		`3`

`fix(x)`	Round toward zero.	`>> fix(13/5)`
		`ans =`
		`2`

`ceil(x)`	Round toward infinity.	`>> ceil(11/5)`
		`ans =`
		`3`

`floor(x)`	Round toward minus infinity.	`>> floor(-9/4)`
		`ans =`
		`-3`

`rem(x,y)`	Returns the remainder after x is divided by y.	`>> rem(13,5)`
		`ans =`
		`3`

Table 1-5: Rounding functions (Continued)

Function	Description	Example

`sign(x)`	Signum function. Returns 1 if $x > 0$, −1 if $x < 0$, and 0 if $x = 0$.	`>> sign(5)`
		`ans =`
		`1`

1.6 DEFINING SCALAR VARIABLES

A variable is a name made of a letter or a combination of several letters (and digits) that is assigned a numerical value. Once a variable is assigned a numerical value, it can be used in mathematical expressions, in functions, and in any MATLAB statements and commands. A variable is actually a name of a memory location. When a new variable is defined, MATLAB allocates an appropriate memory space where the variable's assignment is stored. When the variable is used the stored data is used. If the variable is assigned a new value the content of the memory location is replaced. (In Chapter 1 we consider only variables that are assigned numerical values that are scalars. Assigning and addressing variables that are arrays is discussed in Chapter 2.)

1.6.1 The Assignment Operator

In MATLAB the = sign is called the assignment operator. The assignment operator assigns a value to a variable.

Variable_name = A numerical value, or a computable expression

- The left-hand side of the assignment operator can include only one variable name. The right-hand side can be a number, or a computable expression that can include numbers and/or variables that were previously assigned numerical values. When the **Enter** key is pressed the numerical value of the right-hand side is assigned to the variable, and MATLAB displays the variable and its assigned value in the next two lines.

The following shows how the assignment operator works.

```
>> x=15
x =
    15

>> x=3*x-12
x =
    33
>>
```

The number 15 is assigned to the variable x.

MATLAB displays the variable and its assigned value.

A new value is assigned to x. The new value is 3 times the previous value of x minus 12.

The last statement ($x = 3x - 12$) illustrates the difference between the assignment operator and the equal sign. If in this statement the = sign meant equal, the value of x would be 6 (solving the equation for x).

The use of previously defined variables to define a new variable is demonstrated next.

```
>> a=12
a =
    12
>> B=4
B =
     4
>> C=(a-B)+40-a/B*10
C =
    18
```

Assign 12 to a.

Assign 4 to B.

Assign the value of the expression on the right-hand side to the variable C.

- If a semicolon is typed at the end of the command, then when the **Enter** key is pressed, MATLAB does not display the variable with its assigned value (the variable still exists and is stored in memory).
- If a variable already exists, typing the variable's name and pressing the **Enter** key will display the variable and its value in the next two lines.

As an example, the last demonstration is repeated below using semicolons.

```
>> a=12;
>> B=4;
>> C=(a-B)+40-a/B*10;
>> C
C =
    18
```

The variables a, B, and C are defined but are not displayed since a semicolon is typed at the end of each statement.

The value of the variable C is displayed by typing the name of the variable.

- Several assignments can be typed in the same line. The assignments must be separated with a comma (spaces can be added after the comma). When the **Enter** key is pressed, the assignments are executed from left to right and the variables and their assignments are displayed. A variable is not displayed if a semicolon is typed instead of a comma. For example, the assignments of the variables a, B, and C above can all be done in the same line.

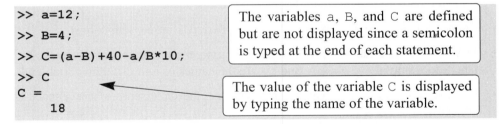

```
>> a=12, B=4; C=(a-B)+40-a/B*10
a =
    12
C =
    18
```

The variable B is not displayed because a semicolon is typed at the end of the assignment.

- A variable that already exists can be reassigned a new value. For example:

```
>> ABB=72;
>> ABB=9;
>> ABB
ABB =
     9
>>
```

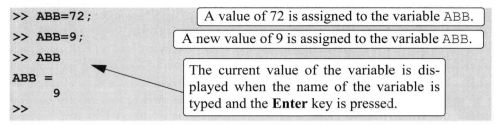

A value of 72 is assigned to the variable ABB.

A new value of 9 is assigned to the variable ABB.

The current value of the variable is displayed when the name of the variable is typed and the **Enter** key is pressed.

- Once a variable is defined it can be used as an argument in functions. For example:

```
>> x=0.75;
>> E=sin(x)^2+cos(x)^2
E =
     1
>>
```

1.6.2 Rules About Variable Names

A variable can be named according to the following rules:

- Must begin with a letter.
- Can be up to 63 characters long.
- Can contain letters, digits, and the underscore character.
- Cannot contain punctuation characters (e.g., period, comma, semicolon).
- MATLAB is case sensitive: it distinguishes between uppercase and lowercase letters. For example, `AA`, `Aa`, `aA`, and `aa` are the names of four different variables.
- No spaces are allowed between characters (use the underscore where a space is desired).
- Avoid using the name of a built-in function for a variable (i.e., avoid using `cos`, `sin`, `exp`, `sqrt`, etc.). Once a function name is used to define a variable, the function cannot be used.

1.6.3 Predefined Variables and Keywords

There are 20 words, called keywords, that are reserved by MATLAB for various purposes and cannot be used as variable names. These words are:

```
break   case   catch  classdef  continue   else   elseif
end   for   function   global   if   otherwise   parfor
persistent  return  spmd   switch   try   while
```

When typed, these words appear in blue. An error message is displayed if the user tries to use a keyword as a variable name. (The keywords can be displayed by typing the command `iskeyword`.)

A number of frequently used variables are already defined when MATLAB is started. Some of the predefined variables are:

`ans` A variable that has the value of the last expression that was not assigned to a specific variable (see Tutorial 1-1). If the user does not assign the value of an expression to a variable, MATLAB automatically stores the result in `ans`.

`pi` The number π.

`eps` The smallest difference between two numbers. Equal to 2^(–52), which is approximately 2.2204e–016.

`inf` Used for infinity.

`i` Defined as $\sqrt{-1}$, which is: 0 + 1.0000i.

`j` Same as i.

`NaN` Stands for Not-a-Number. Used when MATLAB cannot determine a valid numeric value. Example: 0/0.

The predefined variables can be redefined to have any other value. The variables `pi`, `eps`, and `inf`, are usually not redefined since they are frequently used in many applications. Other predefined variables, such as `i` and `j`, are sometime redefined (commonly in association with loops) when complex numbers are not involved in the application.

1.7 USEFUL COMMANDS FOR MANAGING VARIABLES

The following are commands that can be used to eliminate variables or to obtain information about variables that have been created. When these commands are typed in the Command Window and the **Enter** key is pressed, either they provide information, or they perform a task as specified below.

Command	**Outcome**
`clear`	Removes all variables from the memory.
`clear x y z`	Removes only variables `x`, `y`, and `z` from the memory.
`who`	Displays a list of the variables currently in the memory.
`whos`	Displays a list of the variables currently in the memory and their sizes together with information about their bytes and class (see Section 4.1).

1.8 SCRIPT FILES

So far all the commands were typed in the Command Window and were executed when the **Enter** key was pressed. Although every MATLAB command can be executed in this way, using the Command Window to execute a series of commands—especially if they are related to each other (a program)—is not convenient and may be difficult or even impossible. The commands in the Command Window cannot be saved and executed again. In addition, the Command Window is not interactive. This means that every time the **Enter** key is pressed only the last command is executed, and everything executed before is unchanged. If a change or a correction is needed in a command that was previously executed and the results of this command are used in commands that follow, all the commands have to be entered and executed again.

A different (better) way of executing commands with MATLAB is first to create a file with a list of commands (program), save it, and then run (execute) the file. When the file runs, the commands it contains are executed in the order that they are listed. If needed, the commands in the file can be corrected or changed and the file can be saved and run again. Files that are used for this purpose are called script files.

IMPORTANT NOTE: This section covers only the minimum that is required in order to run simple programs. This will allow the student to use script files when practicing the material that is presented in this and the next two chapters (instead of typing repeatedly in the Command Window). Script files are considered again in Chapter 4 where many additional topics that are essential for understanding MATLAB and writing programs in script file are covered.

1.8.1 Notes About Script Files

- A script file is a sequence of MATLAB commands, also called a program.
- When a script file runs (is executed), MATLAB executes the commands in the order they are written just as if they were typed in the Command Window.
- When a script file has a command that generates an output (e.g., assignment of a value to a variable without a semicolon at the end), the output is displayed in the Command Window.
- Using a script file is convenient because it can be edited (corrected or otherwise changed) and executed many times.
- Script files can be typed and edited in any text editor and then pasted into the MATLAB editor.
- Script files are also called M-files because the extension .m is used when they are saved.

1.8.2 Creating and Saving a Script File

In MATLAB script files are created and edited in the Editor/Debugger Window. This window is opened from the Command Window. In the **File** menu, select **New**, and then select **Script**. An open Editor/Debugger Window is shown in Figure 1-6.

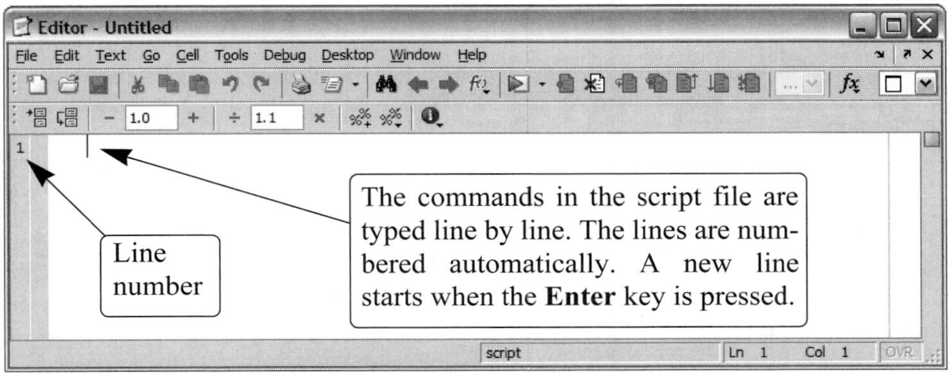

Figure 1-6: The Editor/Debugger Window.

Once the window is open, the commands of the script file are typed line by line. MATLAB automatically numbers a new line every time the **Enter** key is pressed. The commands can also be typed in any text editor or word processor program and then copied and pasted in the Editor/Debugger Window. An example of a short program typed in the Editor/Debugger Window is shown in Figure 1-7. The first few lines in a script file are typically comments (which are not executed since the first character in the line is %) that describe the program written in the script file.

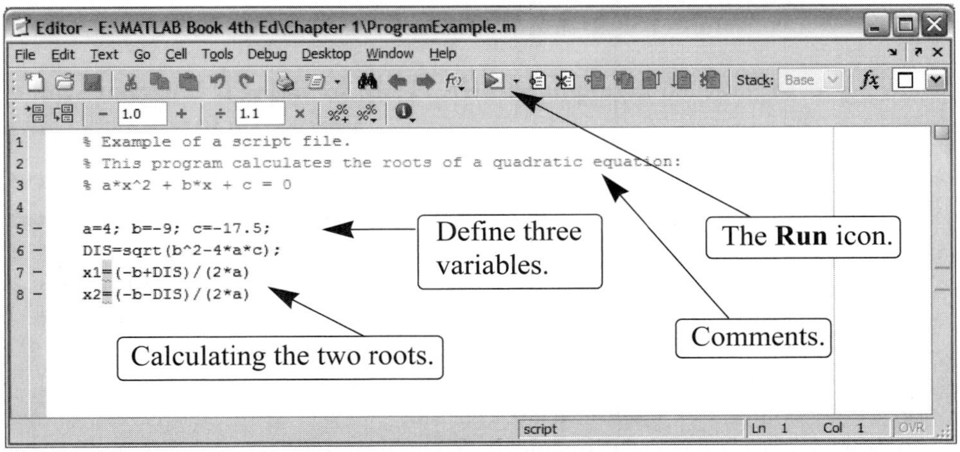

Figure 1-7: A program typed in the Editor/Debugger Window.

Before a script file can be executed it has to be saved. This is done by choosing **Save As**... from the **File** menu, selecting a location (many students save to a flash drive, which appears in the directory as `Drive(F:)` or `(G:)`), and entering a name for the file. When saved, MATLAB adds the extension .m to the name. The rules for naming a script file follow the rules of naming a variable (must begin with a letter, can include digits and underscore, *no spaces*, and up to 63 characters long). The names of user-defined variables, predefined variables, and MATLAB commands or functions should not be used as names of script files.

1.8.3 Running (Executing) a Script File

A script file can be executed either directly from the Editor Window by clicking on the **Run** icon (see Figure 1-7) or by typing the file name in the Command Window and then pressing the **Enter** key. For a file to be executed, MATLAB needs to know where the file is saved. The file will be executed if the folder where the file is saved is the current folder of MATLAB or if the folder is listed in the search path, as explained next.

1.8.4 Current Folder

The current folder is shown in the "Current Folder" field in the desktop toolbar of the Command Window, as shown in Figure 1-8. If an attempt is made to execute a script file by clicking on the **Run** icon (in the Editor Window) when the current folder is not the folder where the script file is saved, then the prompt shown in

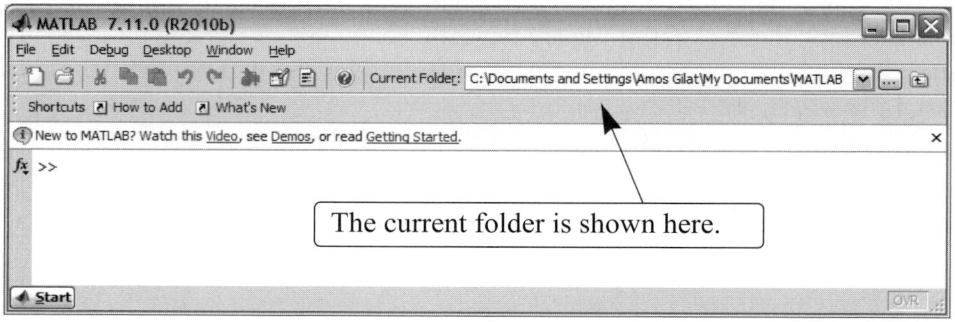

Figure 1-8: The Current folder field in the Command Window.

Figure 1-9 will open. The user can then change the current folder to the folder where the script file is saved, or add it to the search path. Once two or more different current folders are used in a session, it is possible to switch from one to another in the **Current Folder** field in the Command Window. The current folder can also be changed in the Current Folder Window, shown in Figure 1-10, which can be opened from the **Desktop** menu. The Current Folder can be changed by choosing the drive and folder where the file is saved.

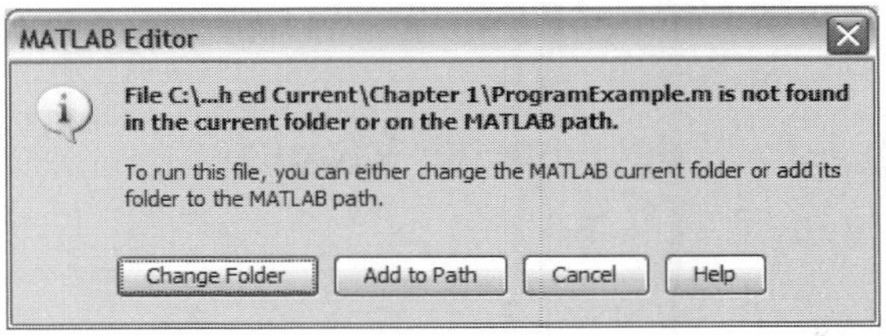

Figure 1-9: Changing the current directory.

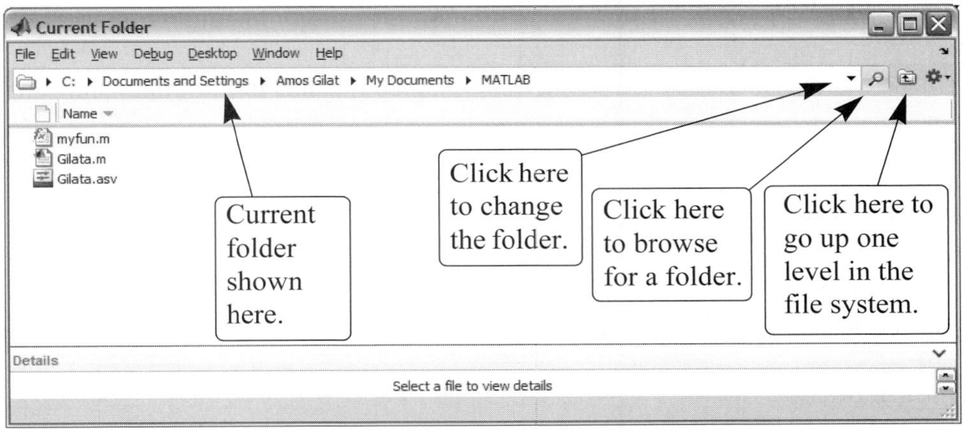

Figure 1-10: The Current Folder Window.

An alternative simple way to change the current folder is to use the `cd` command in the Command Window. To change the current folder to a different drive, type `cd`, space, and then the name of the directory followed by a colon : and press the **Enter** key. For example, to change the current folder to drive F (e.g., the flash drive) type `cd F:`. If the script file is saved in a folder within a drive, the path to that folder has to be specified. This is done by typing the path as a string in the `cd` command. For example, `cd('F:\Chapter 1')` sets the path to the folder Chapter 1 in drive F. The following example shows how the current folder is changed to be drive E. Then the script file from Figure 1-7, which was saved in drive E as ProgramExample.m, is executed by typing the name of the file and pressing the **Enter** key.

```
>> cd E:

>> ProgramExample
x1 =
    3.5000
x2 =
   -1.2500
```

The current directory is changed to drive E.

The script file is executed by typing the name of the file and pressing the **Enter** key.

The output generated by the script file (the roots `x1` and `x2`) is displayed in the Command Window.

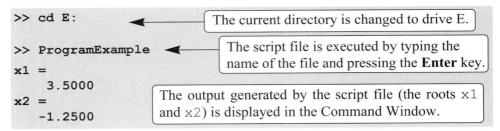

1.9 EXAMPLES OF MATLAB APPLICATIONS

Sample Problem 1-1: Trigonometric identity

A trigonometric identity is given by:

$$\cos^2\frac{x}{2} = \frac{\tan x + \sin x}{2\tan x}$$

Verify that the identity is correct by calculating each side of the equation, substituting $x = \frac{\pi}{5}$.

Solution

The problem is solved by typing the following commands in the Command Window.

```
>> x=pi/5;                          Define x.
>> LHS=cos(x/2)^2                   Calculate the left-hand side.
LHS =
    0.9045
>> RHS=(tan(x)+sin(x))/(2*tan(x))   Calculate the right-hand side.
RHS =
    0.9045
```

Sample Problem 1-2: Geometry and trigonometry

Four circles are placed as shown in the figure. At each point where two circles are in contact they are tangent to each other. Determine the distance between the centers C_2 and C_4.

The radii of the circles are:
$R_1 = 16$ mm, $R_2 = 6.5$mm, $R_3 = 12$ mm, and $R_4 = 9.5$mm.

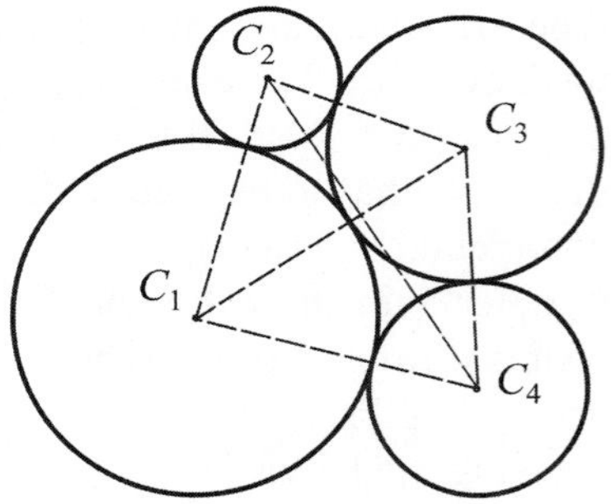

Solution

The lines that connect the centers of the circles create four triangles. In two of the triangles, $\Delta C_1C_2C_3$ and $\Delta C_1C_3C_4$, the lengths of all the sides are known. This information is used to calculate the angles γ_1 and γ_2 in these triangles by using the law of cosines. For example, γ_1 is calculated from:

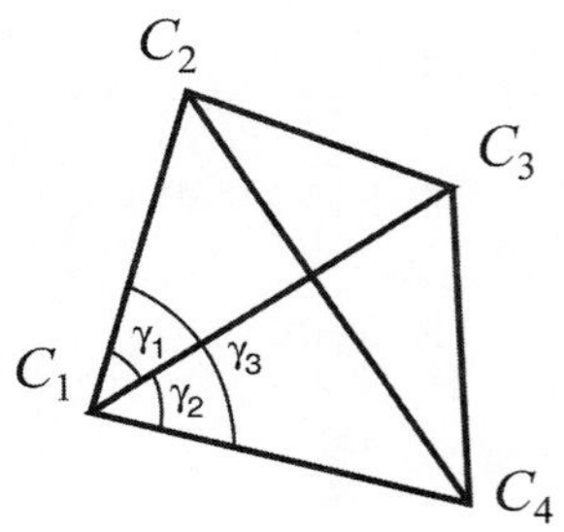

$$(C_2C_3)^2 = (C_1C_2)^2 + (C_1C_3)^2 - 2(C_1C_2)(C_1C_3)\cos\gamma_1$$

Next, the length of the side C_2C_4 is calculated by considering the triangle $\Delta C_1C_2C_4$. This is done, again, by using the law of cosines (the lengths C_1C_2 and C_1C_4 are known and the angle γ_3 is the sum of the angles γ_1 and γ_2).

The problem is solved by writing the following program in a script file:

```
% Solution of Sample Problem 1-2

R1=16; R2=6.5; R3=12; R4=9.5;
C1C2=R1+R2; C1C3=R1+R3; C1C4=R1+R4;
C2C3=R2+R3; C3C4=R3+R4;
Gama1=acos((C1C2^2+C1C3^2-C2C3^2)/(2*C1C2*C1C3));
Gama2=acos((C1C3^2+C1C4^2-C3C4^2)/(2*C1C3*C1C4));
Gama3=Gama1+Gama2;

C2C4=sqrt(C1C2^2+C1C4^2-2*C1C2*C1C4*cos(Gama3))
```

Define the *R*'s.

Calculate the lengths of the sides.

Calculate γ_1, γ_2, and γ_3.

Calculate the length of side C_2C_4.

When the script file is executed, the following (the value of the variable `C2C4`) is displayed in the Command Window:

```
C2C4 =
   33.5051
```

Sample Problem 1-3: Heat transfer

An object with an initial temperature of T_0 that is placed at time $t = 0$ inside a chamber that has a constant temperature of T_s will experience a temperature change according to the equation

$$T = T_s + (T_0 - T_s)e^{-kt}$$

where T is the temperature of the object at time t, and k is a constant. A soda can at a temperature of 120° F (after being left in the car) is placed inside a refrigerator where the temperature is 38°F. Determine, to the nearest degree, the temperature of the can after three hours. Assume $k = 0.45$. First define all of the variables and then calculate the temperature using one MATLAB command.

Solution

The problem is solved by typing the following commands in the Command Window.

```
>> Ts=38;  T0=120; k=0.45; t=3;
>> T=round(Ts+(T0-Ts)*exp(-k*t))
T =
    59
```

Round to the nearest integer.

Sample Problem 1-4: Compounded interest

The balance B of a savings account after t years when a principal P is invested at an annual interest rate r and the interest is compounded n times a year is given by:

$$B = P\left(1 + \frac{r}{n}\right)^{nt} \qquad (1)$$

If the interest is compounded yearly, the balance is given by:

$$B = P(1 + r)^t \qquad (2)$$

Suppose \$5,000 is invested for 17 years in one account where the interest is compounded yearly. In addition, \$5,000 is invested in a second account in which the interest is compounded monthly. In both accounts the interest rate is 8.5%. Use MATLAB to determine how long (in years and months) it would take for the balance in the second account to be the same as the balance of the first account after 17 years.

Solution

Follow these steps:

(*a*) Calculate B for \$5,000 invested in a yearly compounded interest account after 17 years using Equation (2).

(*b*) Calculate t for the B calculated in part (*a*), from the monthly compounded interest formula, Equation (1).

(*c*) Determine the number of years and months that correspond to t.

The problem is solved by writing the following program in a script file:

```
% Solution of Sample Problem 1-4
P=5000;  r=0.085;  ta=17; n=12;
B=P*(1+r)^ta
t=log(B/P)/(n*log(1+r/n))

years=fix(t)
months=ceil((t-years)*12)
```

Step (*a*): Calculate B from Eq. (2).

Step (*b*): Solve Eq. (1) for t, and calculate t.

Step (*c*): Determine the number of years.

Determine the number of months.

When the script file is executed, the following (the values of the variables `B`, `t`, `years`, and `months`) is displayed in the Command Window:

```
>> format short g
B =
        20011
t =
       16.374
years =
    16
months =
     5
```

The values of the variables `B`, `t`, `years`, and `months` are displayed (since a semicolon was not typed at the end of any of the commands that calculate the values).

1.10 PROBLEMS

The following problems can be solved by writing commands in the Command Window, or by writing a program in a script file and then executing the file.

1. Calculate:

   (*a*) $\frac{(14.8^2+6.5^2)}{3.8^2}+\frac{55}{\sqrt{2}+14}$ (*b*) $(-3.5)^3+\frac{e^6}{\ln 524}+206^{1/3}$

2. Calculate:

   (*a*) $\frac{16.5^2(8.4-\sqrt{70})}{4.3^2-17.3}$ (*b*) $\frac{5.2^3-6.4^2+3}{1.6^8-2}+\left(\frac{13.3}{5}\right)^{1.5}$

3. Calculate:

   (*a*) $15\left(\frac{\sqrt{10}+3.7^2}{\log_{10}(1365)+1.9}\right)$ (*b*) $\frac{2.5^3\left(16-\frac{216}{22}\right)}{1.7^4+14}+\sqrt[4]{2050}$

4. Calculate:

   (*a*) $\frac{2.3^2\cdot 1.7}{\sqrt{(1-0.8^2)^2+(2-\sqrt{0.87})^2}}$ (*b*) $2.34+\frac{1}{2}2.7(5.9^2-2.4^2)+9.8\ln 51$

5. Calculate:

(*a*) $\dfrac{\sin\left(\frac{7\pi}{9}\right)}{\cos^2\left(\frac{5}{7}\pi\right)} + \frac{1}{7}\tan\left(\frac{5}{12}\pi\right)$ (*b*) $\dfrac{\tan 64°}{\cos^2 14°} - \dfrac{3\sin 80°}{\sqrt[3]{0.9}} + \dfrac{\cos 55°}{\sin 11°}$

6. Define the variable x as $x = 2.34$, then evaluate:

(*a*) $2x^4 - 6x^3 + 14.8x^2 + 9.1$ (*b*) $\dfrac{e^{2x}}{\sqrt{14 + x^2 - x}}$

7. Define the variable t as $t = 6.8$, then evaluate:

(*a*) $\ln(|t^2 - t^3|)$ (*b*) $\dfrac{75}{2t}\cos(0.8t - 3)$

8. Define the variables x and y as $x = 8.3$ and $y = 2.4$, then evaluate:

(*a*) $x^2 + y^2 - \dfrac{x^2}{y^2}$ (*b*) $\sqrt{xy} - \sqrt{x + y} + \left(\dfrac{x - y}{x - 2y}\right)^2 - \sqrt{\dfrac{x}{y}}$

9. Define the variables a, b, c, and d as:

$a = 13$, $b = 4.2$, $c = (4b)/a$, and $d = \dfrac{abc}{a + b + c}$, then evaluate:

(*a*) $a\dfrac{b}{c + d} + \dfrac{da}{cb} - (a - b^2)(c + d)$ (*b*) $\dfrac{\sqrt{a^2 + b^2}}{(d - c)} + \ln(|b - a + c - d|)$

10. A cube has a side of 18 cm.
(*a*) Determine the radius of a sphere that has the same surface area as the cube.
(*b*) Determine the radius of a sphere that has the same volume as the cube.

11. The perimeter P of an ellipse with semi-minor axes a and b is given approximately by: $P = 2\pi\sqrt{\frac{1}{2}(a^2 + b^2)}$.

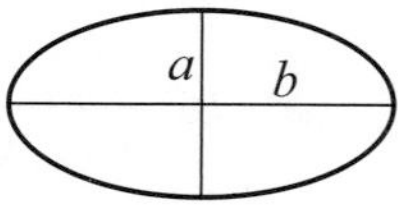

(*a*) Determine the perimeter of an ellipse with $a = 9$ in. and $b = 3$ in.
(*b*) An ellipse with $b = 2a$ has a perimeter of $P = 20$ cm. Determine a and b.

12. Two trigonometric identities are given by:

(*a*) $\sin 4x = 4\sin x\cos x - 8\sin^3 x\cos x$ (*b*) $\cos 2x = \dfrac{1 - \tan^2 x}{1 + \tan^2 x}$

For each part, verify that the identity is correct by calculating the values of the left and right sides of the equation, substituting $x = \frac{\pi}{9}$.

13. Two trigonometric identities are given by:

(*a*) $\tan 4x = \dfrac{4\tan x - 4\tan^3 x}{1 - 6\tan^2 x + \tan^4 x}$ (*b*) $\sin^3 x = \dfrac{1}{4}(3\sin x - \sin 3x)$

For each part, verify that the identity is correct by calculating the values of the left and right sides of the equation, substituting $x = 12°$.

14. Define two variables: *alpha* = 5π/8, and *beta* = π/8. Using these variables, show that the following trigonometric identity is correct by calculating the values of the left and right sides of the equation.

$$\sin\alpha\cos\beta = \frac{1}{2}[\sin(\alpha - \beta) + \sin(\alpha + \beta)]$$

15. Given: $\int \cos^2(ax)dx = \dfrac{1}{2}x - \dfrac{\sin 2ax}{4a}$. Use MATLAB to calculate the following definite integral: $\int_{\frac{\pi}{9}}^{\frac{3\pi}{5}} \cos^2(0.5x)dx$.

16. In the triangle shown $a = 9$ cm, $b = 18$ cm, and $c = 25$ cm. Define a, b, and c as variables, and then:
(*a*) Calculate the angle α (in degrees) by substituting the variables in the Law of Cosines.
(Law of Cosines: $c^2 = a^2 + b^2 - 2ab\cos\gamma$)
(*b*) Calculate the angles β and γ (in degrees) using the Law of Sines.
(*c*) Check that the sum of the angles is 180°.

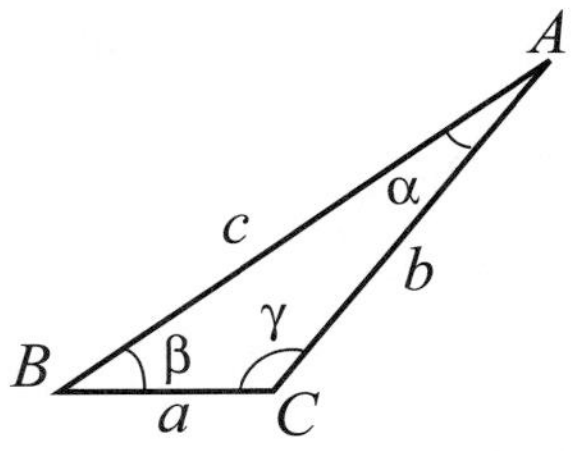

17. In the triangle shown $a = 5$ in., $b = 7$ in., and $\gamma = 25°$. Define a, b, and γ as variables, and then:
(*a*) Calculate the length of c by substituting the variables in the Law of Cosines.
(Law of Cosines: $c^2 = a^2 + b^2 - 2ab\cos\gamma$)
(*b*) Calculate the angles α and β (in degrees) using the Law of Sines.
(*c*) Verify the Law of Tangents by substituting the results from part (*b*) into the right and left sides of the equation.

(Law of Tangents: $\dfrac{a-b}{a+b} = \dfrac{\tan\left[\frac{1}{2}(\alpha - \beta)\right]}{\tan\left[\frac{1}{2}(\alpha + \beta)\right]}$

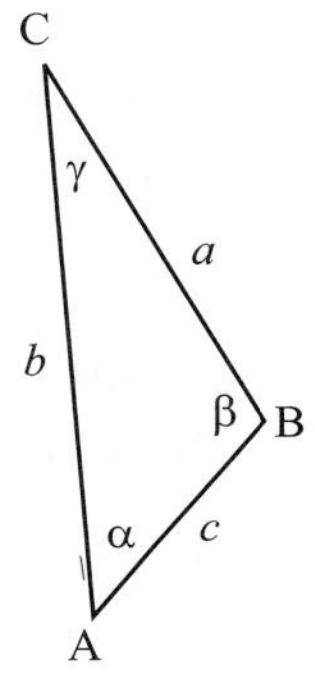

18. For the triangle shown, $a = 200$ mm, $b = 250$ mm, and $c = 300$ mm. Define a, b, and c as variables, and then:

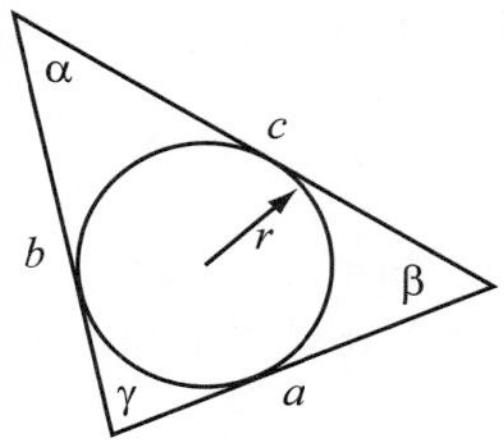

(*a*) Calculate the angle γ (in degrees) by substituting the variables in the Law of Cosines.
(Law of Cosines: $c^2 = a^2 + b^2 - 2ab\cos\gamma$)

(*b*) Calculate the radius r of the circle inscribed in the triangle using the formula $r = \frac{1}{2}(a + b - c)\tan\left(\frac{1}{2}\gamma\right)$.

(*c*) Calculate the radius r of the circle inscribed in the triangle using the formula $r = \frac{\sqrt{s(s-a)(s-b)(s-c)}}{s}$, where $s = \frac{1}{2}(a + b + c)$.

19. In the right triangle shown $a = 16$ cm and $c = 50$ cm. Define a and c as variables, and then:

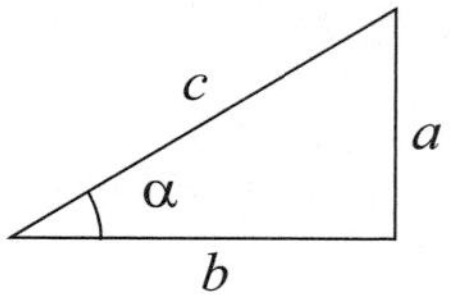

(*a*) Using the Pythagorean Theorem, calculate b by typing one line in the Command Window.

(*b*) Using b from part (*a*) and the `acosd` function, calculate the angle α in degrees by typing one line in the Command Window.

20. The distance d from a point (x_0, y_0, z_0) to a plane $Ax + By + Cz + D = 0$ is given by:

$$d = \frac{|Ax_0 + By_0 + Cz_0 + D|}{\sqrt{A^2 + B^2 + C^2}}$$

Determine the distance of the point $(8, 3, -10)$ from the plane $2x + 23y + 13z - 24 = 0$. First define the variables A, B, C, D, x_0, y_0, and z_0, and then calculate d. (Use the `abs` and `sqrt` functions.)

21. The arc length s of the parabolic segment BOC is given by:

$$s = \frac{1}{2}\sqrt{b^2 + 16a^2} + \frac{b^2}{8a}\ln\left(\frac{4a + \sqrt{b^2 + 16a^2}}{b}\right)$$

Calculate the arc length of a parabola with $a = 12$ in. and $b = 8$ in.

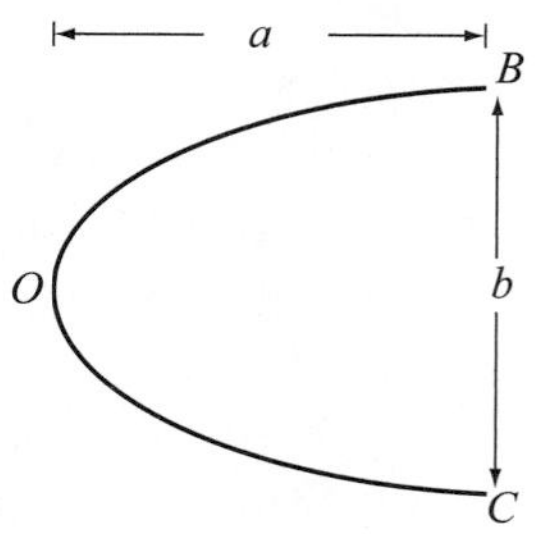

22. Oranges are packed such that 52 are placed in each box. Determine how many boxes are needed to pack 4,000 oranges. Use MATLAB built-in function `ceil`.

23. The voltage difference V_{ab} between points a and b in the Wheatstone bridge circuit is:

$$V_{ab} = V\left(\frac{R_2}{R_1 + R_2} - \frac{R_4}{R_3 + R_4}\right)$$

Calculate the voltage difference when $V = 12$ volts, $R_1 = 120$ ohms, $R_2 = 100$ ohms, $R_3 = 220$ ohms, and $R_4 = 120$ ohms.

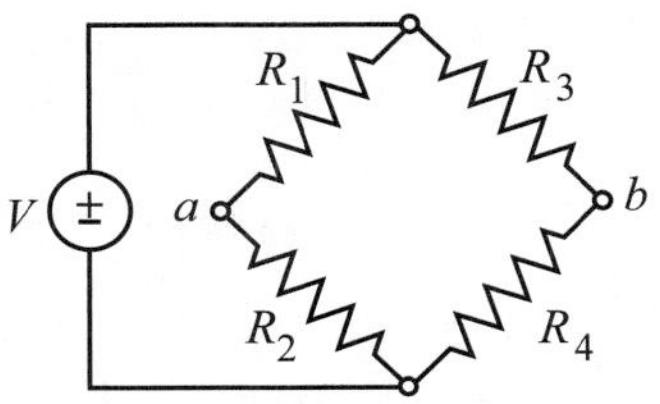

24. The prices of an oak tree and a pine tree are \$54.95 and \$39.95, respectively. Assign the prices to variables named oak and pine, change the display format to `bank`, and calculate the following by typing one command:

(*a*) The total cost of 16 oak trees and 20 pine trees.
(*b*) The same as part (*a*), and add 6.25% sale tax.
(*c*) The same as part (*b*) and round the total cost to the nearest dollar.

25. The resonant frequency f (in Hz) for the circuit shown is given by:

$$f = \frac{1}{2\pi}\sqrt{LC\frac{R_1^2C - L}{R_2^2C - L}}$$

Calculate the resonant frequency when $L = 0.2$ henrys, $R_1 = 1500$ ohms, $R_2 = 1500$ ohms, and $C = 2 \times 10^{-6}$ farads.

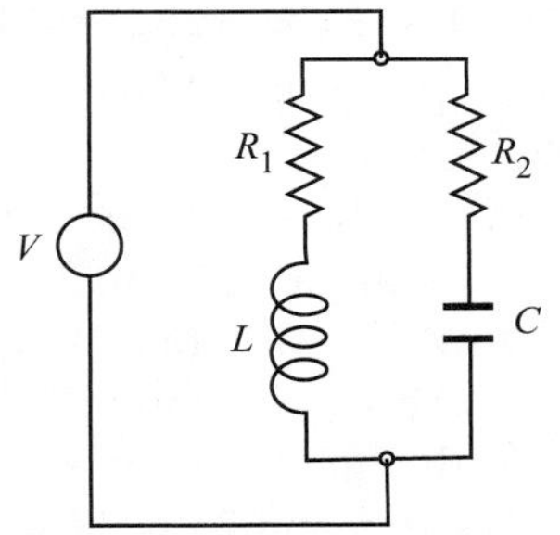

26. The number of combinations $C_{n,r}$ of taking r objects out of n objects is given by:

$$C_{n,r} = \frac{n!}{r!(n-r)!}$$

A deck of poker cards has 52 different cards. Determine how many different combinations are possible for selecting 5 cards from the deck. (Use the built-in function `factorial`.)

27. The formula for changing the base of a logarithm is:

$$\log_a N = \frac{\log_b N}{\log_b a}$$

(*a*) Use MATLAB's function `log(x)` to calculate $\log_4 0.085$.
(*b*) Use MATLAB's function `log10(x)` to calculate $\log_6 1500$.

28. The current I (in amps) t seconds after closing the switch in the circuit shown is:

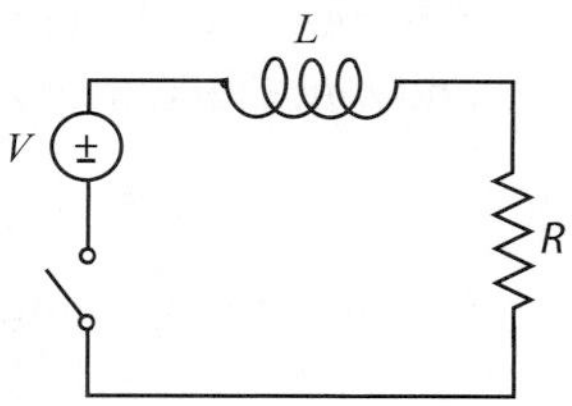

$$I = \frac{V}{R}(1 - e^{-(R/L)t})$$

Given $V = 120$ volts, $R = 240$ ohms, and $L = 0.5$ henrys, calculate the current 0.003 seconds after the switch is closed.

29. Radioactive decay of carbon-14 is used for estimating the age of organic material. The decay is modeled with the exponential function $f(t) = f(0)e^{kt}$, where t is time, $f(0)$ is the amount of material at $t = 0$, $f(t)$ is the amount of material at time t, and k is a constant. Carbon-14 has a half-life of approximately 5,730 years. A sample of paper taken from the Dead Sea Scrolls shows that 78.8% of the initial ($t = 0$) carbon-14 is present. Determine the estimated age of the scrolls. Solve the problem by writing a program in a script file. The program first determines the constant k, then calculates t for $f(t) = 0.788f(0)$, and finally rounds the answer to the nearest year.

30. Fractions can be added by using the smallest common denominator. For example, the smallest common denominator of 1/4 and 1/10 is 20. Use the MATLAB Help Window to find a MATLAB built-in function that determines the least common multiple of two numbers. Then use the function to show that the least common multiple of:
(*a*) 6 and 26 is 78.
(*b*) 6 and 34 is 102.

31. The Moment Magnitude Scale (MMS), denoted M_W, which is used to measure the size of an earthquake, is given by:

$$M_W = \frac{2}{3}\log_{10} M_0 - 10.7$$

where M_0 is the magnitude of the seismic moment in dyne-cm (measure of the energy released during an earthquake). Determine how many times more energy was released from the earthquake in Sumatra, Indonesia ($M_W = 8.5$), in 2007 than the earthquake in San Francisco, California ($M_W = 7.9$), in 1906.

32. According to special relativity, a rod of length L moving at velocity v will shorten by an amount δ, given by:

$$\delta = L\left(1 - \sqrt{1 - \frac{v^2}{c^2}}\right)$$

where c is the speed of light (about 300×10^6 m/s). Calculate how much a rod 2 meter long will contract when traveling at 5,000 m/s.

33. The monthly payment M of a loan amount P for y years and interest rate r can be calculated by the formula:

$$M = \frac{P(r/12)}{1-(1+r/12)^{-12y}}$$

(*a*) Calculate the monthly payment of a \$85,000 loan for 15 years and interest rate of 5.75% ($r = 0.0575$). Define the variables P, r, and y and use them to calculate M.

(*b*) Calculate the total amount needed for paying back the loan.

34. The balance B of a savings account after t years when a principal P is invested at an annual interest rate r and the interest is compounded yearly is given by $B = P(1+r)^t$. If the interest is compounded continuously, the balance is given by $B = Pe^{rt}$. An amount of \$40,000 is invested for 20 years in an account that pays 5.5% interest and the interest is compounded yearly. Use MATLAB to determine how many fewer days it will take to earn the same if the money is invested in an account where the interest is compounded continuously.

35. The temperature dependence of vapor pressure p can be estimated by the Anteing equation:

$$\ln(p) = A - \frac{B}{C+T}$$

where ln is the natural logarithm, p is in mm Hg, T is in kelvins, and A, B, and C are material constants. For toluene ($C_6H_5CH_3$) in the temperature range from 280 to 410 K the material constants are $A = 16.0137$, $B = 3096.52$, and $C = -53.67$. Calculate the vapor pressure of toluene at 315 and 405 K.

36. Sound level L_P in units of decibels (dB) is determined by:

$$L_P = 20\log_{10}\left(\frac{p}{p_0}\right)$$

where p is the sound pressure of the sound, and $p_0 = 20 \times 10^{-6}$ Pa is a reference sound pressure (the sound pressure when $L_P = 0$ dB).

(*a*) The sound pressure of a passing car is 80×10^{-2} Pa. Determine its sound level in decibels.

(*b*) The sound level of a jet engine is 110 decibels. By how many times is the sound pressure of the jet engine larger (louder) than the sound of the passing car?

37. Use the Help Window to find a display format that displays the output as a ratio of integers. For example, the number 3.125 will be displayed as 25/8. Change the display to this format and execute the following operations:

(*a*) $5/8 + 16/6$ (*b*) $1/3 - 11/13 + 2.7^2$

38. The steady-state heat conduction q from a cylindrical solid wall is determined by:

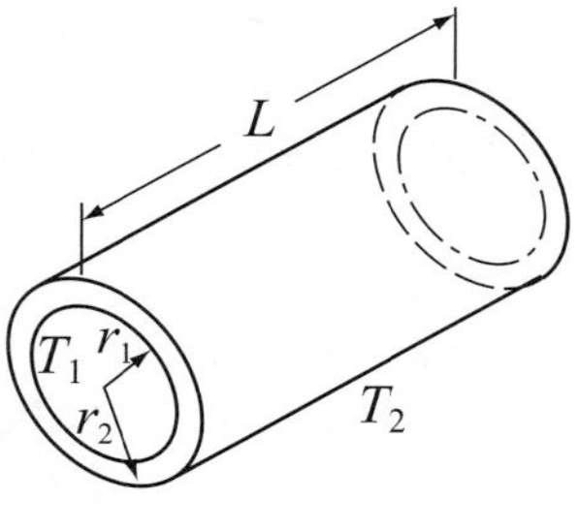

$$q = 2\pi Lk\frac{T_1 - T_2}{\ln\left(\frac{r_2}{r_1}\right)}$$

where k is the thermal conductivity. Calculate q for a copper tube ($k = 401$ Watts/°C/m) of length $L = 300$ cm with an outer radius of $r_2 = 5$ cm and an inner radius of $r_1 = 3$ cm. The external temperature is $T_2 = 20°$ C and the internal temperature is $T_1 = 100°$C.

39. Stirling's approximation for large factorials is given by:

$$n! = \sqrt{2\pi n}\left(\frac{n}{e}\right)^n$$

Use the formula for calculating 20!. Compare the result with the true value obtained with MATLAB's built-in function `factorial` by calculating the error ($Error = (TrueVal - ApproxVal)/TrueVal$).

40. A projectile is launched at an angle θ and speed of V_0. The projectile's travel time t_{travel}, maximum travel distance x_{max}, and maximum height h_{max} are given by:

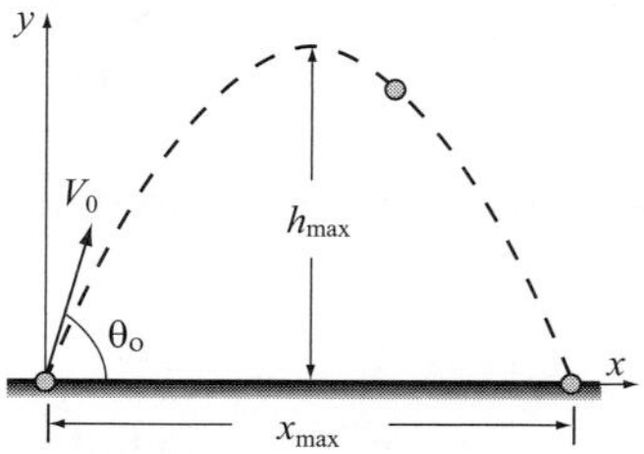

$$t_{travel} = 2\frac{V_0}{g}\sin\theta_0, \quad x_{max} = 2\frac{V_0^2}{g}\sin\theta_0\cos\theta_0,$$

$$h_{max} = 2\frac{V_0^2}{g}\sin^2\theta_0$$

Consider the case where $V_0 = 600$ ft/s and $\theta = 54°$. Define V_0 and θ as MATLAB variables and calculate t_{travel}, x_{max}, and h_{max} ($g = 32.2$ ft/s^2).

Chapter 2
Creating Arrays

The array is a fundamental form that MATLAB uses to store and manipulate data. An array is a list of numbers arranged in rows and/or columns. The simplest array (one-dimensional) is a row or a column of numbers. A more complex array (two-dimensional) is a collection of numbers arranged in rows and columns. One use of arrays is to store information and data, as in a table. In science and engineering, one-dimensional arrays frequently represent vectors, and two-dimensional arrays often represent matrices. This chapter shows how to create and address arrays, and Chapter 3 shows how to use arrays in mathematical operations. In addition to arrays made of numbers, arrays in MATLAB can also be a list of characters, which are called strings. Strings are discussed in Section 2.10.

2.1 Creating a One-Dimensional Array (Vector)

A one-dimensional array is a list of numbers arranged in a row or a column. One example is the representation of the position of a point in space in a three-dimensional Cartesian coordinate system. As shown in Figure 2-1, the position of point *A* is defined by a list of the three numbers 2, 4, and 5, which are the coordinates of the point.

The position of point *A* can be expressed in terms of a position vector:

$$\mathbf{r}_A = 2\mathbf{i} + 4\mathbf{j} + 5\mathbf{k}$$

where **i**, **j**, and **k** are unit vectors in the direction of the *x*, *y*, and *z* axes, respectively. The numbers 2, 4, and 5 can be used to define a row or a column vector.

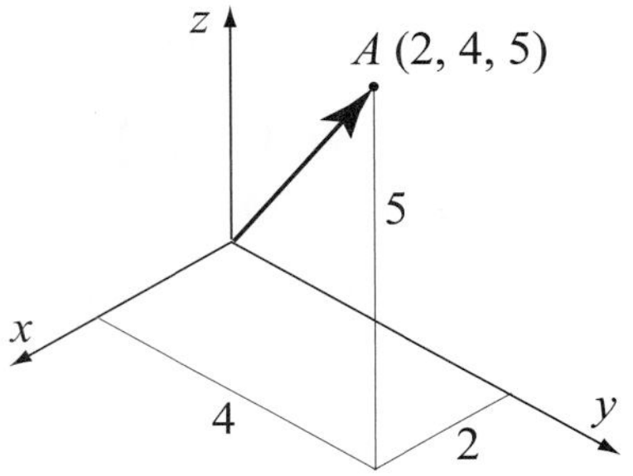

Figure 2-1: Position of a point.

Any list of numbers can be set up as a vector. For example, Table 2-1 contains population growth data that can be used to create two lists of numbers—one of the years and the other of the population values. Each list can be entered as elements in a vector with the numbers placed in a row or in a column.

Table 2-1: Population data

Year	1984	1986	1988	1990	1992	1994	1996
Population (millions)	127	130	136	145	158	178	211

In MATLAB, a vector is created by assigning the elements of the vector to a variable. This can be done in several ways depending on the source of the information that is used for the elements of the vector. When a vector contains specific numbers that are known (like the coordinates of point *A*), the value of each element is entered directly. Each element can also be a mathematical expression that can include predefined variables, numbers, and functions. Often, the elements of a row vector are a series of numbers with constant spacing. In such cases the vector can be created with MATLAB commands. A vector can also be created as the result of mathematical operations as explained in Chapter 3.

Creating a vector from a known list of numbers:

The vector is created by typing the elements (numbers) inside square brackets [ ].

```
variable_name = [ type vector elements ]
```

Row vector: To create a row vector type the elements with a space or a comma between the elements inside the square brackets.

Column vector: To create a column vector type the left square bracket [and then enter the elements with a semicolon between them, or press the **Enter** key after each element. Type the right square bracket] after the last element.

Tutorial 2-1 shows how the data from Table 2-1 and the coordinates of point *A* are used to create row and column vectors.

Tutorial 2-1: Creating vectors from given data.

```
>> yr=[1984 1986 1988 1990 1992 1994 1996]
```

The list of years is assigned to a row vector named `yr`.

```
yr =
    1984      1986      1988      1990      1992      1994      1996
>> pop=[127; 130; 136; 145; 158; 178; 211]
```

The population data is assigned to a column vector named `pop`.

```
pop =
   127
   130
   136
   145
   158
```

Tutorial 2-1: Creating vectors from given data. (Continued)

```
    178
    211
>> pntAH=[2,  4,  5]
pntAH =
     2     4     5
>> pntAV=[2
4
5]
pntAV =
     2
     4
     5
>>
```

The coordinates of point *A* are assigned to a row vector called `pntAH`.

The coordinates of point *A* are assigned to a column vector called `pntAV`. (The **Enter** key is pressed after each element is typed.)

Creating a vector with constant spacing by specifying the first term, the spacing, and the last term:

In a vector with constant spacing the difference between the elements is the same. For example, in the vector $v = 2\ \ 4\ \ 6\ \ 8\ \ 10$, the spacing between the elements is 2. A vector in which the first term is *m*, the spacing is *q*, and the last term is *n* is created by typing:

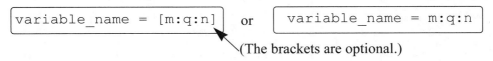

`variable_name = [m:q:n]` or `variable_name = m:q:n`

(The brackets are optional.)

Some examples are:

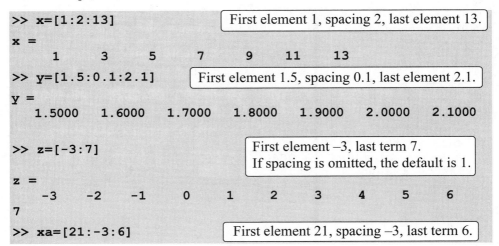

```
>> x=[1:2:13]
x =
     1     3     5     7     9    11    13
>> y=[1.5:0.1:2.1]
y =
    1.5000    1.6000    1.7000    1.8000    1.9000    2.0000    2.1000

>> z=[-3:7]

z =
    -3    -2    -1     0     1     2     3     4     5     6
7
>> xa=[21:-3:6]
```

First element 1, spacing 2, last element 13.

First element 1.5, spacing 0.1, last element 2.1.

First element –3, last term 7. If spacing is omitted, the default is 1.

First element 21, spacing –3, last term 6.

```
xa =
    21    18    15    12     9     6
>>
```

- If the numbers m, q, and n are such that the value of n cannot be obtained by adding q's to m, then (for positive n) the last element in the vector will be the last number that does not exceed n.
- If only two numbers (the first and the last terms) are typed (the spacing is omitted), then the default for the spacing is 1.

Creating a vector with linear (equal) spacing by specifying the first and last terms, and the number of terms:

A vector with *n* elements that are linearly (equally) spaced in which the first element is *xi* and the last element is *xf* can be created by typing the `linspace` command (MATLAB determines the correct spacing):

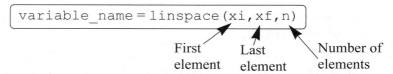

```
variable_name = linspace(xi,xf,n)
```

xi: First element; xf: Last element; n: Number of elements

When the number of elements is omitted, the default is 100. Some examples are:

```
>> va=linspace(0,8,6)
va =
         0    1.6000    3.2000    4.8000    6.4000    8.0000
>> vb=linspace(30,10,11)
vb =
    30    28    26    24    22    20    18    16    14    12    10
>> u=linspace(49.5,0.5)

u =
  Columns 1 through 10
   49.5000   49.0051   48.5101   48.0152   47.5202   47.0253
46.5303   46.0354   45.5404   45.0455
. . . . . . . . . . .
Columns 91 through 100
    4.9545    4.4596    3.9646    3.4697    2.9747    2.4798
1.9848    1.4899    0.9949    0.5000
>>
```

6 elements, first element 0, last element 8.

11 elements, first element 30, last element 10.

First element 49.5, last element 0.5.

When the number of elements is omitted, the default is 100.

100 elements are displayed.

2.2 CREATING A TWO-DIMENSIONAL ARRAY (MATRIX)

A two-dimensional array, also called a matrix, has numbers in rows and columns. Matrices can be used to store information like the arrangement in a table. Matrices play an important role in linear algebra and are used in science and engineering to describe many physical quantities.

In a square matrix the number of rows and the number of columns is equal. For example, the matrix

$$\begin{matrix} 7 & 4 & 9 \\ 3 & 8 & 1 \\ 6 & 5 & 3 \end{matrix} \qquad 3 \times 3 \text{ matrix}$$

is square, with three rows and three columns. In general, the number of rows and columns can be different. For example, the matrix:

$$\begin{matrix} 31 & 26 & 14 & 18 & 5 & 30 \\ 3 & 51 & 20 & 11 & 43 & 65 \\ 28 & 6 & 15 & 61 & 34 & 22 \\ 14 & 58 & 6 & 36 & 93 & 7 \end{matrix} \qquad 4 \times 6 \text{ matrix}$$

has four rows and six columns. A $m \times n$ matrix has m rows and n columns, and m by n is called the size of the matrix.

A matrix is created by assigning the elements of the matrix to a variable. This is done by typing the elements, row by row, inside square brackets [ ]. First type the left bracket [then type the first row, separating the elements with spaces or commas. To type the next row type a semicolon or press **Enter**. Type the right bracket] at the end of the last row.

```
variable_name=[1st row elements; 2nd row elements; 3rd
                    row elements; ... ; last row elements]
```

The elements that are entered can be numbers or mathematical expressions that may include numbers, predefined variables, and functions. *All the rows must have the same number of elements*. If an element is zero, it has to be entered as such. MATLAB displays an error message if an attempt is made to define an incomplete matrix. Examples of matrices defined in different ways are shown in Tutorial 2-2.

Tutorial 2-2: Creating matrices.

```
>> a=[5  35  43;  4  76  81;  21  32  40]
a =
     5    35    43
     4    76    81
    21    32    40
>> b = [7  2  76  33  8
1  98  6  25  6
5  54  68  9  0]
```

A semicolon is typed before a new line is entered.

The **Enter** key is pressed before a new line is entered.

Tutorial 2-2: Creating matrices. (Continued)

```
b =
     7     2    76    33     8
     1    98     6    25     6
     5    54    68     9     0
>> cd=6; e=3; h=4;
>> Mat=[e, cd*h, cos(pi/3); h^2, sqrt(h*h/cd), 14]
Mat =
    3.0000   24.0000    0.5000
   16.0000    1.6330   14.0000
>>
```

Three variables are defined.

Elements are defined by mathematical expressions.

Rows of a matrix can also be entered as vectors using the notation for creating vectors with constant spacing, or the `linspace` command. For example:

```
>> A=[1:2:11; 0:5:25; linspace(10,60,6); 67 2 43 68 4 13]
A =
     1     3     5     7     9    11
     0     5    10    15    20    25
    10    20    30    40    50    60
    67     2    43    68     4    13
>>
```

In this example the first two rows were entered as vectors using the notation of constant spacing, the third row was entered using the `linspace` command, and in the last row the elements were entered individually.

2.2.1 The `zeros`, `ones` and, `eye` Commands

The `zeros(m,n)`, `ones(m,n)`, and `eye(n)` commands can be used to create matrices that have elements with special values. The `zeros(m,n)` and the `ones(m,n)` commands create a matrix with *m* rows and *n* columns in which all elements are the numbers 0 and 1, respectively. The `eye(n)` command creates a square matrix with *n* rows and *n* columns in which the diagonal elements are equal to 1 and the rest of the elements are 0. This matrix is called the identity matrix. Examples are:

```
>> zr=zeros(3,4)
zr =
     0     0     0     0
     0     0     0     0
     0     0     0     0
>> ne=ones(4,3)
```

```
ne =
     1     1     1
     1     1     1
     1     1     1
     1     1     1
>> idn=eye(5)
idn =
     1     0     0     0     0
     0     1     0     0     0
     0     0     1     0     0
     0     0     0     1     0
     0     0     0     0     1
>>
```

Matrices can also be created as a result of mathematical operations with vectors and matrices. This topic is covered in Chapter 3.

2.3 NOTES ABOUT VARIABLES IN MATLAB

- All variables in MATLAB are arrays. A scalar is an array with one element, a vector is an array with one row or one column of elements, and a matrix is an array with elements in rows and columns.
- The variable (scalar, vector, or matrix) is defined by the input when the variable is assigned. There is no need to define the size of the array (single element for a scalar, a row or a column of elements for a vector, or a two-dimensional array of elements for a matrix) before the elements are assigned.
- Once a variable exists—as a scalar, vector, or matrix—it can be changed to any other size, or type, of variable. For example, a scalar can be changed to a vector or a matrix; a vector can be changed to a scalar, a vector of different length, or a matrix; and a matrix can be changed to have a different size, or be reduced to a vector or a scalar. These changes are made by adding or deleting elements. This subject is covered in Sections 2.7 and 2.8.

2.4 THE TRANSPOSE OPERATOR

The transpose operator, when applied to a vector, switches a row (column) vector to a column (row) vector. When applied to a matrix, it switches the rows (columns) to columns (rows). The transpose operator is applied by typing a single quote `'` following the variable to be transposed. Examples are:

```
>> aa=[3  8  1]
aa =
     3     8     1
>> bb=aa'
```

Define a row vector `aa`.

Define a column vector `bb` as the transpose of vector `aa`.

```
bb =
     3
     8
     1
>> C=[2 55 14 8; 21 5 32 11; 41 64 9 1]
C =
     2    55    14     8
    21     5    32    11
    41    64     9     1
>> D=C'
D =
     2    21    41
    55     5    64
    14    32     9
     8    11     1
>>
```

Define a matrix `C` with 3 rows and 4 columns.

Define a matrix `D` as the transpose of matrix `C`. (`D` has 4 rows and 3 columns.)

2.5 *ARRAY ADDRESSING*

Elements in an array (either vector or matrix) can be addressed individually or in subgroups. This is useful when there is a need to redefine only some of the elements, when specific elements are to be used in calculations, or when a subgroup of the elements is used to define a new variable.

2.5.1 *Vector*

The address of an element in a vector is its position in the row (or column). For a vector named `ve`, `ve(k)` refers to the element in position *k*. The first position is 1. For example, if the vector *ve* has nine elements:

ve = 35 46 78 23 5 14 81 3 55

then

ve(4) = 23, *ve*(7) = 81, and *ve*(1) = 35.

A single vector element, *v*(*k*), can be used just as a variable. For example, it is possible to change the value of only one element of a vector by assigning a new value to a specific address. This is done by typing: *v*(*k*) = *value*. A single element can also be used as a variable in a mathematical expression. Examples are:

```
>> VCT=[35 46 78 23 5 14 81 3 55]
VCT =
    35    46    78    23     5    14    81     3    55
>> VCT(4)
```

Define a vector.

Display the fourth element.

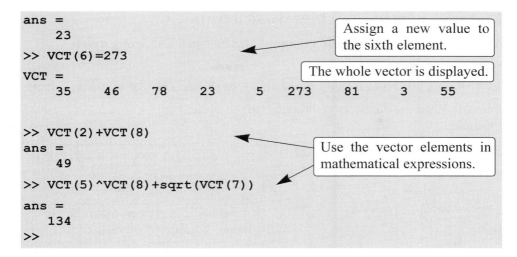

```
ans =
    23
>> VCT(6)=273
VCT =
    35    46    78    23     5   273    81     3    55

>> VCT(2)+VCT(8)
ans =
    49
>> VCT(5)^VCT(8)+sqrt(VCT(7))
ans =
   134
>>
```

Assign a new value to the sixth element.

The whole vector is displayed.

Use the vector elements in mathematical expressions.

2.5.2 Matrix

The address of an element in a matrix is its position, defined by the row number and the column number where it is located. For a matrix assigned to a variable *ma*, *ma*(*k*,*p*) refers to the element in row *k* and column *p*.

For example, if the matrix is: $ma = \begin{bmatrix} 3 & 11 & 6 & 5 \\ 4 & 7 & 10 & 2 \\ 13 & 9 & 0 & 8 \end{bmatrix}$

then *ma*(1,1) = 3 and *ma*(2,3) = 10.

As with vectors, it is possible to change the value of just one element of a matrix by assigning a new value to that element. Also, single elements can be used like variables in mathematical expressions and functions. Some examples are:

```
>> MAT=[3 11 6 5; 4 7 10 2; 13 9 0 8]
MAT =
     3    11     6     5
     4     7    10     2
    13     9     0     8
>> MAT(3,1)=20
MAT =
     3    11     6     5
     4     7    10     2
    20     9     0     8
>> MAT(2,4)-MAT(1,2)
ans =
    -9
```

Create a 3 × 4 matrix.

Assign a new value to the (3,1) element.

Use elements in a mathematical expression.

2.6 *Using a Colon : in Addressing Arrays*

A colon can be used to address a range of elements in a vector or a matrix.

For a vector:

va(:) Refers to all the elements of the vector *va* (either a row or a column vector).

va(*m*:*n*) Refers to elements *m* through *n* of the vector *va*.
Example:

```
>> v=[4 15 8 12 34 2 50 23 11]
v =
     4    15     8    12    34     2    50    23    11
>> u=v(3:7)
u =
     8    12    34     2    50
>>
```

A vector v is created.

A vector u is created from the elements 3 through 7 of vector v.

For a matrix:

A(:,*n*) Refers to the elements in all the rows of column *n* of the matrix *A*.

A(*n*,:) Refers to the elements in all the columns of row *n* of the matrix *A*.

A(:,*m*:*n*) Refers to the elements in all the rows between columns *m* and *n* of the matrix *A*.

A(*m*:*n*,:) Refers to the elements in all the columns between rows *m* and *n* of the matrix *A*.

A(*m*:*n*,*p*:*q*) Refers to the elements in rows *m* through *n* and columns *p* through *q* of the matrix *A*.

The use of the colon symbol in addressing elements of matrices is demonstrated in Tutorial 2-3.

Tutorial 2-3: Using a colon in addressing arrays.

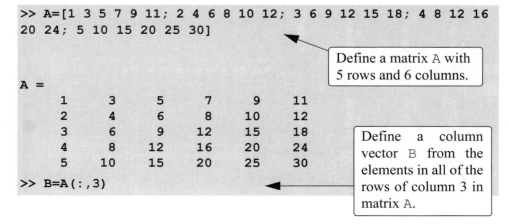

```
>> A=[1 3 5 7 9 11; 2 4 6 8 10 12; 3 6 9 12 15 18; 4 8 12 16
20 24; 5 10 15 20 25 30]

A =
     1     3     5     7     9    11
     2     4     6     8    10    12
     3     6     9    12    15    18
     4     8    12    16    20    24
     5    10    15    20    25    30
>> B=A(:,3)
```

Define a matrix A with 5 rows and 6 columns.

Define a column vector B from the elements in all of the rows of column 3 in matrix A.

Tutorial 2-3: Using a colon in addressing arrays. (Continued)

```
B =
     5
     6
     9
    12
    15
>> C=A(2,:)
C =
     2     4     6     8    10    12
>> E=A(2:4,:)
E =
     2     4     6     8    10    12
     3     6     9    12    15    18
     4     8    12    16    20    24
>> F=A(1:3,2:4)
F =
     3     5     7
     4     6     8
     6     9    12
>>
```

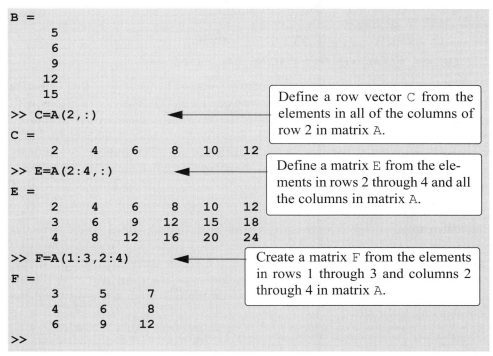

Define a row vector C from the elements in all of the columns of row 2 in matrix A.

Define a matrix E from the elements in rows 2 through 4 and all the columns in matrix A.

Create a matrix F from the elements in rows 1 through 3 and columns 2 through 4 in matrix A.

In Tutorial 2-3 new vectors and matrices are created from existing ones by using a range of elements, or a range of rows and columns (using :). It is possible, however, to select only specific elements, or specific rows and columns of existing variables to create new variables. This is done by typing the selected elements or rows or columns inside brackets, as shown below:

```
>> v=4:3:34
v =
    4     7    10    13    16    19    22    25    28    31    34
>> u=v([3, 5, 7:10])
u =
   10    16    22    25    28    31
>> A=[10:-1:4; ones(1,7); 2:2:14; zeros(1,7)]
A =
   10     9     8     7     6     5     4
    1     1     1     1     1     1     1
    2     4     6     8    10    12    14
    0     0     0     0     0     0     0
>> B = A([1,3],[1,3,5:7])
```

Create a vector v with 11 elements.

Create a vector u from the 3rd, the 5th, and the 7th through 10th elements of v.

Create a 4 × 7 matrix A.

Create a matrix B from the 1st and 3rd rows, and 1st, 3rd, and the 5th through 7th columns of A.

```
B =
    10     8     6     5     4
     2     6    10    12    14
```

2.7 *Adding Elements to Existing Variables*

A variable that exists as a vector, or a matrix, can be changed by adding elements to it (remember that a scalar is a vector with one element). A vector (a matrix with a single row or column) can be changed to have more elements, or it can be changed to be a two-dimensional matrix. Rows and/or columns can also be added to an existing matrix to obtain a matrix of different size. The addition of elements can be done by simply assigning values to the additional elements, or by appending existing variables.

Adding elements to a vector:

Elements can be added to an existing vector by assigning values to the new elements. For example, if a vector has 4 elements, the vector can be made longer by assigning values to elements 5, 6, and so on. If a vector has n elements and a new value is assigned to an element with an address of $n+2$ or larger, MATLAB assigns zeros to the elements that are between the last original element and the new element. Examples:

```
>> DF=1:4                          Define vector DF with 4 elements.
DF =
     1     2     3     4
>> DF(5:10)=10:5:35                Adding 6 elements starting with the 5th.
DF =
     1     2     3     4    10    15    20    25    30    35
>> AD=[5  7  2]                    Define vector AD with 3 elements.
AD =
     5     7     2
>> AD(8)=4                         Assign a value to the 8th element.
AD =                               MATLAB assigns zeros to
     5   7   2   0   0   0   0   4 the 4th through 7th elements.
>> AR(5)=24                        Assign a value to the 5th element of a new vector.
AR =                               MATLAB assigns zeros to the
     0     0     0     0    24     1st through 4th elements.
>>
```

Elements can also be added to a vector by appending existing vectors. Two examples are:

```
>> RE=[3  8 1  24];                Define vector RE with 4 elements.
```

```
>> GT=4:3:16;
>> KNH=[RE  GT]
KNH =
    3    8    1   24    4    7   10   13   16
>> KNV=[RE'; GT']
KNV =
     3
     8
     1
    24
     4
     7
    10
    13
    16
```

Define vector GT with 5 elements.

Define a new vector KNH by appending RE and GT.

Create a new column vector KNV by appending RE′ and GT′.

Adding elements to a matrix:

Rows and/or columns can be added to an existing matrix by assigning values to the new rows or columns. This can be done by assigning new values, or by appending existing variables. This must be done carefully since the size of the added rows or columns must fit the existing matrix. Examples are:

```
>> E=[1 2 3 4; 5 6 7 8]
E =
     1     2     3     4
     5     6     7     8
>> E(3,:)=[10:4:22]

E =
     1     2     3     4
     5     6     7     8
    10    14    18    22
>> K=eye(3)
K =
     1     0     0
     0     1     0
     0     0     1
>> G=[E  K]
G =
     1     2     3     4     1     0     0
     5     6     7     8     0     1     0
    10    14    18    22     0     0     1
```

Define a 2×4 matrix E.

Add the vector 10 14 18 22 as the third row of E.

Define a 3×3 matrix K.

Append matrix K to matrix E. The numbers of rows in E and K must be the same.

If a matrix has a size of $m \times n$, and a new value is assigned to an element with an address beyond the size of the matrix, MATLAB increases the size of the matrix to include the new element. Zeros are assigned to the other elements that are added. Examples:

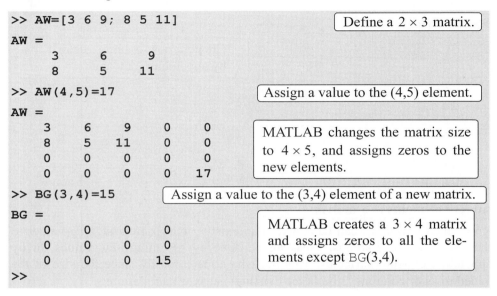

```
>> AW=[3 6 9; 8 5 11]                    Define a 2 × 3 matrix.
AW =
     3     6     9
     8     5    11
>> AW(4,5)=17                            Assign a value to the (4,5) element.
AW =
    3    6    9    0    0               MATLAB changes the matrix size
    8    5   11    0    0               to 4 × 5, and assigns zeros to the
    0    0    0    0    0               new elements.
    0    0    0    0   17
>> BG(3,4)=15                  Assign a value to the (3,4) element of a new matrix.
BG =
    0    0    0    0                    MATLAB creates a 3 × 4 matrix
    0    0    0    0                    and assigns zeros to all the ele-
    0    0    0   15                    ments except BG(3,4).
>>
```

2.8 *Deleting Elements*

An element, or a range of elements, of an existing variable can be deleted by reassigning nothing to these elements. This is done by using square brackets with nothing typed in between them. By deleting elements a vector can be made shorter and a matrix can be made to have a smaller size. Examples are:

```
>> kt=[2 8 40 65 3 55 23 15 75 80]         Define a vector
kt =                                         with 10 elements.
    2    8   40   65    3   55   23   15   75   80
>> kt(6)=[]            <----------          Eliminate the 6th element.
kt =
    2    8  40  65   3  23  15  75  80       The vector now
                                             has 9 elements.
>> kt(3:6)=[]        <----------            Eliminate elements 3 through 6.
kt =
    2    8   15   75   80                   The vector now has 5 elements.
>> mtr=[5 78 4 24 9; 4 0 36 60 12; 56 13 5 89 3]

                                             Define a 3 × 5 matrix.
```

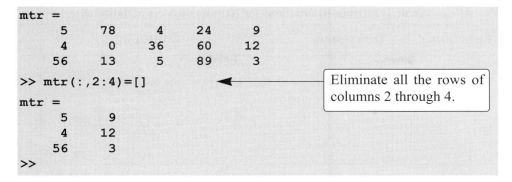

```
mtr =
     5    78     4    24     9
     4     0    36    60    12
    56    13     5    89     3
>> mtr(:,2:4)=[]
mtr =
     5     9
     4    12
    56     3
>>
```

Eliminate all the rows of columns 2 through 4.

2.9 Built-in Functions for Handling Arrays

MATLAB has many built-in functions for managing and handling arrays. Some of these are listed below:

Table 2-2: Built-in functions for handling arrays

Function	Description	Example

`length(A)`	Returns the number of elements in the vector `A`.	`>> A=[5 9 2 4];`
		`>> length(A)`
		`ans =`
		`4`

`size(A)`	Returns a row vector `[m,n]`, where `m` and `n` are the size $m \times n$ of the array `A`.	`>> A=[6 1 4 0 12; 5 19 6 8 2]`
		`A =`
		`6 1 4 0 12`
		`5 19 6 8 2`
		`>> size(A)`
		`ans =`
		`2 5`

`reshape(A, m,n)`	Creates a `m` by `n` matrix from the elements of matrix `A`. The elements are taken column after column. Matrix `A` must have `m` times `n` elements.	`>> A=[5 1 6; 8 0 2]`
		`A =`
		`5 1 6`
		`8 0 2`
		`>> B = reshape(A,3,2)`
		`B =`
		`5 0`
		`8 6`
		`1 2`

Table 2-2: Built-in functions for handling arrays (Continued)

Function	Description	Example

`diag(v)`	When `v` is a vector, creates a square matrix with the elements of `v` in the diagonal.	`>> v=[7 4 2];`
		`>> A=diag(v)`
		`A =`
		`7 0 0`
		`0 4 0`
		`0 0 2`

`diag(A)`	When `A` is a matrix, creates a vector from the diagonal elements of `A`.	`>> A=[1 2 3; 4 5 6; 7 8 9]`
		`A =`
		`1 2 3`
		`4 5 6`
		`7 8 9`
		`>> vec=diag(A)`
		`vec =`
		`1`
		`5`
		`9`

Additional built-in functions for manipulation of arrays are described in the Help Window. In this window select "Functions by Category," then "Mathematics," and then "Arrays and Matrices."

Sample Problem 2-1: Create a matrix

Using the ones and zeros commands, create a 4×5 matrix in which the first two rows are 0s and the next two rows are 1s.

Solution

```
>> A(1:2,:)=zeros(2,5)
A =
     0     0     0     0     0
     0     0     0     0     0
>> A(3:4,:)=ones(2,5)
A =
     0     0     0     0     0
     0     0     0     0     0
     1     1     1     1     1
     1     1     1     1     1
```

First, create a 2×5 matrix with 0s.

Add rows 3 and 4 with 1s.

A different solution to the problem is:

```
>> A=[zeros(2,5);ones(2,5)]
A =
     0     0     0     0     0
     0     0     0     0     0
     1     1     1     1     1
     1     1     1     1     1
```

Create a 4×5 matrix from two 2×5 matrices.

Sample Problem 2-2: Create a matrix

Create a 6×6 matrix in which the middle two rows and the middle two columns are 1s, and the rest of the entries are 0s.

Solution

```
>> AR=zeros(6,6)
AR =
     0     0     0     0     0     0
     0     0     0     0     0     0
     0     0     0     0     0     0
     0     0     0     0     0     0
     0     0     0     0     0     0
     0     0     0     0     0     0
>> AR(3:4,:)=ones(2,6)
AR =
     0     0     0     0     0     0
     0     0     0     0     0     0
     1     1     1     1     1     1
     1     1     1     1     1     1
     0     0     0     0     0     0
     0     0     0     0     0     0
>> AR(:,3:4)=ones(6,2)
AR =
     0     0     1     1     0     0
     0     0     1     1     0     0
     1     1     1     1     1     1
     1     1     1     1     1     1
     0     0     1     1     0     0
     0     0     1     1     0     0
```

First, create a 6×6 matrix with 0s.

Reassign the number 1 to the 3rd and 4th rows.

Reassign the number 1 to the 3rd and 4th columns.

Sample Problem 2-3: Matrix manipulation

Given are a 5×6 matrix A, a 3×6 matrix B, and a 9-element vector v.

$$A = \begin{bmatrix} 2 & 5 & 8 & 11 & 14 & 17 \\ 3 & 6 & 9 & 12 & 15 & 18 \\ 4 & 7 & 10 & 13 & 16 & 19 \\ 5 & 8 & 11 & 14 & 17 & 20 \\ 6 & 9 & 12 & 15 & 18 & 21 \end{bmatrix} \qquad B = \begin{bmatrix} 5 & 10 & 15 & 20 & 25 & 30 \\ 30 & 35 & 40 & 45 & 50 & 55 \\ 55 & 60 & 65 & 70 & 75 & 80 \end{bmatrix}$$

$$v = \begin{bmatrix} 99 & 98 & 97 & 96 & 95 & 94 & 93 & 92 & 91 \end{bmatrix}$$

Create the three arrays in the Command Window, and then, by writing one command, replace the last four columns of the first and third rows of A with the first four columns of the first two rows of B, the last four columns of the fourth row of A with the elements 5 through 8 of v, and the last four columns of the fifth row of A with columns 3 through 5 of the third row of B.

Solution

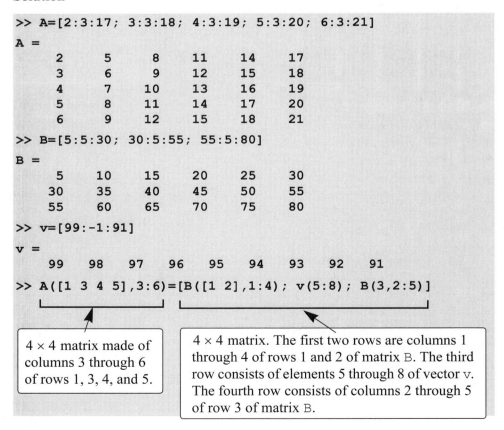

```
>> A=[2:3:17; 3:3:18; 4:3:19; 5:3:20; 6:3:21]
A =
     2     5     8    11    14    17
     3     6     9    12    15    18
     4     7    10    13    16    19
     5     8    11    14    17    20
     6     9    12    15    18    21
>> B=[5:5:30; 30:5:55; 55:5:80]
B =
     5    10    15    20    25    30
    30    35    40    45    50    55
    55    60    65    70    75    80
>> v=[99:-1:91]
v =
    99    98    97    96    95    94    93    92    91
>> A([1 3 4 5],3:6)=[B([1 2],1:4); v(5:8); B(3,2:5)]
```

4 × 4 matrix made of columns 3 through 6 of rows 1, 3, 4, and 5.

4 × 4 matrix. The first two rows are columns 1 through 4 of rows 1 and 2 of matrix `B`. The third row consists of elements 5 through 8 of vector `v`. The fourth row consists of columns 2 through 5 of row 3 of matrix `B`.

```
A =
     2     5     5    10    15    20
     3     6     9    12    15    18
     4     7    30    35    40    45
     5     8    95    94    93    92
     6     9    60    65    70    75
```

2.10 *STRINGS AND STRINGS AS VARIABLES*

- A string is an array of characters. It is created by typing the characters within single quotes.
- Strings can include letters, digits, other symbols, and spaces.
- Examples of strings: 'ad ef ', '3%fr2', '{edcba:21!', 'MATLAB'.
- A string that contains a single quote is created by typing two single quotes within the string.
- When a string is being typed in, the color of the text on the screen changes to maroon when the first single quote is typed. When the single quote at the end of the string is typed, the color of the string changes to purple.

Strings have several different uses in MATLAB. They are used in output commands to display text messages (Chapter 4), in formatting commands of plots (Chapter 5), and as input arguments of some functions (Chapter 7). More details are given in these chapters when strings are used for these purposes.

- When strings are being used in formatting plots (labels to axes, title, and text notes), characters within the string can be formatted to have a specified font, size, position (uppercase, lowercase), color, etc. See Chapter 5 for details.

Strings can also be assigned to variables by simply typing the string on the right side of the assignment operator, as shown in the examples below:

```
>> a='FRty 8'
a =
FRty 8
>> B='My name is John Smith'
B =
My name is John Smith
>>
```

When a variable is defined as a string, the characters of the string are stored in an array just as numbers are. Each character, including a space, is an element in the array. This means that a one-line string is a row vector in which the number of elements is equal to the number of characters. The elements of the vectors are addressed by position. For example, in the vector B that was defined above the 4th element is the letter n, the 12th element is J, and so on.

```
>> B(4)
ans =
n
>> B(12)
ans =
J
```

As with a vector that contains numbers, it is also possible to change specific elements by addressing them directly. For example, in the vector B above the name John can be changed to Bill by:

```
>> B(12:15)='Bill'
B =
My name is Bill Smith
>>
```

Using a colon to assign new characters to elements 12 through 15 in the vector B.

Strings can also be placed in a matrix. As with numbers, this is done by typing a semicolon ; (or pressing the **Enter** key) at the end of each row. Each row must be typed as a string, which means that it must be enclosed in single quotes. In addition, as with a numerical matrix, all rows must have the same number of elements. This requirement can cause problems when the intention is to create rows with specific wording. Rows can be made to have the same number of elements by adding spaces.

MATLAB has a built-in function named `char` that creates an array with rows having the same number of characters from an input of rows not all of the same length. MATLAB makes the length of all the rows equal to that of the longest row by adding spaces at the end of the short lines. In the `char` function, the rows are entered as strings separated by a comma according to the following format:

```
variable_name=char('string 1','string 2','string 3')
```

For example:

```
>> Info=char('Student Name:','John Smith','Grade:','A+')
Info =
Student Name:
John Smith
Grade:
A+
>>
```

A variable named `Info` is assigned four rows of strings, each with different length.

The function `char` creates an array with four rows with the same length as the longest row by adding empty spaces to the shorter lines.

A variable can be defined as either a number or a string made up of the same digits. For example, as shown below, x is defined to be the number 536, and y is defined to be a string made up of the digits 536.

```
>> x=536
x =
   536
>> y='536'
y =
536
>>
```

The two variables are not the same even though they appear identical on the screen. Note that the characters 536 in the line below the x= are indented, while the characters 536 in the line below the y= are not indented. The variable x can be used in mathematical expressions, while the variable y cannot.

2.11 PROBLEMS

1. Create a row vector that has the following elements: 3, $4 \cdot 2.55$, $68/16$, 45, $\sqrt[3]{110}$, $\cos 25°$, and 0.05.

2. Create a row vector that has the following elements: $\frac{54}{3 + 4.2^2}$, 32, $6.3^2 - 7.2^2$, 54, $e^{3.7}$, and $\sin 66° + \cos\frac{3\pi}{8}$.

3. Create a column vector that has the following elements: 25.5, $\frac{(14\tan 58°)}{(2.1^2 + 11)}$, $6!$, 2.7^4, 0.0375, and $\pi/5$.

4. Create a column vector that has the following elements: $\frac{32}{3.2^2}$, $\sin^2 35°$, 6.1, $\ln 29^2$, 0.00552, $\ln^2 29$, and 133.

5. Define the variables $x = 0.85$, $y = 12.5$, and then use them to create a column vector that has the following elements: y, y^x, $\ln(y/x)$, $y \cdot x$, and $x + y$.

6. Define the variables $a = 3.5$, $b = -6.4$, and then use them to create a row vector that has the following elements: a, a^2, a/b, $a \cdot b$, and $\sqrt{a}$.

7. Create a row vector in which the first element is 2 and the last element is 37, with an increment of 5 between the elements (2, 7, 12, … , 37).

8. Create a row vector with 9 equally spaced elements in which the first element is 81 and the last element is 12.

9. Create a column vector in which the first element is 22.5, the elements decrease with increments of –2.5, and the last element is 0. (A column vector can be created by the transpose of a row vector.)

10. Create a column vector with 15 equally spaced elements in which the first element is –21 and the last element is 12.

11. Using the colon symbol, create a row vector (assign it to a variable named `same`) with seven elements that are all –3.

12. Use a single command to create a row vector (assign it to a variable named `a`) with 9 elements such that the last element is 7.5 and the rest of the elements are 0s. Do not type the vector explicitly.

13. Use a single command to create a row vector (assign it to a variable named `b`) with 19 elements such that
    `b = 1 2 3 4 5 6 7 8 9 10 9 8 7 6 5 4 3 2 1`
    Do not type the vector explicitly.

14. Create a vector (name it `vecA`) that has 14 elements of which the first is 49, the increment is –3, and the last element is 10. Then, using the colon symbol, create a new vector (call it `vecB`) that has 8 elements. The first 4 elements are the first 4 elements of the vector `vecA`, and the last 4 are the last 4 elements of the vector `vecA`.

15. Create a vector (name it `vecC`) that has 16 elements of which the first is 13, the increment is 4 and the last element is 73. Then create the following two vectors:
    (*a*) A vector (name it `Codd`) that contains all the elements with odd index of `vecCodd` (*vecCodd*(1), *vecCodd*(3), etc; i.e., `Codd = 13 21 29 ... 69`).
    (*b*) A vector (name it `Ceven`) that contains all the elements with even index of `vecCodd` (*vecCodd*(2), *vecCodd*(4), etc; i.e., `Codd = 17 25 33 ... 73`).
    In both parts use vectors of odd and even numbers for the index of `Codd` and `Ceven`, respectively. Do not type the vectors explicitly.

16. Create the following matrix by using vector notation for creating vectors with constant spacing and/or the `linspace` command. Do not type individual elements explicitly.

$$A = \begin{bmatrix} 0 & 5 & 10 & 15 & 20 & 25 & 30 \\ 600 & 500 & 400 & 300 & 200 & 100 & 0 \\ 0 & 0.8333 & 1.6667 & 2.5 & 3.3333 & 4.1667 & 5 \end{bmatrix}$$

17. Create the following matrix by using vector notation for creating vectors with constant spacing and/or the `linspace` command. Do not type individual elements explicitly.

$$B = \begin{bmatrix} 1 & 0 & 3 \\ 2 & 0 & 3 \\ 3 & 0 & 3 \\ 4 & 0 & 3 \\ 5 & 0 & 3 \end{bmatrix}$$

18. Using the colon symbol, create a 4×6 matrix (assign it to a variable named `Anine`) in which all the elements are the number 9.

19. Create the following matrix by typing one command. Do not type individual elements explicitly.

$$C = \begin{bmatrix} 0 & 0 & 0 & 0 & 0 \\ 0 & 0 & 0 & 0 & 0 \\ 0 & 0 & 0 & 0 & 8 \end{bmatrix}$$

20. Create the following matrix by typing one command. Do not type individual elements explicitly.

$$D = \begin{bmatrix} 0 & 0 & 0 & 0 & 0 \\ 0 & 0 & 0 & 6 & 6 \\ 0 & 0 & 0 & 6 & 6 \end{bmatrix}$$

21. Create the following matrix by typing one command. Do not type individual elements explicitly.

$$E = \begin{bmatrix} 0 & 0 & 0 & 0 & 0 \\ 0 & 0 & 1 & 2 & 3 \\ 0 & 0 & 4 & 5 & 6 \\ 0 & 0 & 7 & 8 & 9 \end{bmatrix}$$

22. Create the following matrix by typing one command. Do not type individual elements explicitly.

$$F = \begin{bmatrix} 0 & 0 & 0 & 0 & 0 \\ 0 & 0 & 1 & 10 & 20 \\ 0 & 0 & 2 & 8 & 26 \\ 0 & 0 & 3 & 6 & 32 \end{bmatrix}$$

23. Create three row vectors:

$$a = \begin{bmatrix} 7 & 2 & -3 & 1 & 0 \end{bmatrix},\ b = \begin{bmatrix} -3 & 10 & 0 & 7 & -2 \end{bmatrix},\ c = \begin{bmatrix} 1 & 0 & 4 & -6 & 5 \end{bmatrix}$$

(*a*) Use the three vectors in a MATLAB command to create a 3×5 matrix in which the rows are the vectors *a*, *b*, and *c*.

(*b*) Use the three vectors in a MATLAB command to create a 5×3 matrix in which the columns are the vectors *a*, *b*, and *c*.

24. Create three row vectors:

$$a = \begin{bmatrix} 7 & 2 & -3 & 1 & 0 \end{bmatrix},\ b = \begin{bmatrix} -3 & 10 & 0 & 7 & -2 \end{bmatrix},\ c = \begin{bmatrix} 1 & 0 & 4 & -6 & 5 \end{bmatrix}$$

(*a*) Use the three vectors in a MATLAB command to create a 3×3 matrix such that the first, second, and third rows consist of the first three elements of the vectors *a*, *b*, and *c*, respectively.

(*b*) Use the three vectors in a MATLAB command to create a 3×3 matrix such that the first, second, and third columns consist of the last three elements of the vectors *a*, *b*, and *c*, respectively.

25. Create two row vectors:

$$a = \begin{bmatrix} -4 & 10 & 0.5 & 1.8 & -2.3 & 7 \end{bmatrix},\ b = \begin{bmatrix} 0.7 & 9 & -5 & 3 & -0.6 & 12 \end{bmatrix}$$

(*a*) Use the two vectors in a MATLAB command to create a 2×4 matrix such that the first row consists of elements 2 through 5 of vector *a*, and the second row consists of elements 3 through 6 of vector *b*.

(*b*) Use the two vectors in a MATLAB command to create a 3×4 matrix such that the first column consists of elements 2 through 4 of vector *a*, the second column consists of elements 4 through 6 of vector *a*, the third column consists of elements 1 through 3 of vector *b*, and the fourth column consists of elements 3 through 5 of vector *b*.

26. By hand (pencil and paper) write what will be displayed if the following commands are executed by MATLAB. Check your answers by executing the commands with MATLAB. (Parts (*b*), (*c*), and (*d*) use the vector that was defined in part (*a*).)

(*a*) `a=9:-3:0` (*b*) `b=[a a]` or `b=[a,a]` (*c*) `c=[a;a]`
(*d*) `d=[a' a']` or `d=[a',a']` (*e*) `e=[[a; a; a; a] a']`

27. The following vector is defined in MATLAB:

$$v = \begin{bmatrix} 15 & 0 & 6 & -2 & 3 & -5 & 4 & 9 & 1.8 & -0.35 & 7 \end{bmatrix}$$

By hand (pencil and paper) write what will be displayed if the following commands are executed by MATLAB. Check your answers by executing the commands with MATLAB.

(*a*) `a=v(2:5)` (*b*) `b=v([1,3:7,11])` (*c*) `c=v([10,2,9,4])`

28. The following vector is defined in MATLAB:

$$v = \begin{bmatrix} 15 & 0 & 6 & -2 & 3 & -5 & 4 & 9 & 1.8 & -0.35 & 7 \end{bmatrix}$$

By hand (pencil and paper) write what will be displayed if the following commands are executed by MATLAB. Check your answers by executing the commands with MATLAB.

(*a*) `a=[v([2 7:10]);v([3,5:7,2])]`
(*b*) `b=[v([3:5,8])' v([10 6 4 1])' v(7:-1:4)']`

29. Create the following matrix A.

$$A = \begin{bmatrix} 1 & 2 & 3 & 4 & 5 & 6 \\ 7 & 8 & 9 & 10 & 11 & 12 \\ 13 & 14 & 15 & 16 & 17 & 18 \end{bmatrix}$$

Use the matrix A to:

(*a*) Create a six-element row vector named `ha` that contains the elements of the first row of A.
(*b*) Create a three-element row vector named `hb` that contains the elements of the sixth column of A.
(*c*) Create a six-element row vector named `hc` that contains the first three elements of the second row of A and the last three element of the third row of A.

30. Create the following matrix B.

$$B = \begin{bmatrix} 18 & 17 & 16 & 15 & 14 & 13 \\ 12 & 11 & 10 & 9 & 8 & 7 \\ 6 & 5 & 4 & 3 & 2 & 1 \end{bmatrix}$$

Use the matrix B to:

(*a*) Create a six-element column vector named `va` that contains the elements of the second and fifth columns of B.
(*b*) Create a seven-element column vector named `vb` that contains elements 3 through 6 of the third row of B and the elements of the second column of B.
(*c*) Create a nine-element column vector named `vc` that contains the elements of the second, fourth, and sixth columns of B.

31. Create the following vector C.

$$C = [0.7\ 1.9\ 3.1\ 4.3\ 5.5\ 6.7\ 7.9\ 9.1\ 10.3\ 11.5\ 12.7\ 13.9\ 15.1\ 16.3\ 17.5]$$

Then use MATLAB's built-in `reshape` function and the transpose operation to create the following matrix D from the vector C:

$$D = \begin{bmatrix} 0.7 & 1.9 & 3.1 & 4.3 & 5.5 \\ 6.7 & 7.9 & 9.1 & 10.3 & 11.5 \\ 12.7 & 13.9 & 15.1 & 16.3 & 17.5 \end{bmatrix}$$

Use the matrix D to:

(*a*) Create a nine-element column vector named `ua` that contains the elements of the first, third, and fourth columns of D.

(*b*) Create an eight-element raw vector named `ub` that contains the elements of the second row of D and the third column of D.

(*c*) Create a six-element row vector named `uc` that contains the first three elements of the first row of D and the last three elements of the last row of D.

32. Create the following matrix E.

$$E = \begin{bmatrix} 0 & 0 & 0 & 0 & 2 & 2 & 2 \\ 0.7 & 0.6 & 0.5 & 0.4 & 0.3 & 0.2 & 0.1 \\ 2 & 4 & 6 & 8 & 10 & 12 & 14 \\ 22 & 19 & 16 & 13 & 10 & 7 & 4 \end{bmatrix}$$

(*a*) Create a 2×5 matrix F from the second and fourth rows, and the third through the seventh columns of matrix E.

(*b*) Create a 4×3 matrix G from all rows and the third through fifth columns of matrix E.

33. Create the following matrix H.

$$H = \begin{bmatrix} 1.7 & 1.6 & 1.5 & 1.4 & 1.3 & 1.2 \\ 22 & 24 & 26 & 28 & 30 & 32 \\ 9 & 8 & 7 & 6 & 5 & 4 \end{bmatrix}$$

(*a*) Create a 2×4 matrix G such that its first row includes the first two elements and the last two elements of the first row of H, and the second row of G includes the second through the fifth elements of the third row of H.

(*b*) Create a 3×3 matrix K such that the first, second, and third rows are the first, fourth, and sixth columns of matrix H.

34. The following matrix is defined in MATLAB:

$$M = \begin{bmatrix} 3 & 5 & 7 & 9 & 11 & 13 \\ 15 & 14 & 13 & 12 & 11 & 10 \\ 1 & 2 & 3 & 1 & 2 & 3 \end{bmatrix}$$

By hand (pencil and paper) write what will be displayed if the following commands are executed by MATLAB. Check your answers by executing the commands with MATLAB.

a) `A=M([1,2],[2,4,5])` *b*) `B=M(:,[1:3,6])`
c) `C=M([1,3],:)` *d*) `D=M([2,3],5)`

35. The following matrix is defined in MATLAB:

$$N = \begin{bmatrix} 33 & 21 & 9 & 14 & 30 \\ 30 & 18 & 6 & 18 & 34 \\ 27 & 15 & 6 & 22 & 38 \\ 24 & 12 & 10 & 26 & 42 \end{bmatrix}$$

By hand (pencil and paper) write what will be displayed if the following commands are executed by MATLAB. Check your answers by executing the commands with MATLAB.

(*a*) `A=[N(1,1:4)',N(2,2:5)']`
(*b*) `B=[N(:,3)' N(3,:)]`
(*c*) `C(3:4,5:6)=N(2:3,4:5)`

36. By hand (pencil and paper) write what will be displayed if the following commands are executed by MATLAB. Check your answers by executing the commands with MATLAB.

```
v=1:3:34
M=reshape(v,3,4)
M(2,:)=[]
M(:,3)=[]
N=ones(size(M))
```

37. Using the `zeros`, `ones`, and `eye` commands, create the following arrays:

(*a*) $\begin{bmatrix} 1 & 1 \\ 1 & 1 \\ 0 & 0 \\ 0 & 0 \end{bmatrix}$ (*b*) $\begin{bmatrix} 1 & 0 & 0 & 1 & 1 & 1 \\ 0 & 1 & 0 & 1 & 1 & 1 \\ 0 & 0 & 1 & 1 & 1 & 1 \end{bmatrix}$ (*c*) $\begin{bmatrix} 1 & 1 & 1 & 1 \\ 1 & 1 & 1 & 1 \\ 0 & 0 & 0 & 0 \\ 1 & 1 & 1 & 1 \end{bmatrix}$

38. Using the `zeros`, `ones`, and `eye` commands create the following arrays:

$$(a)\quad \begin{bmatrix} 1 & 0 & 0 & 1 & 1 \\ 0 & 1 & 0 & 1 & 1 \end{bmatrix} \qquad (b)\quad \begin{bmatrix} 0 & 0 & 1 & 1 \\ 0 & 0 & 1 & 1 \\ 0 & 0 & 0 & 0 \\ 1 & 1 & 1 & 1 \end{bmatrix} \qquad (c)\quad \begin{bmatrix} 1 & 1 & 0 & 0 & 1 \\ 1 & 1 & 0 & 0 & 0 \\ 1 & 1 & 0 & 0 & 0 \\ 1 & 1 & 0 & 0 & 0 \end{bmatrix}$$

39. Use the `eye` command to create the array *A* shown on the left below. Then use the colon to address elements in the arrays and the `eye` command to change *A* to match the array shown on the right.

$$A = \begin{bmatrix} 1 & 0 & 0 & 0 & 0 & 0 \\ 0 & 1 & 0 & 0 & 0 & 0 \\ 0 & 0 & 1 & 0 & 0 & 0 \\ 0 & 0 & 0 & 1 & 0 & 0 \\ 0 & 0 & 0 & 0 & 1 & 0 \\ 0 & 0 & 0 & 0 & 0 & 1 \end{bmatrix} \qquad A = \begin{bmatrix} 1 & 0 & 0 & 1 & 0 & 0 \\ 0 & 1 & 0 & 0 & 1 & 0 \\ 0 & 0 & 1 & 0 & 0 & 1 \\ 1 & 0 & 0 & 1 & 0 & 0 \\ 0 & 1 & 0 & 0 & 1 & 0 \\ 0 & 0 & 1 & 0 & 0 & 1 \end{bmatrix}$$

40. Create a 2×2 matrix *A* in which all the elements are 1. Then reassign *A* to itself (several times) such that *A* will become:

$$A = \begin{bmatrix} 1 & 1 & 0 & 0 & 1 & 1 & 0 & 0 \\ 1 & 1 & 0 & 0 & 1 & 1 & 0 & 0 \\ 0 & 0 & 1 & 1 & 0 & 0 & 1 & 1 \\ 0 & 0 & 1 & 1 & 0 & 0 & 1 & 1 \end{bmatrix}$$

Chapter 3
Mathematical Operations with Arrays

Once variables are created in MATLAB they can be used in a wide variety of mathematical operations. In Chapter 1 the variables that were used in mathematical operations were all defined as scalars. This means that they were all 1×1 arrays (arrays with one row and one column that have only one element) and the mathematical operations were done with single numbers. Arrays, however, can be one-dimensional (arrays with one row, or with one column), two-dimensional (arrays with multiple rows and columns), and even of higher dimensions. In these cases the mathematical operations are more complex. MATLAB, as its name indicates, is designed to carry out advanced array operations that have many applications in science and engineering. This chapter presents the basic, most common mathematical operations that MATLAB performs using arrays.

Addition and subtraction are relatively simple operations and are covered first, in Section 3.1. The other basic operations—multiplication, division, and exponentiation—can be done in MATLAB in two different ways. One way, which uses the standard symbols (*, /, and ^), follows the rules of linear algebra and is presented in Sections 3.2 and 3.3. The second way, which is called element-by-element operations, is covered in Section 3.4. These operations use the symbols .*, ./, and .^ (a period is typed in front of the standard operation symbol). In addition, in both types of calculations, MATLAB has left division operators (.\ or \ ), which are also explained in Sections 3.3 and 3.4.

A Note to First-Time Users of MATLAB:

Although matrix operations are presented first and element-by-element operations next, the order can be reversed since the two are independent of each other. It is expected that almost every MATLAB user has some knowledge of matrix operations and linear algebra, and thus will be able to follow the material covered in Sections 3.2 and 3.3 without any difficulty. Some readers, however, might prefer to read Section 3.4 first. MATLAB can be used with element-by-element operations in numerous applications that do not require linear algebra multiplication (or division) operations.

3.1 ADDITION AND SUBTRACTION

The operations + (addition) and – (subtraction) can be used to add (subtract) arrays of identical size (the same numbers of rows and columns) and to add (subtract) a scalar to an array. When two arrays are involved the sum, or the difference, of the arrays is obtained by adding, or subtracting, their corresponding elements.

In general, if A and B are two arrays (for example, 2×3 matrices),

$$A = \begin{bmatrix} A_{11} & A_{12} & A_{13} \\ A_{21} & A_{22} & A_{23} \end{bmatrix} \quad \text{and} \quad B = \begin{bmatrix} B_{11} & B_{12} & B_{13} \\ B_{21} & B_{22} & B_{23} \end{bmatrix}$$

then the matrix that is obtained by adding A and B is:

$$\begin{bmatrix} (A_{11}+B_{11}) & (A_{12}+B_{12}) & (A_{13}+B_{13}) \\ (A_{21}+B_{21}) & (A_{22}+B_{22}) & (A_{23}+B_{23}) \end{bmatrix}$$

Examples are:

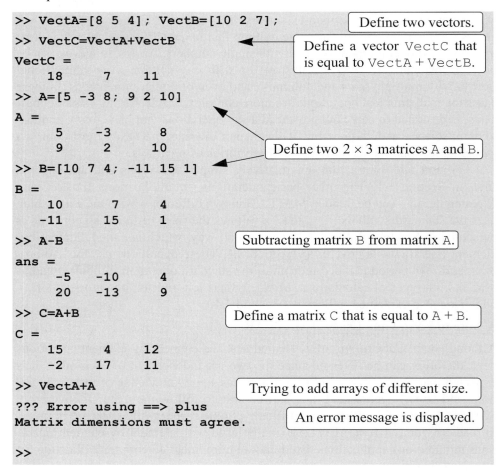

```
>> VectA=[8 5 4]; VectB=[10 2 7];
>> VectC=VectA+VectB
VectC =
    18     7    11
>> A=[5 -3 8; 9 2 10]
A =
     5    -3     8
     9     2    10
>> B=[10 7 4; -11 15 1]
B =
    10     7     4
   -11    15     1
>> A-B
ans =
    -5   -10     4
    20   -13     9
>> C=A+B
C =
    15     4    12
    -2    17    11
>> VectA+A
??? Error using ==> plus
Matrix dimensions must agree.

>>
```

Define two vectors.

Define a vector VectC that is equal to VectA + VectB.

Define two 2 × 3 matrices A and B.

Subtracting matrix B from matrix A.

Define a matrix C that is equal to A + B.

Trying to add arrays of different size.

An error message is displayed.

When a scalar (number) is added to (or subtracted from) an array, the scalar is added to (or subtracted from) all the elements of the array. Examples are:

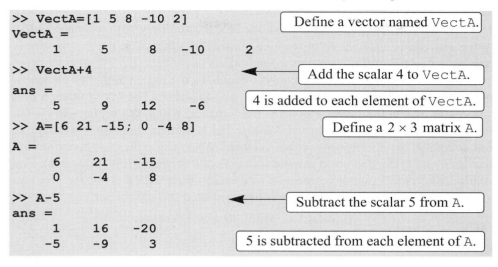

```
>> VectA=[1 5 8 -10 2]                     Define a vector named VectA.
VectA =
     1     5     8   -10     2
>> VectA+4                                 Add the scalar 4 to VectA.
ans =
     5     9    12    -6                   4 is added to each element of VectA.
>> A=[6 21 -15; 0 -4 8]                    Define a 2 × 3 matrix A.
A =
     6    21   -15
     0    -4     8
>> A-5                                     Subtract the scalar 5 from A.
ans =
     1    16   -20
    -5    -9     3                         5 is subtracted from each element of A.
```

3.2 *ARRAY MULTIPLICATION*

The multiplication operation * is executed by MATLAB according to the rules of linear algebra. This means that if *A* and *B* are two matrices, the operation *A*B* can be carried out only if the number of columns in matrix *A* is equal to the number of rows in matrix *B*. The result is a matrix that has the same number of rows as *A* and the same number of columns as *B*. For example, if *A* is a 4×3 matrix and *B* is a 3×2 matrix:

$$A = \begin{bmatrix} A_{11} & A_{12} & A_{13} \\ A_{21} & A_{22} & A_{23} \\ A_{31} & A_{32} & A_{33} \\ A_{41} & A_{42} & A_{43} \end{bmatrix} \quad \text{and} \quad B = \begin{bmatrix} B_{11} & B_{12} \\ B_{21} & B_{22} \\ B_{31} & B_{32} \end{bmatrix}$$

then the matrix that is obtained with the operation *A*B* has dimensions 4×2 with the elements:

$$\begin{bmatrix} (A_{11}B_{11} + A_{12}B_{21} + A_{13}B_{31}) & (A_{11}B_{12} + A_{12}B_{22} + A_{13}B_{32}) \\ (A_{21}B_{11} + A_{22}B_{21} + A_{23}B_{31}) & (A_{21}B_{12} + A_{22}B_{22} + A_{23}B_{32}) \\ (A_{31}B_{11} + A_{32}B_{21} + A_{33}B_{31}) & (A_{31}B_{12} + A_{32}B_{22} + A_{33}B_{32}) \\ (A_{41}B_{11} + A_{42}B_{21} + A_{43}B_{31}) & (A_{41}B_{12} + A_{42}B_{22} + A_{43}B_{32}) \end{bmatrix}$$

A numerical example is:

$$\begin{bmatrix} 1 & 4 & 3 \\ 2 & 6 & 1 \\ 5 & 2 & 8 \end{bmatrix} \begin{bmatrix} 5 & 4 \\ 1 & 3 \\ 2 & 6 \end{bmatrix} = \begin{bmatrix} (1 \cdot 5 + 4 \cdot 1 + 3 \cdot 2) & (1 \cdot 4 + 4 \cdot 3 + 3 \cdot 6) \\ (2 \cdot 5 + 6 \cdot 1 + 1 \cdot 2) & (2 \cdot 4 + 6 \cdot 3 + 1 \cdot 6) \\ (5 \cdot 5 + 2 \cdot 1 + 8 \cdot 2) & (5 \cdot 4 + 2 \cdot 3 + 8 \cdot 6) \end{bmatrix} = \begin{bmatrix} 15 & 34 \\ 18 & 32 \\ 43 & 74 \end{bmatrix}$$

The product of the multiplication of two square matrices (they must be of the same size) is a square matrix of the same size. However, the multiplication of matrices is not commutative. This means that if A and B are both $n \times n$, then $A*B \neq B*A$. Also, the power operation can be executed only with a square matrix (since $A*A$ can be carried out only if the number of columns in the first matrix is equal to the number of rows in the second matrix).

Two vectors can be multiplied only if they have the same number of elements, and one is a row vector and the other is a column vector. The multiplication of a row vector by a column vector gives a 1×1 matrix, which is a scalar. This is the dot product of two vectors. (MATLAB also has a built-in function, `dot(a,b)`, that computes the dot product of two vectors.) When using the `dot` function, the vectors `a` and `b` can each be a row vector or a column vector (see Table 3-1). The multiplication of a column vector by a row vector, each with n elements, gives an $n \times n$ matrix. Multiplication of array is demonstrated in Tutorial 3-1,

Tutorial 3-1: Multiplication of arrays.

```
>> A=[1 4 2; 5 7 3; 9 1 6; 4 2 8]
A =
     1     4     2
     5     7     3
     9     1     6
     4     2     8
>> B=[6 1; 2 5; 7 3]
B =
     6     1
     2     5
     7     3
>> C=A*B
C =
    28    27
    65    49
    98    32
    84    38
>> D=B*A
??? Error using ==> *
Inner matrix dimensions must agree.

>> F=[1 3; 5 7]
F =
     1     3
     5     7
>> G=[4 2; 1 6]
```

Define a 4×3 matrix `A`.

Define a 3×2 matrix `B`.

Multiply matrix `A` by matrix `B` and assign the result to variable `C`.

Trying to multiply `B` by `A`, `B*A`, gives an error since the number of columns in `B` is 2 and the number of rows in `A` is 4.

Define two 2×2 matrices `F` and `G`.

Tutorial 3-1: Multiplication of arrays. (Continued)

```
G =
     4     2
     1     6
>> F*G                                   Multiply F*G
ans =
     7    20
    27    52
>> G*F                                   Multiply G*F
ans =                                    Note that the answer for G*F is not the
    14    26                             same as the answer for F*G.
    31    45
>> AV=[2 5 1]                            Define a three-element row vector AV.
AV =
     2     5     1
>> BV=[3; 1; 4]                          Define a three-element column vector BV.
BV =
     3
     1
     4
>> AV*BV                                 Multiply AV by BV. The answer is a scalar.
ans =                                    (Dot product of two vectors.)
    15
>> BV*AV                                 Multiply BV by AV. The
ans =                                    answer is a 3 × 3 matrix.
     6    15     3
     2     5     1
     8    20     4
>>
```

When an array is multiplied by a number (actually a number is a 1×1 array), each element in the array is multiplied by the number. For example:

```
>> A=[2 5 7 0; 10 1 3 4; 6 2 11 5]       Define a 3 × 4 matrix A.
A =
     2     5     7     0
    10     1     3     4
     6     2    11     5
>> b=3                                   Assign the number 3 to the variable b.
b =
     3
```

```
>> b*A

ans =
     6    15    21     0
    30     3     9    12
    18     6    33    15
>> C=A*5
C =
    10    25    35     0
    50     5    15    20
    30    10    55    25
```

Multiply the matrix A by b. This can be done by either typing b*A or A*b.

Multiply the matrix A by 5 and assign the result to a new variable C. (Typing C = 5*A gives the same result.)

Linear algebra rules of array multiplication provide a convenient way for writing a system of linear equations. For example, the system of three equations with three unknowns

$$\begin{aligned} A_{11}x_1 + A_{12}x_2 + A_{13}x_3 &= B_1 \\ A_{21}x_1 + A_{22}x_2 + A_{23}x_3 &= B_2 \\ A_{31}x_1 + A_{32}x_2 + A_{33}x_3 &= B_3 \end{aligned}$$

can be written in a matrix form as

$$\begin{bmatrix} A_{11} & A_{12} & A_{13} \\ A_{21} & A_{22} & A_{23} \\ A_{31} & A_{32} & A_{33} \end{bmatrix} \begin{bmatrix} x_1 \\ x_2 \\ x_3 \end{bmatrix} = \begin{bmatrix} B_1 \\ B_2 \\ B_3 \end{bmatrix}$$

and in matrix notation as

$$AX = B \quad \text{where } A = \begin{bmatrix} A_{11} & A_{12} & A_{13} \\ A_{21} & A_{22} & A_{23} \\ A_{31} & A_{32} & A_{33} \end{bmatrix},\ X = \begin{bmatrix} x_1 \\ x_2 \\ x_3 \end{bmatrix},\ \text{and } B = \begin{bmatrix} B_1 \\ B_2 \\ B_3 \end{bmatrix}.$$

3.3 *Array Division*

The division operation is also associated with the rules of linear algebra. This operation is more complex and only a brief explanation is given below. A full explanation can be found in books on linear algebra.

The division operation can be explained with the help of the identity matrix and the inverse operation.

Identity matrix:

The identity matrix is a square matrix in which the diagonal elements are 1s, and the rest of the elements are 0s. As was shown in Section 2.2.1, an identity matrix can be created in MATLAB with the eye command. When the identity matrix multiplies another matrix (or vector), that matrix (or vector) is unchanged (the multiplication has to be done according to the rules of linear algebra). This is equivalent to multiplying a scalar by 1. For example:

$$\begin{bmatrix} 7 & 3 & 8 \\ 4 & 11 & 5 \end{bmatrix}\begin{bmatrix} 1 & 0 & 0 \\ 0 & 1 & 0 \\ 0 & 0 & 1 \end{bmatrix} = \begin{bmatrix} 7 & 3 & 8 \\ 4 & 11 & 5 \end{bmatrix} \text{ or } \begin{bmatrix} 1 & 0 & 0 \\ 0 & 1 & 0 \\ 0 & 0 & 1 \end{bmatrix}\begin{bmatrix} 8 \\ 2 \\ 15 \end{bmatrix} = \begin{bmatrix} 8 \\ 2 \\ 15 \end{bmatrix} \text{ or } \begin{bmatrix} 6 & 2 & 9 \\ 1 & 8 & 3 \\ 7 & 4 & 5 \end{bmatrix}\begin{bmatrix} 1 & 0 & 0 \\ 0 & 1 & 0 \\ 0 & 0 & 1 \end{bmatrix} = \begin{bmatrix} 6 & 2 & 9 \\ 1 & 8 & 3 \\ 7 & 4 & 5 \end{bmatrix}$$

If a matrix A is square, it can be multiplied by the identity matrix, I, from the left or from the right:

$$AI = IA = A$$

Inverse of a matrix:

The matrix B is the inverse of the matrix A if, when the two matrices are multiplied, the product is the identity matrix. Both matrices must be square and the multiplication order can be BA or AB.

$$BA = AB = I$$

Obviously B is the inverse of A, and A is the inverse of B. For example:

$$\begin{bmatrix} 2 & 1 & 4 \\ 4 & 1 & 8 \\ 2 & -1 & 3 \end{bmatrix}\begin{bmatrix} 5.5 & -3.5 & 2 \\ 2 & -1 & 0 \\ -3 & 2 & 1 \end{bmatrix} = \begin{bmatrix} 5.5 & -3.5 & 2 \\ 2 & -1 & 0 \\ -3 & 2 & 1 \end{bmatrix}\begin{bmatrix} 2 & 1 & 4 \\ 4 & 1 & 8 \\ 2 & -1 & 3 \end{bmatrix} = \begin{bmatrix} 1 & 0 & 0 \\ 0 & 1 & 0 \\ 0 & 0 & 1 \end{bmatrix}$$

The inverse of a matrix A is typically written as A^{-1}. In MATLAB the inverse of a matrix can be obtained either by raising A to the power of –1, A^{-1}, or with the `inv(A)` function. Multiplying the matrices above with MATLAB is shown below.

```
>> A=[2 1 4; 4 1 8; 2 -1 3]
A =
     2     1     4
     4     1     8
     2    -1     3
>> B=inv(A)
B =
    5.5000   -3.5000    2.0000
    2.0000   -1.0000         0
   -3.0000    2.0000   -1.0000
>> A*B
ans =
     1     0     0
     0     1     0
     0     0     1
```

Creating the matrix A.

Use the inv function to find the inverse of A and assign it to B.

Multiplication of A and B gives the identity matrix.

```
>> A*A^-1
ans =
     1     0     0
     0     1     0
     0     0     1
```

Use the power –1 to find the inverse of A. Multiplying it by A gives the identity matrix.

Not every matrix has an inverse. A matrix has an inverse only if it is square and its determinant is not equal to zero.

Determinants:

A determinant is a function associated with square matrices. A short review on determinants is given below. For a more detailed coverage refer to books on linear algebra.

The determinant is a function that associates with each square matrix A a number, called the determinant of the matrix. The determinant is typically denoted by det(A) or $|A|$. The determinant is calculated according to specific rules. For a second-order 2×2 matrix the rule is:

$$|A| = \begin{vmatrix} a_{11} & a_{12} \\ a_{21} & a_{22} \end{vmatrix} = a_{11}a_{22} - a_{12}a_{21}, \text{ for example, } \begin{vmatrix} 6 & 5 \\ 3 & 9 \end{vmatrix} = 6 \cdot 9 - 5 \cdot 3 = 39$$

The determinant of a square matrix can be calculated with the `det` command (see Table 3-1).

Array division:

MATLAB has two types of array division, right division and left division.

Left division, \ :

Left division is used to solve the matrix equation $AX = B$. In this equation X and B are column vectors. This equation can be solved by multiplying, on the left, both sides by the inverse of A:

$$A^{-1}AX = A^{-1}B$$

The left-hand side of this equation is X since

$$A^{-1}AX = IX = X$$

So the solution of $AX = B$ is:

$$X = A^{-1}B$$

In MATLAB the last equation can be written by using the left division character:

$$X = A\backslash B$$

It should be pointed out here that although the last two operations appear to give the same result, the method by which MATLAB calculates X is different. In the first, MATLAB calculates A^{-1} and then uses it to multiply B. In the second (left division), the solution X is obtained numerically with a method that is based on Gauss elimination. The left division method is recommended for solving a set of linear equations because the calculation of the inverse may be less accurate than the Gauss elimination method when large matrices are involved.

Right division, / :

The right division is used to solve the matrix equation $XC = D$. In this equation X and D are row vectors. This equation can be solved by multiplying, on the right, both sides by the inverse of C:

$$X \cdot CC^{-1} = D \cdot C^{-1}$$

which gives

$$X = D \cdot C^{-1}$$

In MATLAB the last equation can be written using the right division character:

$$X = D/C$$

The following example demonstrates the use of the left and right division, and the `inv` function to solve a set of linear equations.

Sample Problem 3-1: Solving three linear equations (array division)

Use matrix operations to solve the following system of linear equations.

$$4x - 2y + 6z = 8$$
$$2x + 8y + 2z = 4$$
$$6x + 10y + 3z = 0$$

Solution

Using the rules of linear algebra demonstrated earlier, the above system of equations can be written in the matrix form $AX = B$ or in the form $XC = D$:

$$\begin{bmatrix} 4 & -2 & 6 \\ 2 & 8 & 2 \\ 6 & 10 & 3 \end{bmatrix} \begin{bmatrix} x \\ y \\ z \end{bmatrix} = \begin{bmatrix} 8 \\ 4 \\ 0 \end{bmatrix} \quad \text{or} \quad \begin{bmatrix} x & y & z \end{bmatrix} \begin{bmatrix} 4 & 2 & 6 \\ -2 & 8 & 10 \\ 6 & 2 & 3 \end{bmatrix} = \begin{bmatrix} 8 & 4 & 0 \end{bmatrix}$$

Solutions for both forms are shown below:

```
>> A=[4 -2 6; 2 8 2; 6 10 3];
>> B=[8; 4; 0];
>> X=A\B
X =
   -1.8049
    0.2927
    2.6341
>> Xb=inv(A)*B
Xb =
   -1.8049
    0.2927
    2.6341
```

Solving the form $AX = B$.

Solving by using left division: $X = A \backslash B$.

Solving by using the inverse of A: $X = A^{-1}B$.

```
>> C=[4 2 6; -2 8 10; 6 2 3];
>> D=[8 4 0];
>> Xc=D/C
Xc =
   -1.8049    0.2927    2.6341
>> Xd=D*inv(C)
Xd =
   -1.8049    0.2927    2.6341
```

Solving the form $XC = D$.

Solving by using right division: $X = D/C$.

Solving by using the inverse of C: $X = D \cdot C^{-1}$.

3.4 *ELEMENT-BY-ELEMENT OPERATIONS*

In Sections 3.2 and 3.3 it was shown that when the regular symbols for multiplication and division (* and /) are used with arrays, the mathematical operations follow the rules of linear algebra. There are, however, many situations that require element-by-element operations. These operations are carried out on each of the elements of the array (or arrays). Addition and subtraction are by definition already element-by-element operations since when two arrays are added (or subtracted) the operation is executed with the elements that are in the same position in the arrays. Element-by-element operations can be done only with arrays of the same size.

Element-by-element multiplication, division, or exponentiation of two vectors or matrices is entered in MATLAB by typing a period in front of the arithmetic operator.

Symbol	Description	Symbol	Description
.*	Multiplication	./	Right division
.^	Exponentiation	.\	Left Division

If two vectors a and b are $a = [a_1\ a_2\ a_3\ a_4]$ and $b = [b_1\ b_2\ b_3\ b_4]$, then element-by-element multiplication, division, and exponentiation of the two vectors gives:

$$a\ .*\ b = [a_1b_1\ a_2b_2\ a_3b_3\ a_4b_4]$$

$$a\ ./\ b = [a_1/b_1\ a_2/b_2\ a_3/b_3\ a_4/b_4]$$

$$a\ .\wedge\ b = [(a_1)^{b_1}\ (a_2)^{b_2}\ (a_3)^{b_3}\ (a_4)^{b_4}]$$

If two matrices A and B are

$$A = \begin{bmatrix} A_{11} & A_{12} & A_{13} \\ A_{21} & A_{22} & A_{23} \\ A_{31} & A_{32} & A_{33} \end{bmatrix} \quad \text{and} \quad B = \begin{bmatrix} B_{11} & B_{12} & B_{13} \\ B_{21} & B_{22} & B_{23} \\ B_{31} & B_{32} & B_{33} \end{bmatrix}$$

then element-by-element multiplication and division of the two matrices give:

$$A\ .\!*\ B = \begin{bmatrix} A_{11}B_{11} & A_{12}B_{12} & A_{13}B_{13} \\ A_{21}B_{21} & A_{22}B_{22} & A_{23}B_{23} \\ A_{31}B_{31} & A_{32}B_{32} & A_{33}B_{33} \end{bmatrix} \qquad A\ ./\ B = \begin{bmatrix} A_{11}/B_{11} & A_{12}/B_{12} & A_{13}/B_{13} \\ A_{21}/B_{21} & A_{22}/B_{22} & A_{23}/B_{23} \\ A_{31}/B_{31} & A_{32}/B_{32} & A_{33}/B_{33} \end{bmatrix}$$

Element-by-element exponentiation of matrix A gives:

$$\mathrm{A}\ .\hat{}\ n = \begin{bmatrix} (A_{11})^n & (A_{12})^n & (A_{13})^n \\ (A_{21})^n & (A_{22})^n & (A_{23})^n \\ (A_{31})^n & (A_{32})^n & (A_{33})^n \end{bmatrix}$$

Element-by-element multiplication, division, and exponentiation are demonstrated in Tutorial 3-2.

Tutorial 3-2: Element-by-element operations.

```
>> A=[2 6 3; 5 8 4]
A =
     2     6     3
     5     8     4
>> B=[1 4 10; 3 2 7]
B =
     1     4    10
     3     2     7
>> A.*B
ans =
     2    24    30
    15    16    28
>> C=A./B
C =
    2.0000    1.5000    0.3000
    1.6667    4.0000    0.5714
```

Define a 2 × 3 array A.

Define a 2 × 3 array B.

Element-by-element multiplication of array A by B.

Element-by-element division of array A by B. The result is assigned to variable C.

Tutorial 3-2: Element-by-element operations. (Continued)

```
>> B.^3

ans =
     1    64   1000
    27     8    343
```

Element-by-element exponentiation of array B. The result is an array in which each term is the corresponding term in B raised to the power of 3.

```
>> A*B

??? Error using ==> *
Inner matrix dimensions must agree.
```

Trying to multiply A*B gives an error since A and B cannot be multiplied according to linear algebra rules. (The number of columns in A is not equal to the number of rows in B.)

Element-by-element calculations are very useful for calculating the value of a function at many values of its argument. This is done by first defining a vector that contains values of the independent variable, and then using this vector in element-by-element computations to create a vector in which each element is the corresponding value of the function. One example is:

```
>> x=[1:8]
x =
   1   2   3   4   5   6   7   8
>> y=x.^2-4*x
y =
  -3  -4  -3   0   5  12  21  32
>>
```

Create a vector x with eight elements.

Vector x is used in element-by-element calculations of the elements of vector y.

In the example above $y = x^2 - 4x$. Element-by-element operation is needed when x is squared. Each element in the vector y is the value of y that is obtained when the value of the corresponding element of the vector x is substituted in the equation. Another example is:

```
>> z=[1:2:11]
z =
    1     3     5     7     9    11
>> y=(z.^3 + 5*z)./(4*z.^2 - 10)

 y =
   -1.0000    1.6154    1.6667    2.0323    2.4650    2.9241
```

Create a vector z with six elements.

Vector z is used in element-by-element calculations of the elements of vector y.

In the last example $y = \frac{z^3 + 5z}{4z^2 - 10}$. Element-by-element operations are used in this example three times: to calculate z^3 and z^2, and to divide the numerator by the denominator.

3.5 USING ARRAYS IN MATLAB BUILT-IN MATH FUNCTIONS

The built-in functions in MATLAB are written such that when the argument (input) is an array, the operation that is defined by the function is executed on each element of the array. (One can think of the operation as element-by-element application of the function.) The result (output) from such an operation is an array in which each element is calculated by entering the corresponding element of the argument (input) array into the function. For example, if a vector with seven elements is substituted in the function `cos(x)`, the result is a vector with seven elements in which each element is the cosine of the corresponding element in `x`. This is shown below.

```
>> x=[0:pi/6:pi]
x =
     0    0.5236    1.0472    1.5708    2.0944    2.6180    3.1416
>>y=cos(x)
y =
    1.0000    0.8660    0.5000    0.0000   -0.5000   -0.8660   -1.0000
>>
```

An example in which the argument variable is a matrix is:

```
>> d=[1 4 9; 16 25 36; 49 64 81]          Creating a 3 × 3 array.
d =
     1     4     9
    16    25    36
    49    64    81
>> h=sqrt(d)
h =
     1     2     3
     4     5     6
     7     8     9
```

h is a 3 × 3 array in which each element is the square root of the corresponding element in array d.

The feature of MATLAB in which arrays can be used as arguments in functions is called vectorization.

3.6 BUILT-IN FUNCTIONS FOR ANALYZING ARRAYS

MATLAB has many built-in functions for analyzing arrays. Table 3-1 lists some of these functions.

Table 3-1: Built-in array functions

Function	Description	Example

`mean(A)`	If `A` is a vector, returns the mean value of the elements of the vector.	`>> A=[5 9 2 4];`
		`>> mean(A)`
		`ans =`
		`     5`

`C=max(A)`	If `A` is a vector, `C` is the largest element in `A`. If `A` is a matrix, `C` is a row vector containing the largest element of each column of `A`.	`>> A=[5 9 2 4 11 6 11 1];`
		`>> C=max(A)`
		`C =`
		`    11`

`[d,n]=max(A)`	If `A` is a vector, `d` is the largest element in `A`, and `n` is the position of the element (the first if several have the max value).	`>> [d,n]=max(A)`
		`d =`
		`    11`
		`n =`
		`     5`

`min(A)`
 	
`[d,n]=min(A)`
	The same as `max(A)`, but for the smallest element.
	
	The same as `[d,n]= max(A)`, but for the smallest element.
		`>> A=[5 9 2 4];`
		`>> min(A)`
		`ans =`
		`     2`

`sum(A)`	If `A` is a vector, returns the sum of the elements of the vector.	`>> A=[5 9 2 4];`
		`>> sum(A)`
		`ans =`
		`    20`

`sort(A)`	If `A` is a vector, arranges the elements of the vector in ascending order.	`>> A=[5 9 2 4];`
		`>> sort(A)`
		`ans =`
		`     2     4     5     9`

`median(A)`	If `A` is a vector, returns the median value of the elements of the vector.	`>> A=[5 9 2 4];`
		`>> median(A)`
		`ans =`
		`    4.5000`

Table 3-1: Built-in array functions (Continued)

Function	Description	Example

`std(A)`	If `A` is a vector, returns the standard deviation of the elements of the vector.	`>> A=[5 9 2 4];`
		`>> std(A)`
		`ans =`
		`    2.9439`

`det(A)`	Returns the determinant of a square matrix `A`.	`>> A=[2 4; 3 5];`
		`>> det(A)`
		`ans =`
		`    -2`

`dot(a,b)`	Calculates the scalar (dot) product of two vectors `a` and `b`. The vectors can each be row or column vectors.	`>> a=[1 2 3];`
		`>> b=[3 4 5];`
		`>> dot(a,b)`
		`ans =`
		`    26`

`cross(a,b)`	Calculates the cross product of two vectors `a` and `b`, (`axb`). The two vectors must have each three elements.	`>> a=[1 3 2];`
		`>> b=[2 4 1];`
		`>> cross(a,b)`
		`ans =`
		`    -5    3    -2`

`inv(A)`	Returns the inverse of a square matrix `A`.	`>> A=[2 -2 1; 3 2 -1; 2 -3 2];`
		`>> inv(A)`
		`ans =`
		`   0.2000   0.2000        0`
		`  -1.6000   0.4000   1.0000`
		`  -2.6000   0.4000   2.0000`

3.7 GENERATION OF RANDOM NUMBERS

Simulations of many physical processes and engineering applications frequently require using a number (or a set of numbers) with a random value. MATLAB has three commands—`rand`, `randn`, and `randi`—that can be used to assign random numbers to variables.

The `rand` command:

The `rand` command generates uniformly distributed random numbers with values between 0 and 1. The command can be used to assign these numbers to a scalar, a vector, or a matrix, as shown in Table 3-2.

Table 3-2: The `rand` command

Command	Description	Example

`rand`	Generates a single random number between 0 and 1.	`>> rand`
		`ans =`
		`0.2311`

`rand(1,n)`	Generates an `n`-element row vector of random numbers between 0 and 1.	`>> a=rand(1,4)`
		`a =`
		`0.6068 0.4860 0.8913 0.7621`

`rand(n)`	Generates an `n` × `n` matrix with random numbers between 0 and 1.	`>> b=rand(3)`
		`b =`
		`0.4565 0.4447 0.9218`
		`0.0185 0.6154 0.7382`
		`0.8214 0.7919 0.1763`

`rand(m,n)`	Generates an `m` × `n` matrix with random numbers between 0 and 1.	`>> c=rand(2,4)`
		`c =`
		`0.4057 0.9169 0.8936 0.3529`
		`0.9355 0.4103 0.0579 0.8132`

`randperm(n)`	Generates a row vector with `n` elements that are random permutation of integers 1 through `n`.	`>> randperm(8)`
		`ans =`
		`8 2 7 4 3 6 5 1`

Sometimes there is a need for random numbers that are distributed in an interval other than (0,1), or for numbers that are integers only. This can be done using mathematical operations with the `rand` function. Random numbers that are distributed in a range (a,b) can be obtained by multiplying `rand` by $(b - a)$ and adding the product to a:

$$(b-a)*\mathtt{rand}+a$$

For example, a vector of 10 elements with random values between –5 and 10 can be created by ($a = -5$, $b = 10$):

```
>> v=15*rand(1,10)-5
v =
   -1.8640    0.6973    6.7499    5.2127    1.9164    3.5174
6.9132   -4.1123    4.0430   -4.2460
```

The `randi` command:

The `randi` command generates uniformly distributed random integer. The command can be used to assign these numbers to a scalar, a vector, or a matrix, as shown in Table 3-3.

Table 3-3: The `randi` command

Command	Description	Example
`randi(imax)` (imax is an integer)	Generates a single random number between 1 and imax.	`>> a=randi(15)` `a =` `9`
`randi(imax, n)`	Generates an n × n matrix with random integers between 1 and imax.	`>> b=randi(15,3)` `b =` `4 8 11` `14 3 8` `1 15 8`
`randi(imax, m,n)`	Generates an m × n matrix with random integers between 1 and imax.	`>> c=randi(15,2,4)` `c =` `1 1 8 13` `11 2 2 13`

The range of the random integers can be set to be between any two integers by typing `[imin imax]` instead of `imax`. For example, a 3 × 4 matrix with random integers between 50 and 90 is created by:

```
>> d=randi([50 90],3,4)
d =
    57    82    71    75
    66    52    67    61
    84    66    76    67
```

The `randn` command:

The `randn` command generates normally distributed numbers with mean 0 and standard deviation of 1. The command can be used to generate a single number, a vector, or a matrix in the same way as the `rand` command. For example, a 3 × 4 matrix is created by:

```
>> d=randn(3,4)
d =
  -0.4326    0.2877    1.1892    0.1746
  -1.6656   -1.1465   -0.0376   -0.1867
   0.1253    1.1909    0.3273    0.7258
```

The mean and standard deviation of the numbers can be changed by mathematical operations to have any values. This is done by multiplying the number generated by the `randn` function by the desired standard deviation, and adding the desired mean. For example, a vector of six numbers with a mean of 50 and standard deviation of 6 is generated by:

```
>> v=4*randn(1,6)+50
v =
   42.7785   57.4344   47.5819   50.4134   52.2527   50.4544
```

Integers of normally distributed numbers can be obtained by using the `round` function.

```
>> w=round(4*randn(1,6)+50)
w =
    51    49    46    49    50    44
```

3.8 EXAMPLES OF MATLAB APPLICATIONS

Sample Problem 3-2: Equivalent force system (addition of vectors)

Three forces are applied to a bracket as shown. Determine the total (equivalent) force applied to the bracket.

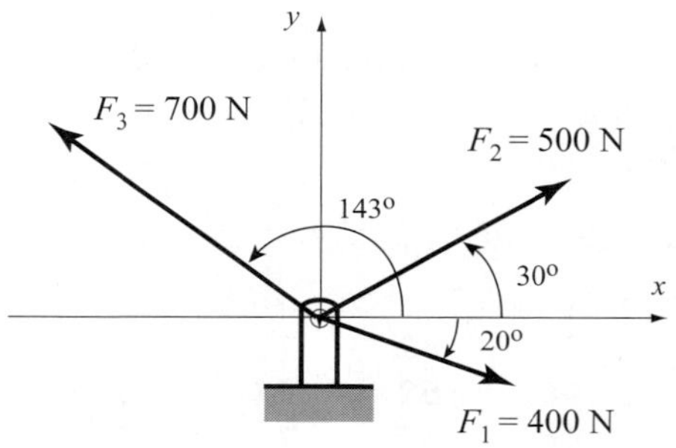

Solution

A force is a vector (a physical quantity that has a magnitude and direction). In a Cartesian coordinate system a two-dimensional vector **F** can be written as:

$$\mathbf{F} = F_x\mathbf{i} + F_y\mathbf{j} = F\cos\theta\mathbf{i} + F\sin\theta\mathbf{j} = F(\cos\theta\mathbf{i} + \sin\theta\mathbf{j})$$

where F is the magnitude of the force and θ is its angle relative to the x axis, F_x and F_y are the components of **F** in the directions of the x and y axes, respectively, and **i** and **j** are unit vectors in these directions. If F_x and F_y are known, then F and θ can be determined by:

$$F = \sqrt{F_x^2 + F_y^2} \quad \text{and} \quad \tan\theta = \frac{F_y}{F_x}$$

The total (equivalent) force applied on the bracket is obtained by adding the forces that are acting on the bracket. The MATLAB solution below follows three steps:

- Write each force as a vector with two elements, where the first element is the x component of the vector and the second element is the y component.
- Determine the vector form of the equivalent force by adding the vectors.
- Determine the magnitude and direction of the equivalent force.

The problem is solved in the following script file.

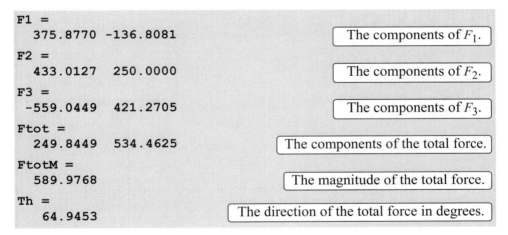

```
% Sample Problem 3-2 solution (script file)
clear
F1M=400; F2M=500; F3M=700;              Define variables with the
                                        magnitude of each vector.
Th1=-20; Th2=30; Th3=143;
                     Define variables with the angle of each vector.
F1=F1M*[cosd(Th1) sind(Th1)]
F2=F2M*[cosd(Th2) sind(Th2)]        Define the three vectors.
F3=F3M*[cosd(Th3) sind(Th3)]
Ftot=F1+F2+F3                       Calculate the total force vector.
FtotM=sqrt(Ftot(1)^2+Ftot(2)^2)     Calculate the magnitude of
Th=atand(Ftot(2)/Ftot(1))           the total force vector.
                  Calculate the angle of the total force vector.
```

When the program is executed, the following is displayed in the Command Window:

```
F1 =
  375.8770 -136.8081               The components of F1.
F2 =
  433.0127  250.0000               The components of F2.
F3 =
 -559.0449  421.2705               The components of F3.
Ftot =
  249.8449  534.4625               The components of the total force.
FtotM =
  589.9768                         The magnitude of the total force.
Th =
   64.9453                  The direction of the total force in degrees.
```

The equivalent force has a magnitude of 589.98 N, and is directed $64.95°$ (ccw) relative to the x axis. In vector notation the force is $\mathbf{F} = 249.84\mathbf{i} + 534.46\mathbf{j}$ N.

Sample Problem 3-3: Friction experiment (element-by-element calculations)

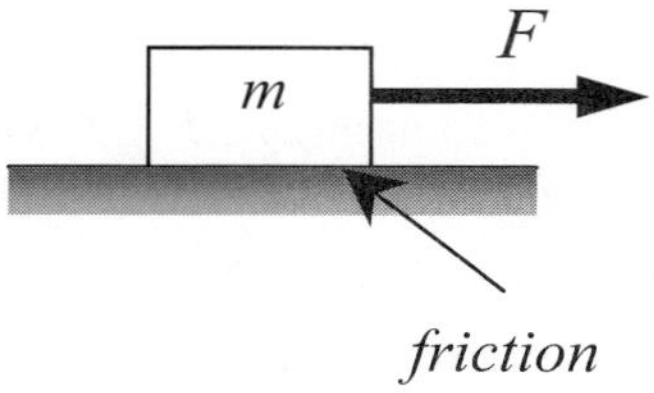

The coefficient of friction, μ, can be determined in an experiment by measuring the force *F* required to move a mass *m*. When *F* is measured and *m* is known, the coefficient of friction can be calculated by:

$$\mu = F/(mg) \quad (g = 9.81 \text{ m/s}^2).$$

Results from measuring *F* in six tests are given in the table below. Determine the coefficient of friction in each test, and the average from all tests.

Test	1	2	3	4	5	6
Mass *m* (kg)	2	4	5	10	20	50
Force *F* (N)	12.5	23.5	30	61	117	294

Solution

A solution using MATLAB commands in the Command Window is shown below.

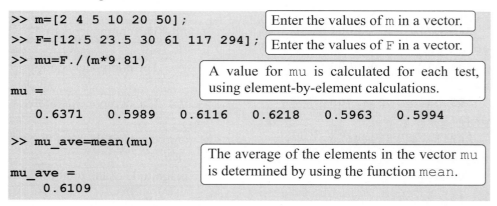

```
>> m=[2 4 5 10 20 50];
>> F=[12.5 23.5 30 61 117 294];
>> mu=F./(m*9.81)

mu =

    0.6371    0.5989    0.6116    0.6218    0.5963    0.5994

>> mu_ave=mean(mu)

mu_ave =
    0.6109
```

Enter the values of `m` in a vector.

Enter the values of `F` in a vector.

A value for `mu` is calculated for each test, using element-by-element calculations.

The average of the elements in the vector `mu` is determined by using the function `mean`.

Sample Problem 3-4: Electrical resistive network analysis (solving a system of linear equations)

The electrical circuit shown consists of resistors and voltage sources. Determine the current in each resistor using the mesh current method, which is based on Kirchhoff's voltage law.

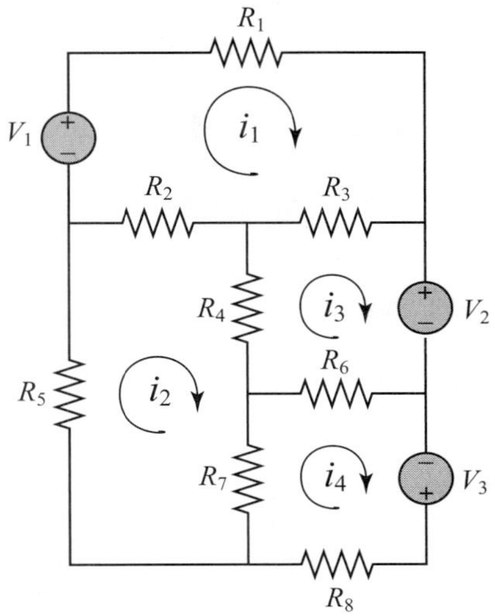

$V_1 = 20$ V, $V_2 = 12$ V, $V_3 = 40$ V
$R_1 = 18\,\Omega$, $R_2 = 10\,\Omega$, $R_3 = 16\,\Omega$
$R_4 = 6\,\Omega$, $R_5 = 15\,\Omega$, $R_6 = 8\,\Omega$
$R_7 = 12\,\Omega$, $R_8 = 14\,\Omega$

Solution

Kirchhoff's voltage law states that the sum of the voltage around a closed circuit is zero. In the mesh current method a current is first assigned for each mesh (i_1, i_2, i_3, i_4 in the figure). Then Kirchhoff's voltage law is applied for each mesh. This results in a system of linear equations for the currents (in this case four equations). The solution gives the values of the mesh currents. The current in a resistor that belongs to two meshes is the sum of the currents in the corresponding meshes. It is convenient to assume that all the currents are in the same direction (clockwise in this case). In the equation for each mesh, the voltage source is positive if the current flows to the – pole, and the voltage of a resistor is negative for current in the direction of the mesh current.

The equations for the four meshes in the current problem are:

$$V_1 - R_1 i_1 - R_3(i_1 - i_3) - R_2(i_1 - i_2) = 0$$
$$-R_5 i_2 - R_2(i_2 - i_1) - R_4(i_2 - i_3) - R_7(i_2 - i_4) = 0$$
$$-V_2 - R_6(i_3 - i_4) - R_4(i_3 - i_2) - R_3(i_3 - i_1) = 0$$
$$V_3 - R_8 i_4 - R_7(i_4 - i_2) - R_6(i_4 - i_3) = 0$$

The four equations can be rewritten in matrix form $[A][x] = [B]$:

$$\begin{bmatrix} -(R_1 + R_2 + R_3) & R_2 & R_3 & 0 \\ R_2 & -(R_2 + R_4 + R_5 + R_7) & R_4 & R_7 \\ R_3 & R_4 & -(R_3 + R_4 + R_6) & R_6 \\ 0 & R_7 & R_6 & -(R_6 + R_7 + R_8) \end{bmatrix} \begin{bmatrix} i_1 \\ i_2 \\ i_3 \\ i_4 \end{bmatrix} = \begin{bmatrix} -V_1 \\ 0 \\ V_2 \\ -V_3 \end{bmatrix}$$

The problem is solved in the following program, written in a script file:

```
V1=20; V2=12; V3=40;
R1=18; R2=10; R3=16; R4=6;
R5=15; R6=8; R7=12; R8=14;
A=[-(R1+R2+R3) R2 R3 0
R2 -(R2+R4+R5+R7) R4 R7
R3 R4 -(R3+R4+R6) R6
0 R7 R6 -(R6+R7+R8)]
>> B=[-V1; 0; V2; -V3]
>> I=A\B
```

Define variables with the values of the V's and R's.

Create the matrix A.

Create the vector B.

Solve for the currents by using left division.

When the script file is executed, the following is displayed in the Command Window:

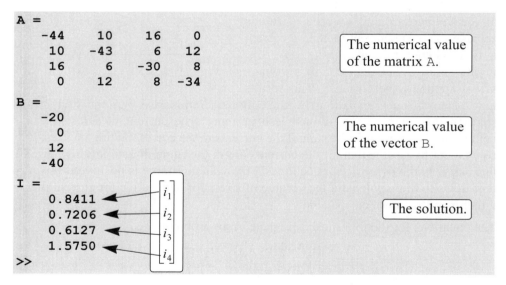

```
A =
   -44    10    16     0
    10   -43     6    12
    16     6   -30     8
     0    12     8   -34
B =
   -20
     0
    12
   -40
I =
    0.8411
    0.7206
    0.6127
    1.5750
>>
```

The numerical value of the matrix A.

The numerical value of the vector B.

The solution. (i_1, i_2, i_3, i_4)

The last column vector gives the current in each mesh. The currents in the resistors R_1, R_5, and R_8 are $i_1 = 0.8411$ A, $i_2 = 0.7206$ A, and $i_4 = 1.5750$ A, respectively. The other resistors belong to two meshes and their current is the sum of the currents in the meshes.

The current in resistor R_2 is $i_1 - i_2 = 0.1205$ A.

The current in resistor R_3 is $i_1 - i_3 = 0.2284$ A.

The current in resistor R_4 is $i_2 - i_3 = 0.1079$ A.

The current in resistor R_6 is $i_4 - i_3 = 0.9623$ A.

The current in resistor R_7 is $i_4 - i_2 = 0.8544$ A.

Sample Problem 3-5: Motion of two particles

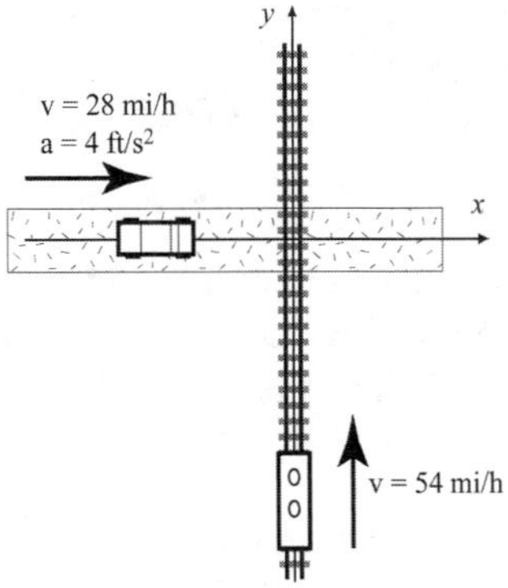

A train and a car are approaching a road crossing. At time $t = 0$ the train is 400 ft south of the crossing traveling north at a constant speed of 54 mi/h. At the same time the car is 200 ft west of the crossing traveling east at a speed of 28 mi/h and accelerating at 4 ft/s^2. Determine the positions of the train and the car, the distance between them, and the speed of the train relative to the car every second for the next 10 seconds.

To show the results, create an 11×6 matrix in which each row has the time in the first column and the train position, car position, distance between the train and the car, car speed, and the speed of the train relative to the car, in the next five columns, respectively.

Solution

The position of an object that moves along a straight line at a constant acceleration is given by $s = s_o + v_o t + \frac{1}{2}at^2$ where s_o and v_o are the position and velocity at $t = 0$, and a is the acceleration. Applying this equation to the train and the car gives:

$$y = -400 + v_{otrain}t \qquad \text{(train)}$$

$$x = -200 + v_{ocar}t + \frac{1}{2}a_{car}t^2 \qquad \text{(car)}$$

The distance between the car and the train is: $d = \sqrt{x^2 + y^2}$.
The velocity of the train is constant and in vector notation is $\mathbf{v}_{train} = v_{otrain}\mathbf{j}$. The car is accelerating and its velocity at time t is given by $\mathbf{v}_{car} = (v_{ocar} + a_{car}t)\mathbf{i}$. The velocity of the train relative to the car, $\mathbf{v}_{t/c}$, is given by $\mathbf{v}_{t/c} = \mathbf{v}_{train} - \mathbf{v}_{car} = -(v_{ocar} + a_{car}t)\mathbf{i} + v_{otrain}\mathbf{j}$. The magnitude (speed) of this velocity is the length of the vector.

The problem is solved in the following program, written in a script file. First a vector `t` with 11 elements for the time from 0 to 10 s is created, then the positions of the train and the car, the distance between them, and the speed of the train relative to the car at each time element are calculated.

```
v0train=54*5280/3600; v0car=28*5280/3600; acar=4;
```
Create variables for the initial velocities (in ft/s) and the acceleration.

```
t=0:10;
```
Create the vector `t`.

```
y=-400+v0train*t;
x=-200+v0car*t+0.5*acar*t.^2;
```
Calculate the train and car positions.

```
d=sqrt(x.^2+y.^2);
```
Calculate the distance between the train and car.

```
vcar=v0car+acar*t;                    Calculate the car's velocity.
speed_trainRcar=sqrt(vcar.^2+v0train^2);
                    Calculate the speed of the train relative to the car.
table=[t' y' x' d' vcar' speed_trainRcar']
                                Create a table (see note below).
```

Note: In the commands above, `table` is the name of the variable that is a matrix containing the data to be displayed.

When the script file is executed, the following is displayed in the Command Window:

```
table =
         0 -400.0000 -200.0000  447.2136   41.0667   89.2139
    1.0000 -320.8000 -156.9333  357.1284   45.0667   91.1243
    2.0000 -241.6000 -109.8667  265.4077   49.0667   93.1675
    3.0000 -162.4000  -58.8000  172.7171   53.0667   95.3347
    4.0000  -83.2000   -3.7333   83.2837   57.0667   97.6178
    5.0000   -4.0000   55.3333   55.4777   61.0667  100.0089
    6.0000   75.2000  118.4000  140.2626   65.0667  102.5003
    7.0000  154.4000  185.4667  241.3239   69.0667  105.0849
    8.0000  233.6000  256.5333  346.9558   73.0667  107.7561
    9.0000  312.8000  331.6000  455.8535   77.0667  110.5075
   10.0000  392.0000  410.6667  567.7245   81.0667  113.3333
```

Time (s)	Train position (ft)	Car position (ft)	Car-train distance (ft)	Car speed (ft/s)	Train speed relative to the car (ft/s)

In this problem the results (numbers) are displayed by MATLAB without any text. Instructions on how to add text to output generated by MATLAB are presented in Chapter 4.

3.9 PROBLEMS

Note: *Additional problems for practicing mathematical operations with arrays are provided at the end of Chapter 4.*

1. For the function $y = x^3 - 2x^2 + x$, calculate the value of y for the following values of x using element-by-element operations: $-2, -1, 0, 1, 2, 3, 4$.

2. For the function $y = \frac{x^2 - 2}{x + 4}$, calculate the value of y for the following values of x using element-by-element operations: $-3, -2, -1, 0, 1, 2, 3$.

3. For the function $y = \frac{(x-3)(x^2+3)}{x^2}$, calculate the value of y for the following values of x using element-by-element operations: 1, 2, 3, 4, 5, 6, 7 .

4. For the function $y = \frac{20t^{2/3}}{t+1} - \frac{(t+1)^2}{e^{(0.3t+5)}} + \frac{2}{t+1}$, calculate the value of y for the following values of t using element-by-element operations: 0, 1, 2, 3, 4, 5, 6, 7, 8 .

5. A ball that is dropped on the floor bounces back up many times, reaching a lower height after each bounce. When the ball impacts the floor its rebound velocity is 0.85 times the impact velocity. The velocity v with which a ball hits the floor after being dropped from a height h is given by $v = \sqrt{2gh}$, where $g = 9.81$ m/s^2. The time between successive bounces is given by $t = v/g$, where v is the upward velocity after the last impact. Consider a ball that is dropped from a height of 2 m. Determine the times at which the ball hits the floor for the first eight bounces. Set $t = 0$ when the ball hits the floor for the first time. (Calculate the velocity of the ball when it hits the floor for the first time. Derive a formula for the time of the following hits as a function of the bounce number. Then create a vector $n = 1, 2, \ldots, 8$ and use the formula (use element-by element operations) to calculate a vector with the values of t for each n.) Display the results in a two-column table where the values of n and t are displayed in the first and second columns, respectively.

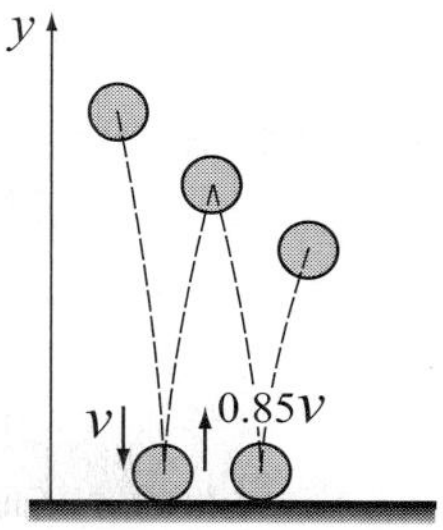

6. An aluminum sphere ($r = 0.2$ cm) is dropped in a glass cylinder filled with glycerin. The velocity of the sphere as a function of time $v(t)$ can be modeled by the equation

$$v(t) = \sqrt{\frac{V(\rho_{al} - \rho_{gl})g}{k}} \tanh\left(\frac{\sqrt{V(\rho_{al} - \rho_{gl})gk}}{V\rho_{al}} t\right)$$

where V is the volume of the sphere, $g = 9.81$ m/s^2 is the gravitational acceleration, $k = 0.0018$ is a constant, and $\rho_{al} = 2700$ kg/m^3 and $\rho_{gl} = 1260$ kg/m^3 are the density of aluminum and glycerin, respectively. Determine the velocity of the sphere for t = 0, 0.05, 0.1, 0.15, 0.2, 0.25, 0.3, and 0.35 s. Note that initially the velocity increases rapidly, but then, due to the resistance of the glycerin, the velocity increases more gradually. Eventually the velocity approaches a limit that is called the terminal velocity.

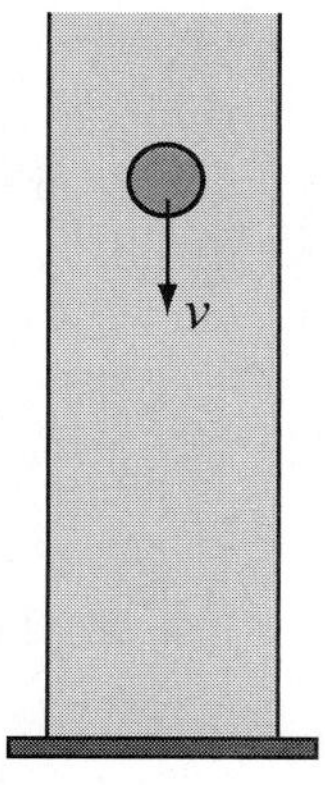

7. The current i (in amps) t seconds after closing the switch in the circuit shown is given by:

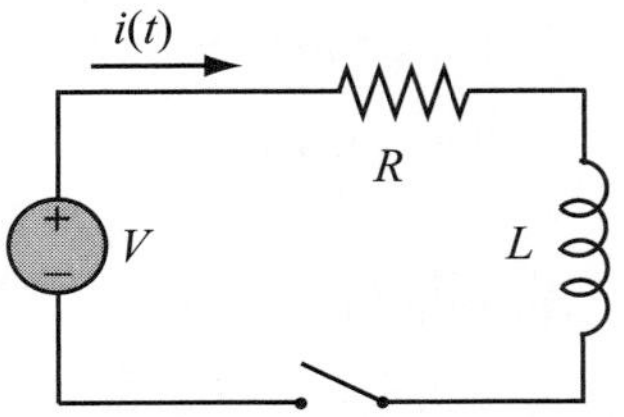

$$i(t) = \frac{V}{R}(1 - e^{-(R/L)\,t})$$

Consider the case where $V = 120$ volts, $R = 120$ ohms and $L = 0.1$ henry.

(*a*) Find the time t_m required for the current to reach 1% of its initial value, then use `linspace` to create a vector t having 10 elements with the first element 0 and maximum value t_m.

(*b*) Calculate the current i for each value of t from part (*a*).

8. The length $|\mathbf{u}|$ (magnitude) of a vector $\mathbf{u} = x\mathbf{i} + y\mathbf{j} + z\mathbf{k}$ is given by $|\mathbf{u}| = \sqrt{x^2 + y^2 + z^2}$. Given the vector $\mathbf{u} = 23.5\mathbf{i} - 17\mathbf{j} + 6\mathbf{k}$, determine its length two ways:

(*a*) Define the vector in MATLAB, and then write a mathematical expression that uses the components of the vector.

(*b*) Define the vector in MATLAB, then use element-by element operations to create a new vector with elements that are the squares of the elements of the original vector. Then use MATLAB built-in functions `sum` and `sqrt` to calculate the length. All of these steps can be written in one command.

9. The unit vector $\mathbf{u}_n$ in the direction of the vector $\mathbf{u} = x\mathbf{i} + y\mathbf{j} + z\mathbf{k}$ is given by $\mathbf{u}_n = \frac{x\mathbf{i} + y\mathbf{j} + z\mathbf{k}}{\sqrt{x^2 + y^2 + z^2}}$. Determine the unit vector of the vector $\mathbf{u} = -8\mathbf{i} - 14\mathbf{j} + 25\mathbf{k}$ by writing one MATLAB command.

10. The following two vectors are defined in MATLAB:

$$v = [3, -2, 4] \qquad u = [5, 3, -1]$$

By hand (pencil and paper) write what will be displayed if the following commands are executed by MATLAB. Check your answers by executing the commands with MATLAB.

(*a*) `v.*u` (*b*) `v*u'` (*c*) `v'*u`

11. Two vectors are given:

$$\mathbf{u} = -3\mathbf{i} + 8\mathbf{j} - 2\mathbf{k} \quad \text{and} \quad \mathbf{v} = 6.5\mathbf{i} - 5\mathbf{j} - 4\mathbf{k}$$

Use MATLAB to calculate the dot product $\mathbf{u} \cdot \mathbf{v}$ of the vectors in three ways:

(*a*) Write an expression using element-by-element calculation and the MATLAB built-in function `sum`.

(*b*) Define $\mathbf{u}$ as a row vector and $\mathbf{v}$ as a column vector, and then use matrix multiplication.

(*c*) Use the MATLAB built-in function `dot`.

12. Define the vector $v = [2\ \ 4\ \ 6\ \ 8\ \ 10]$. Then use the vector in a mathematical expression to create the following vectors:
(a) $a = \left[\frac{1}{2}\ \ \frac{1}{4}\ \ \frac{1}{6}\ \ \frac{1}{8}\ \ \frac{1}{10}\right]$ (b) $b = \left[\frac{1}{2^2}\ \ \frac{1}{4^2}\ \ \frac{1}{6^2}\ \ \frac{1}{8^2}\ \ \frac{1}{10^2}\right]$
(c) $c = [1\ \ 2\ \ 3\ \ 4\ \ 5]$ (d) $d = [1\ \ 1\ \ 1\ \ 1\ \ 1]$

13. Define the vector $v = [5\ \ 4\ \ 3\ \ 2\ \ 1]$. Then use the vector in a mathematical expression to create the following vectors:
(a) $a = [5^2\ \ 4^2\ \ 3^2\ \ 2^2\ \ 1^2]$ (b) $b = [5^5\ \ 4^4\ \ 3^3\ \ 2^2\ \ 1^1]$
(c) $c = [25\ \ 20\ \ 15\ \ 10\ \ 5]$ (d) $d = [4\ \ 3\ \ 2\ \ 1\ \ 0]$

14. Define x and y as the vectors $x = [1, 3, 5, 7, 9]$ and $y = [2, 5, 8, 11, 14$. Then use them in the following expressions to calculate z using element-by-element calculations.
(a) $z = \dfrac{xy^2}{x+y}$ (b) $z = x(x^2 - y) - (x - y)^2$

15. Define p and w as scalars, $p = 2.3$ and define $w = 5.67$, and, t, x, and y as the vectors $t = [1, 2, 3, 4, 5]$, $x = [2.8, 2.5, 2.2, 1.9, 1.6]$, and $y = [4, 7, 10, 13, 17]$. Then use these variables to calculate the following expressions using element-by-element calculations for the vectors.
(a) $T = \dfrac{p(x+y)^2}{y}w$ (b) $S = \dfrac{p(x+y)^2}{yw} + \dfrac{wxt}{py}$

16. The area of the parallelogram shown can be calculated by $|\mathbf{r}_{AB} \times \mathbf{r}_{AC}|$. Use the following steps in a script file to calculate the area:
Define the position of points A, B, and C as vectors $A = [2, 0]$, $B = [10, 3]$, and $C = [4, 6]$.
Determine the vectors $\mathbf{r}_{AB}$ and $\mathbf{r}_{AC}$ from the points.
Determine the area by using MATLAB's built-in functions `cross`, `sum`, and `sqrt`.

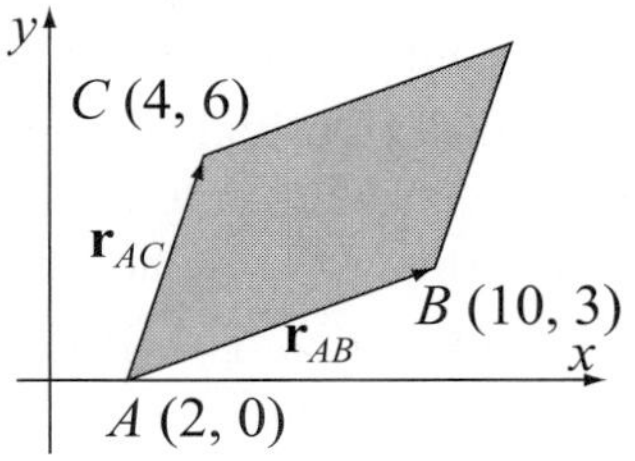

17. Define the vectors:

$$\mathbf{u} = -2\mathbf{i} + 6\mathbf{j} + 5\mathbf{k},\ \mathbf{v} = 5\mathbf{i} - 1\mathbf{j} + 3\mathbf{k}, \text{ and } \mathbf{w} = 4\mathbf{i} + 7\mathbf{j} - 2\mathbf{k}$$

Use the vectors to verify the identity:

$$\mathbf{u} \times (\mathbf{v} \times \mathbf{w}) = \mathbf{v}(\mathbf{u} \cdot \mathbf{w}) - \mathbf{w}(\mathbf{u} \cdot \mathbf{v})$$

Using MATLAB's built-in functions `cross` and `abs`, calculate the value of the left and right sides of the identity.

18. The dot product can be used for determining the angle between two vectors:

$$\theta = \cos^{-1}\left(\frac{\mathbf{r}_1 \cdot \mathbf{r}_2}{|\mathbf{r}_1||\mathbf{r}_2|}\right)$$

Use MATLAB's built-in functions `cosd`, `sqrt`, and `dot` to find the angle (in degrees) between $\mathbf{r}_1 = 3\mathbf{i} - 2\mathbf{j} + \mathbf{k}$ and $\mathbf{r}_2 = 1\mathbf{i} + 2\mathbf{j} - 4\mathbf{k}$.
Recall that $|\mathbf{r}| = \sqrt{\mathbf{r} \cdot \mathbf{r}}$.

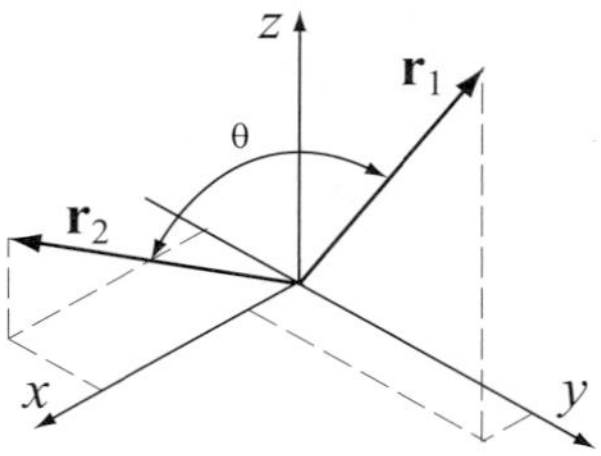

19. The position as a function of time $(x(t), y(t))$ of a projectile fired with a speed of v_0 at an angle α is given by

$$x(t) = v_0 \cos\alpha \cdot t \qquad y(t) = v_0 \sin\alpha \cdot t - \frac{1}{2}gt^2$$

where $g = 9.81$ m/s^2. The polar coordinates of the projectile at time t are $(r(t), \theta(t))$, where $r(t) = \sqrt{x(t)^2 + y(t)^2}$ and $\tan\theta = \frac{y(t)}{x(t)}$. Consider the case where $v_0 = 162$ m/s and $\theta = 70°$. Determine $r(t)$ and $\theta(t)$ for $t = 1, 6, 11, \ldots, 31$s.

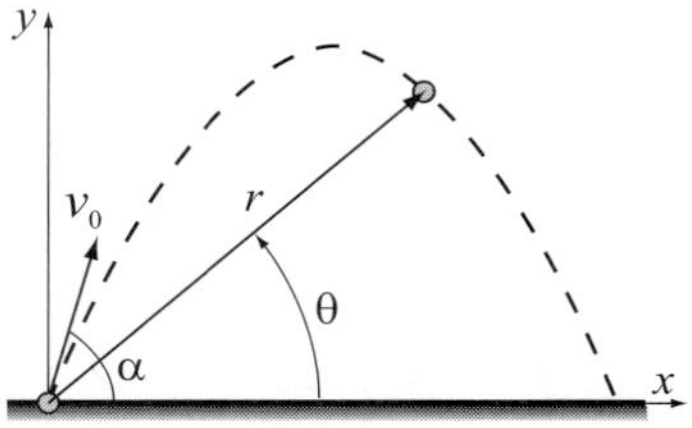

20. Two projectiles, *A* and *B*, are shot at the same instant from the same spot. Projectile *A* is shot at a speed of 560 m/s at an angle of 43° and projectile *B* is shot at a speed of 680 m/s at an angle of 50°. Determine which projectile will hit the ground first. Then take the flying time t_f of that projectile and divide it into ten increments by creating a vector t with 11 equally spaced elements (the first element is 0, the last is t_f). At each time t calculate the position vector $\mathbf{r}_{AB}$ between the two projectiles. Display the results in a three-column matrix where the first column is t and the second and third columns are the corresponding x and y components of $\mathbf{r}_{AB}$.

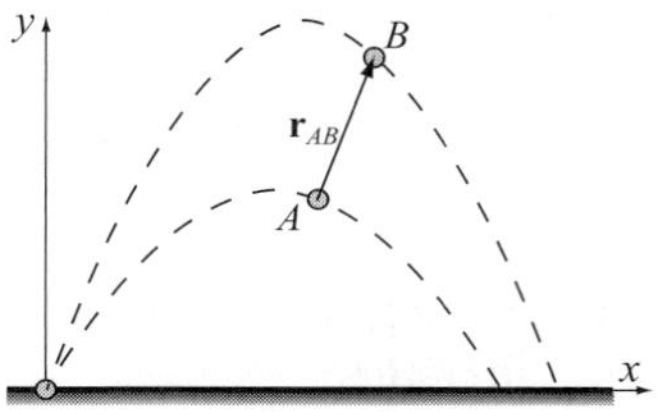

21. Show that $\lim_{x \to 0} \frac{\sin x}{x} = 1$.
Do this by first creating a vector x that has the elements 1.5, 1.0, 0.5, 0.1, 0.01, 0.001, and 0.00001. Then, create a new vector y in which each element is determined from the elements of x by $\frac{\sin x}{x}$. Compare the elements of y with the value 1 (use format long to display the numbers).

22. Show that $\lim_{x \to 1} \frac{x^2 - 1}{x - 1} = 2$.

Do this by first creating a vector x that has the elements: 5, 3, 2, 1.5, 1.1, 1.001, and 1.00001. Then, create a new vector y in which each element is determined from the elements of x by $\frac{x^2 - 1}{x - 1}$. Compare the elements of y with the value 2 (use format long to display the numbers).

23. Use MATLAB to show that the sum of the infinite series $\sum_{n=1}^{\infty} \frac{1}{2^n} = \frac{1}{2} + \frac{1}{2^2} + \frac{1}{2^3} + \ldots$ converges to 1. Do it by computing the sum for:

(*a*) $n = 10$ (*b*) $n = 20$

(*c*) $n = 30$ (*c*) $n = 40$

For each part create a vector `n` in which the first element is 1, the increment is 1, and the last term is 10, 20, 30, or 40. Then use element-by-element calculations to create a vector in which the elements are $\frac{1}{2^n}$. Finally, use the MATLAB built-in function `sum` to add the terms of the series. Compare the values obtained in parts (*a*), (*b*), (*c*), and (*d*) with the value of 1. (Don't forget to type semicolons at the end of commands that otherwise will display large vectors.)

24. Use MATLAB to show that the sum of the infinite series $\sqrt{12}\sum_{n=0}^{\infty} \frac{(-3)^{-n}}{2n+1}$ is equal to π. Do this by computing the sum for:

(*a*) $n = 10$ (*b*) $n = 20$ (*c*) $n = 50$

For each part create a vector `n` in which the first element is 0, the increment is 1 and the last term is 10, 50, or 100. Then, use element-by-element calculation to create a vector in which the elements are $\frac{(-3)^{-n}}{2n+1}$. Finally, use the function `sum` to add the terms of the series and multiply the result by $\sqrt{12}$. Compare the values obtained in parts (*a*), (*b*), and (*c*) to the value of π in MATLAB.

25. Fisheries commonly estimate the growth of a fish population using the von Bertalanffy growth law:

$$L = L_{max}(1 - e^{-K(t+\tau)})$$

where L_{max} is the maximum length, K is a rate constant, and τ is a time constant. These constants vary with the species of fish. Assuming L_{max} = 58 cm, K = 0.45 years^{-1}, and τ = 0.65 years, calculate the length of a fish at 0, 1, 2, 3, 4, and 5 years of age.

26. The path of a projectile fired with an initial speed v_0 at an angle θ is described by the equation

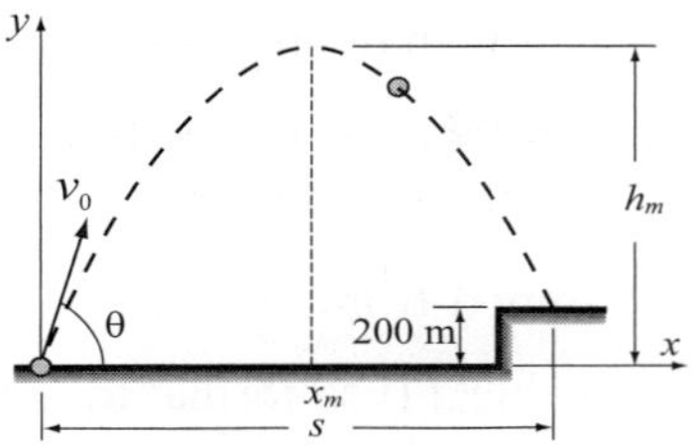

$$y = x\tan\theta - \frac{g}{2v_0^2\cos^2\theta}x^2$$

where $g = 9.81\text{ m/s}^2$. Consider the case where $\theta = 75°$ and $v_0 = 110\text{ m/s}$. Write a MATLAB script that does the following: calculates the distance s traveled by the projectile, creates a vector x with 100 elements such that the first element is 0 and the last is s, calculates the value of y for each value of x, finds the maximum height h_m that the projectile reaches (use MATLAB built-in function `max`) and the distance x_{hm} where the maximum height is reached. When the script is executed only the values of h_m and x_{hm} are displayed.

27. Create the following three matrices:

$$A = \begin{bmatrix} 2 & 4 & -1 \\ 3 & 1 & -5 \\ 0 & 1 & 4 \end{bmatrix} \qquad B = \begin{bmatrix} -2 & 5 & 0 \\ -3 & 2 & 7 \\ -1 & 6 & 9 \end{bmatrix} \qquad C = \begin{bmatrix} 0 & 3 & 5 \\ 2 & 1 & 0 \\ 4 & 6 & -3 \end{bmatrix}$$

(*a*) Calculate $A + B$ and $B + A$ to show that addition of matrices is commutative.

(*b*) Calculate $A + (B + C)$ and $(A + B) + C$ to show that addition of matrices is associative.

(*c*) Calculate $5(A + C)$ and $5A + 5C$ to show that, when matrices are multiplied by a scalar, the multiplication is distributive.

(*d*) Calculate $A*(B + C)$ and $A*B + A*C$ to show that matrix multiplication is distributive.

28. Use the matrices *A*, *B*, and *C* from the previous problem to answer the following:

(*a*) Does $A*B = B*A$?
(*b*) Does $A*(B*C) = (A*B)*C$?
(*c*) Does $(A*B)^t = B^t*A^t$? (^{t} means transpose)
(*d*) Does $(A + B)^t = A^t + B^t$?

29. Create a 4×4 matrix having random integer values between 1 and 10. Call the matrix *A* and using MATLAB perform the following operations. For each part explain the operation.

(*a*) $A * A$
(*b*) $A .*A$
(*c*) $A \backslash A$
(*d*) $A .\backslash A$
(*e*) det(*A*)
(*e*) inv(*A*)

30. The mechanical power output P in a contracting muscle is given by

$$P = Tv = \frac{kvT_0\left(1 - \dfrac{v}{v_{max}}\right)}{k + \dfrac{v}{v_{max}}}$$

where T is the muscle tension, v is the shortening velocity (max of v_{max}), T_0 is the isometric tension (i.e., tension at zero velocity), and k is a non-dimensional constant that ranges between 0.15 and 0.25 for most muscles. The equation can be written in non-dimensional form:

$$p = \frac{ku(1-u)}{k+u}$$

where $p = (Tv)/(T_0 v_{max})$, and $u = v/v_{max}$. A figure with $k = 0.25$ is shown here.

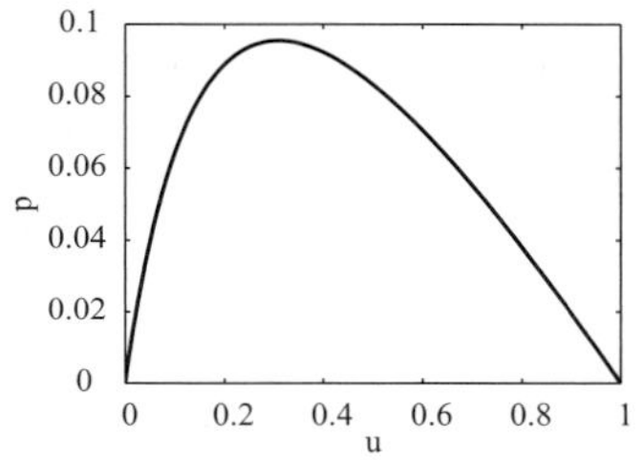

(*a*) Create a vector u ranging from 0 to 1 with increments of 0.05.

(*b*) Using $k = 0.25$, calculate the value of p for each value of u.

(*c*) Using MATLAB built-in function `max`, find the maximum value of p.

(*d*) Repeat the first three steps with increments of 0.01 and calculate the percent relative error, defined by $E = \left|\dfrac{p_{max_{0.01}} - p_{max_{0.05}}}{p_{max_{0.05}}}\right| \times 100$.

31. Solve the following system of three linear equations:

$$\begin{aligned} 3x - 2y + 5z &= 7.5 \\ -4.5 + 2y + 3z &= 5.5 \\ 5x + y - 2.5z &= 4.5 \end{aligned}$$

32. Solve the following system of five linear equations:

$$\begin{aligned} 3u + 1.5v + w + 0.5x + 4y &= -11.75 \\ -2u + v + 4w - 3.5x + 2y &= 19 \\ 6u - 3v + 2w + 2.5x + y &= -23 \\ u + 4v - 3w + 0.5x - 2y &= -1.5 \\ 3u + 2v - w + 1.5x - 3y &= -3.5 \end{aligned}$$

33. A juice company manufactures one-gallon bottles of three types of juice blends using orange, pineapple, and mango juice. The blends have the following compositions:

1 gallon orange blend: 3 quarts of orange juice, 0.75 quart of pineapple juice, 0.25 quart of mango juice.

1 gallon pineapple blend: 1 quart of orange juice, 2.5 quarts of pineapple juice, 0.5 quart of mango juice.

1 gallon mango blend: 0.5 quart of orange juice, 0.5 quart of pineapple juice, 3 quarts of mango juice.
How many gallons of each blend can be manufactured if 7,600 gallons of orange juice, 4,900 gallons of pineapple juice, and 3,500 gallons mango juice are available? Write a system of linear equations and solve.

34. The electrical circuit shown consists of resistors and voltage sources. Determine the current in each resistor, using the mesh current method based on Kirchhoff's voltage law (see Sample Problem 3-4).

$V_1 = 12\text{ V},\ V_2 = 24\text{ V}$
$R_1 = 20\,\Omega,\ R_2 = 12\,\Omega,\ R_3 = 8\ \Omega$
$R_4 = 6\,\Omega,\ R_5 = 10\,\Omega$

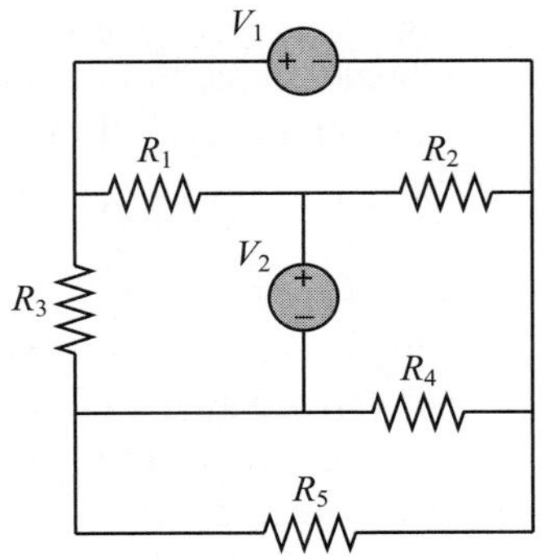

35. The electrical circuit shown consists of resistors and voltage sources. Determine the current in each resistor, using the mesh current method based on Kirchhoff's voltage law (see Sample Problem 3-4).

$V_1 = 12\text{ V},\ V_2 = 24\text{ V}$
$R_1 = 20\,\Omega,\ R_2 = 12\,\Omega,\ R_3 = 8\ \Omega$
$R_4 = 6\,\Omega,\ R_5 = 10\,\Omega$

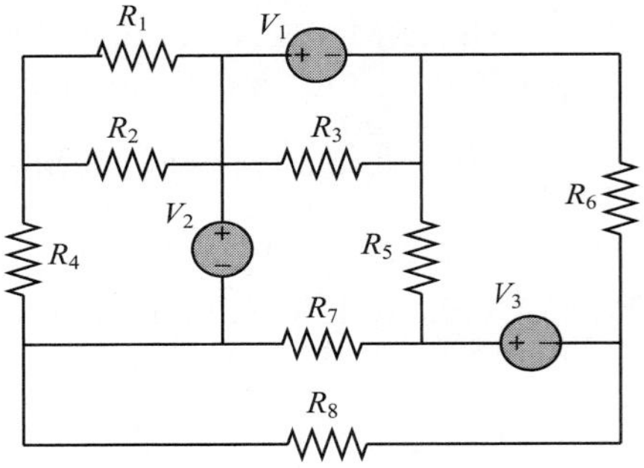

Chapter 4
Using Script Files and Managing Data

A script file (see Section 1.8) is a list of MATLAB commands, called a program, that is saved in a file. When the script file is executed (run), MATLAB executes the commands. Section 1.8 describes how to create, save, and run a simple script file in which the commands are executed in the order in which they are listed, and in which all the variables are defined within the script file. The present chapter gives more details of how to input data to a script file, how data is stored in MATLAB, various ways to display and save data that is created in script files, and how to exchange data between MATLAB and other applications. (How to write more advanced programs where commands are not necessarily executed in a simple order is covered in Chapter 6.)

In general, variables can be defined (created) in several ways. As shown in Chapter 2, variables can be defined implicitly by assigning values to a variable name. Variables can also be assigned values by the output of a function. In addition, variables can be defined with data that is imported from files outside MATLAB. Once defined (either in the Command Window or when a script file is executed) the variables are stored in MATLAB's Workspace.

Variables that reside in the workspace can be displayed in various ways, saved, or exported to applications outside MATLAB. Similarly, data from files outside MATLAB can be imported to the workspace and then used in MATLAB.

Section 4.1 explains how MATLAB stores data in the workspace and how the user can see the data that is stored. Section 4.2 shows how variables that are used in script files can be defined in the Command Window and/or in script files. Section 4.3 shows how to output data that is generated when script files are executed. Section 4.4 explains how the variables in the workspace can be saved and then retrieved, and Section 4.5 shows how to import and export data from and to applications outside MATLAB.

4.1 THE MATLAB WORKSPACE AND THE WORKSPACE WINDOW

The MATLAB workspace consists of the set of variables (named arrays) that are defined and stored during a MATLAB session. It includes variables that have been defined in the Command Window and variables defined when script files are executed. This means that the Command Window and script files share the same memory zone within the computer. This implies that once a variable is in the workspace, it is recognized and can be used, and it can be reassigned new values, in both the Command Window and script files. As will be explained in Chapter 7 (Section 7.3), there is another type of file in MATLAB, called a function file, where variables can also be defined. These variables, however, are normally not shared with other parts of the program since they use a separate workspace.

Recall from Chapter 1 that the `who` command displays a list of the variables currently in the workspace. The `whos` command displays a list of the variables currently in the workspace and information about their size, bytes, and class. An example is shown below.

```
>> 'Variables in memory'
ans =
Variables in memory
>> a = 7;
>> E = 3;
>> d = [5,  a+E,  4,  E^2]
d =
     5    10     4     9
>> g = [a, a^2,  13;  a*E,  1,  a^E]
g =
      7    49    13
     21     1   343
>> who
Your variables are:
E    a    ans   d    g
>> whos
  Name      Size            Bytes  Class     Attributes

  E         1x1                 8  double
  a         1x1                 8  double
  ans       1x19               38  char
  d         1x4                32  double
  g         2x3                48  double

>>
```

Typing a string.

The string is assigned to `ans`.

Creating the variables `a`, `E`, `d`, and `g`.

The `who` command displays the variables currently in the workspace.

The `whos` command displays the variables currently in the workspace, and information about their size and other information.

The variables currently in memory can also be viewed in the Workspace Window. If not open, this window can be opened by selecting **Workspace** in the **Desktop** menu. Figure 4-1 shows the Workspace Window that corresponds to the variables defined above. The variables that are displayed in the Workspace Window can also be edited (changed). Double-clicking on a variable opens the Variable Editor Window, where the content of the variable is displayed in a table. For example, Figure 4-2 shows the Variable Editor Window that opens when the variable `g` in Figure 4-1 is double-clicked.

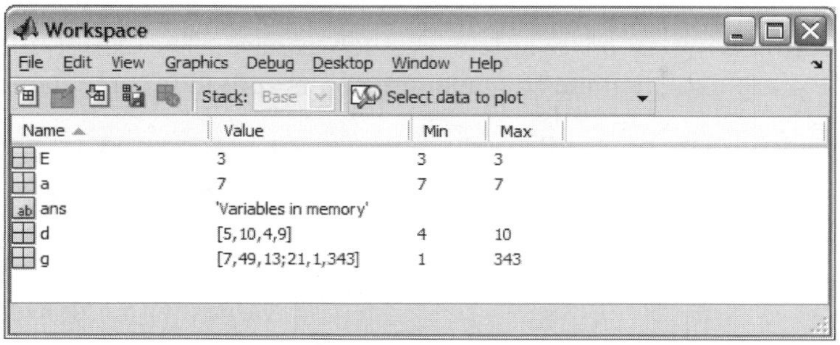

Figure 4-1: The Workspace Window.

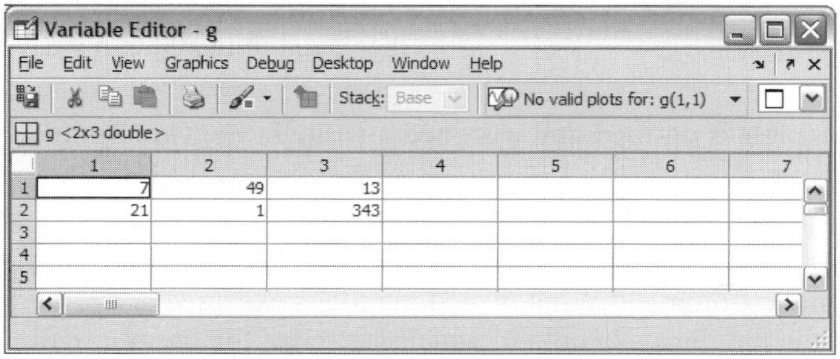

Figure 4-2: The Variable Editor Window.

The elements in the Variable Editor Window can be edited. The variables in the Workspace Window can be deleted by selecting them, and then either pressing the **delete** key on the keyboard or selecting **delete** from the **edit** menu. This has the same effect as entering the command `clear variable_name` in the Command Window.

4.2 *INPUT TO A SCRIPT FILE*

When a script file is executed, the variables that are used in the calculations within the file must have assigned values. In other words, the variables must be in the workspace. The assignment of a value to a variable can be done in three ways, depending on where and how the variable is defined.

1. The variable is defined and assigned a value in the script file.

In this case the assignment of a value to the variable is part of the script file. If the user wants to run the file with a different variable value, the file must be edited and the assignment of the variable changed. Then, after the file is saved, it can be executed again.

The following is an example of such a case. The script file (saved as Chapter4Example2) calculates the average points scored in three games.

```
% This script file calculates the average points scored in three games.
% The assignment of the values of the points is part of the script file.
game1=75;
game2=93;
game3=68;
ave_points=(game1+game2+game3)/3
```

The variables are assigned values within the script file.

The display in the Command Window when the script file is executed is:

```
>> Chapter4Example2

ave_points =
   78.6667
>>
```

The script file is executed by typing the name of the file.

The variable `ave_points` with its value is displayed in the Command Window.

2. The variable is defined and assigned a value in the Command Window.

In this case the assignment of a value to the variable is done in the Command Window. (Recall that the variable is recognized in the script file.) If the user wants to run the script file with a different value for the variable, the new value is assigned in the Command Window and the file is executed again.

For the previous example in which the script file has a program that calculates the average of points scored in three games, the script file (saved as Chapter4Example3) is:

```
% This script file calculates the average points scored in three games.
% The assignment of the values of the points to the variables
% game1, game2, and game3 is done in the Command Window.

ave_points=(game1+game2+game3)/3
```

The Command Window for running this file is:

```
>> game1 = 67;
>> game2 = 90;
>> game3 = 81;
```

The variables are assigned values in the Command Window.

```
>> Chapter4Example3

ave_points =
   79.3333

>> game1 = 87;
>> game2 = 70;
>> game3 = 50;
>> Chapter4Example3

ave_points =
    69
>>
```

The script file is executed.

The output from the script file is displayed in the Command Window.

New values are assigned to the variables.

The script file is executed again.

The output from the script file is displayed in the Command Window.

3. The variable is defined in the script file, but a specific value is entered in the Command Window when the script file is executed.

In this case the variable is defined in the script file, and when the file is executed, the user is prompted to assign a value to the variable in the Command Window. This is done by using the `input` command for creating the variable.

The form of the `input` command is:

```
variable_name = input('string with a message that
                      is displayed in the Command Window')
```

When the `input` command is executed as the script file runs, the string is displayed in the Command Window. The string is a message prompting the user to enter a value that is assigned to the variable. The user types the value and presses the **Enter** key. This assigns the value to the variable. As with any variable, the variable and its assigned value will be displayed in the Command Window unless a semicolon is typed at the very end of the `input` command. A script file that uses the `input` command to enter the points scored in each game to the program that calculates the average of the scores is shown below.

```
% This script file calculates the average of points scored in three games.
% The points from each game are assigned to the variables by
% using the input command.
game1=input('Enter the points scored in the first game ');
game2=input('Enter the points scored in the second game ');
game3=input('Enter the points scored in the third game ');
ave_points=(game1+game2+game3)/3
```

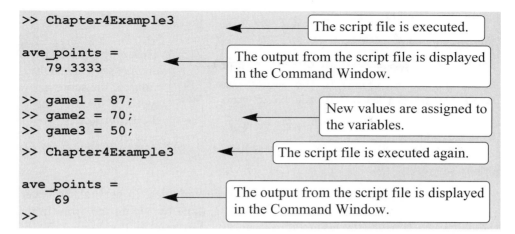

The following shows the Command Window when this script file (saved as

Chapter4Example4) is executed.

```
>> Chapter4Example4
Enter the points scored in the first game   67

Enter the points scored in the second game   91
Enter the points scored in the third game   70


ave_points =
    76
>>
```

The computer displays the message. Then the value of the score is typed by the user and the **Enter** key is pressed.

In this example scalars are assigned to the variables. In general, however, vectors and arrays can also be assigned. This is done by typing the array in the same way that it is usually assigned to a variable (left bracket, then typing row by row, and a right bracket).

The `input` command can also be used to assign a string to a variable. This can be done in one of two ways. One way is to use the command in the same form as shown above, and when the prompt message appears the string is typed between two single quotes in the same way that a string is assigned to a variable without the `input` command. The second way is to use an option in the `input` command that defines the characters that are entered as a string. The form of the command is:

```
variable_name = input('prompt message','s')
```

where the `'s'` inside the command defines the characters that will be entered as a string. In this case when the prompt message appears, the text is typed in without the single quotes, but it is assigned to the variable as a string. An example where the `input` command is used with this option is included in Sample Problem 6-4.

4.3 OUTPUT COMMANDS

As discussed before, MATLAB automatically generates a display when some commands are executed. For example, when a variable is assigned a value, or the name of a previously assigned variable is typed and the **Enter** key is pressed, MATLAB displays the variable and its value. This type of output is not displayed if a semicolon is typed at the end of the command. In addition to this automatic display, MATLAB has several commands that can be used to generate displays. The displays can be messages that provide information, numerical data, and plots. Two commands that are frequently used to generate output are `disp` and `fprintf`. The `disp` command displays the output on the screen, while the `fprintf` command can be used to display the output on the screen or to save the output to a file. The commands can be used in the Command Window, in a script file, and, as will be shown later, in a function file. When these commands are used in a script file, the display output that they generate is displayed in the Command Window.

4.3.1 The disp Command

The disp command is used to display the elements of a variable without displaying the name of the variable, and to display text. The format of the disp command is:

```
disp(name of a variable)   or   disp('text as string')
```

- Every time the disp command is executed, the display it generates appears in a new line. One example is:

```
>> abc = [5  9  1;  7  2  4];     A 2 × 3 array is assigned to variable abc.
>> disp(abc)                      The disp command is used to display the abc array.
     5     9     1
     7     2     4                The array is displayed without its name.

>> disp('The problem has no solution.')
                                  The disp command is used
                                  to display a message.
The problem has no solution.
>>
```

The next example shows the use of the disp command in the script file that calculates the average points scored in three games.

```
% This script file calculates the average points scored in three games.
% The points from each game are assigned to the variables by
% using the input command.
% The disp command is used to display the output.

game1=input('Enter the points scored in the first game    ');
game2=input('Enter the points scored in the second game   ');
game3=input('Enter the points scored in the third game    ');
ave_points=(game1+game2+game3)/3;
disp(' ')                                   Display empty line.
disp('The average of points scored in a game is:')   Display text.
disp(' ')                                   Display empty line.
disp(ave_points)       Display the value of the variable ave_points.
```

When this file (saved as Chapter4Example5) is executed, the display in the

Command Window is:

```
>> Chapter4Example5
Enter the points scored in the first game    89
Enter the points scored in the second game   60
Enter the points scored in the third game    82
                                             An empty line is displayed.
The average of points scored in a game is:   The text line is displayed.
                                             An empty line is displayed.
   77                  The value of the variable ave_points is displayed.
```

- Only one variable can be displayed in a disp command. If elements of two variables need to be displayed together, a new variable (that contains the elements to be displayed) must first be defined and then displayed.

In many situations it is nice to display output (numbers) in a table. This can be done by first defining a variable that is an array with the numbers and then using the disp command to display the array. Headings to the columns can also be created with the disp command. Since in the disp command the user cannot control the format (the width of the columns and the distance between the columns) of the display of the array, the position of the headings has to be aligned with the columns by adding spaces. As an example, the script file below shows how to display the population data from Chapter 2 in a table.

```
yr=[1984 1986 1988 1990 1992 1994 1996];     The population data is
pop=[127 130 136 145 158 178 211];           entered in two row vectors.
tableYP(:,1)=yr';     yr is entered as the first column in the array tableYP.
tableYP(:,2)=pop';    pop is entered as the second column in the array tableYP.
disp('          YEAR     POPULATION')     Display heading (first line).
disp('                   (MILLIONS)')     Display heading (second line).
disp(' ')                                 Display an empty line.
disp(tableYP)                             Display the array tableYP.
```

When this script file (saved as PopTable) is executed, the display in the Command Window is:

```
>> PopTable
          YEAR     POPULATION             Headings are displayed.
                   (MILLIONS)
                                          An empty line is displayed.
          1984        127
          1986        130
```

```
        1988        136
        1990        145
        1992        158
        1994        178
        1996        211
```

The `tableYP` array is displayed.

Another example of displaying a table is shown in Sample Problem 4-3. Tables can also be created and displayed with the `fprintf` command, which is explained in the next section.

4.3.2 The `fprintf` *Command*

The `fprintf` command can be used to display output (text and data) on the screen or to save it to a file. With this command (unlike with the `disp` command) the output can be formatted. For example, text and numerical values of variables can be intermixed and displayed in the same line. In addition, the format of the numbers can be controlled.

With many available options, the `fprintf` command can be long and complicated. To avoid confusion, the command is presented gradually. First, this section shows how to use the command to display text messages, then how to mix numerical data and text, next how to format the display of numbers, and finally how to save the output to a file.

Using the `fprintf` command to display text:

To display text, the `fprintf` command has the form:

```
fprintf('text typed in as a string')
```

For example:

```
fprintf('The problem, as entered, has no solution. Please check the
input data.')
```

If this line is part of a script file, then when the line is executed, the following is displayed in the Command Window:

```
The problem, as entered, has no solution. Please check the input data.
```

With the `fprintf` command it is possible to start a new line in the middle of the string. This is done by inserting `\n` before the character that will start the new line. For example, inserting `\n` after the first sentence in the previous example gives:

```
fprintf('The problem, as entered, has no solution.\nPlease
check the input data.')
```

When this line executes, the display in the Command Window is:

```
The problem, as entered, has no solution.
Please check the input data.
```

The `\n` is called an escape character. It is used to control the display. Other escape characters that can be inserted within the string are:

`\b` Backspace.
`\t` Horizontal tab.

When a program has more than one `fprintf` command, the display generated is continuous (the `fprintf` command does not automatically start a new line). This is true even if there are other commands between the `fprintf` commands. An example is the following script file:

```
fprintf('The problem, as entered, has no solution. Please check the
input data.')
x = 6; d = 19 + 5*x;
fprintf('Try to run the program later.')
y = d + x;
fprintf('Use different input values.')
```

When this file is executed the display in the Command Window is:

```
The problem, as entered, has no solution. Please check the
input data.Try to run the program later.Use different input
values.
```

To start a new line with the `fprintf` command, `\n` must be typed at the start of the string.

Using the `fprintf` command to display a mix of text and numerical data:

To display a mix of text and a number (value of a variable), the `fprintf` command has the form:

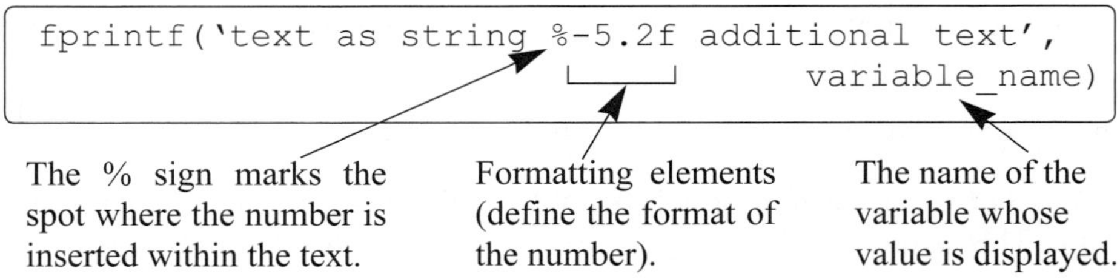
fprintf('text as string %-5.2f additional text',
variable_name)
The % sign marks the spot where the number is inserted within the text.
Formatting elements (define the format of the number).
The name of the variable whose value is displayed.

The formatting elements are:

–5.2f

Flag (optional). → –

Field width and precision (optional). → 5.2

Conversion character (required). → f

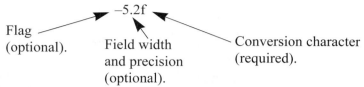

The flag, which is optional, can be one of the following three characters:

Character used for flag	Description
– (minus sign)	Left-justifies the number within the field.
+ (plus sign)	Prints a sign character (+ or –) in front of the number.
0 (zero)	Adds zeros if the number is shorter than the field.

The field width and precision (5.2 in the previous example) are optional. The first number (5 in the example) is the field width, which specifies the minimum number of digits in the display. If the number to be displayed is shorter than the field width, spaces or zeros are added in front of the number. The precision is the second number (2 in the example). It specifies the number of digits to be displayed to the right of the decimal point.

The last element in the formatting elements, which is required, is the conversion character, which specifies the notation in which the number is displayed. Some of the common notations are:

e	Exponential notation using lower-case e (e.g., 1.709098e+001).
E	Exponential notation using upper-case E (e.g., 1.709098E+001).
f	Fixed-point notation (e.g., 17.090980).
g	The shorter of e or f notations.
G	The shorter of E or f notations.
i	Integer.

Information about additional notation is available in the help menu of MATLAB. As an example, the `fprintf` command with a mix of text and a number is used in the script file that calculates the average points scored in three games.

```
% This script file calculates the average points scored in three games.
% The values are assigned to the variables by using the input command.
% The fprintf command is used to display the output.
game(1) = input('Enter the points scored in the first game    ');
game(2) = input('Enter the points scored in the second game   ');
game(3) = input('Enter the points scored in the third game    ');
ave_points = mean(game);
```

```
fprintf('An average of %f points was scored in the three games.',ave_points)
```

Text

% marks the position of the number.

Additional text.

The name of the variable whose value is displayed.

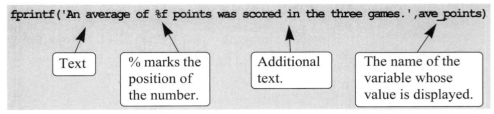

Notice that, besides using the `fprintf` command, this file differs from the ones shown earlier in the chapter in that the scores are stored in the first three elements of a vector named `game`, and the average of the scores is calculated by using the `mean` function. The Command Window where the script file above (saved as Chapter4Example6) was run is shown below.

```
>> Chapter4Example6
Enter the points scored in the first game     75
Enter the points scored in the second game    60
Enter the points scored in the third game     81
An average of 72.000000 points was scored in the three games.
>>
```

The display generated by the `fprintf` command combines text and a number (value of a variable).

With the `fprintf` command it is possible to insert more than one number (value of a variable) within the text. This is done by typing %g (or % followed by any formatting elements) at the places in the text where the numbers are to be inserted. Then, after the string argument of the command (following the comma), the names of the variables are typed in the order in which they are inserted in the text. In general the command looks like:

```
fprintf('..text...%g...%g...%f...',variable1,variable2,variable3)
```

An example is shown in the following script file:

```
% This program calculates the distance a projectile flies,
% given its initial velocity and the angle at which it is shot.
% the fprintf command is used to display a mix of text and numbers.

v=1584;   % Initial velocity (km/h)
theta=30;   % Angle (degrees)
vms=v*1000/3600;
t=vms*sind(30)/9.81;
d=vms*cosd(30)*2*t/1000;
```

Changing velocity units to m/s.

Calculating the time to highest point.

Calculating max distance.

```
fprintf('A projectile shot at %3.2f degrees with a velocity
of %4.2f km/h will travel a distance of %g km.\n',theta,v,d)
```

When this script file (saved as Chapter4Example7) is executed, the display in the Command Window is:

```
>> Chapter4Example7
A projectile shot at 30.00 degrees with a velocity of
1584.00 km/h will travel a distance of 17.091 km.
>>
```

Additional remarks about the `fprintf` command:

- To place a single quotation mark in the displayed text, type two single quotation marks in the string inside the command.
- The `fprintf` command is vectorized. This means that when a variable that is a vector or a matrix is included in the command, the command repeats itself until all the elements are displayed. If the variable is a matrix, the data is used column by column.

For example, the script file below creates a 2×5 matrix `T` in which the first row contains the numbers 1 through 5, and the second row shows the corresponding square roots.

```
x=1:5;                                    Create a vector x.
y=sqrt(x);                                Create a vector y.
T=[x; y]          Create 2 x 5 matrix T, first row is x, second row is y.
fprintf('If the number is: %i, its square root is: %f\n',T)
          The fprintf command displays two numbers from T in every line.
```

When this script file is executed, the display in the Command Window is:

```
T =
    1.0000    2.0000    3.0000    4.0000    5.0000     The 2 x 5 matrix T.
    1.0000    1.4142    1.7321    2.0000    2.2361
If the number is: 1, its square root is: 1.000000     The fprintf
If the number is: 2, its square root is: 1.414214     command repeats
If the number is: 3, its square root is: 1.732051     five times, using
If the number is: 4, its square root is: 2.000000     the numbers from
If the number is: 5, its square root is: 2.236068     the matrix T col-
                                                      umn after column.
```

Using the `fprintf` command to save output to a file:

In addition to displaying output in the Command Window, the `fprintf` command can be used for writing the output to a file when it is necessary to save the output. The data that is saved can subsequently be displayed or used in MATLAB and in other applications.

Writing output to a file requires three steps:

a) Opening a file using the `fopen` command.
b) Writing the output to the open file using the `fprintf` command.
c) Closing the file using the `fclose` command.

Step *a*:

Before data can be written to a file, the file must be opened. This is done with the `fopen` command, which creates a new file or opens an existing file. The `fopen` command has the form:

```
fid=fopen('file_name','permission')
```

`fid` is a variable called the file identifier. A scalar value is assigned to `fid` when `fopen` is executed. The file name is written (including its extension) within single quotes as a string. The permission is a code (also written as a string) that tells how the file is opened. Some of the more common permission codes are:

- `'r'` Open file for reading (default).
- `'w'` Open file for writing. If the file already exists, its content is deleted. If the file does not exist, a new file is created.
- `'a'` Same as `'w'`, except that if the file exists the written data is appended to the end of the file.
- `'r+'` Open file for reading and writing.
- `'w+'` Open file for writing and writing. If the file already exists, its content is deleted. If the file does not exists, a new file is created.
- `'a+'` Same as `'w+'`, except that if the file exists the written data is appended to the end of the file.

If a permission code is not included in the command, the file opens with the default code `'r'`. Additional permission codes are described in the help menu.

Step *b*:

Once the file is open, the `fprintf` command can be used to write output to the file. The `fprintf` command is used in exactly the same way as it is used to display output in the Command Window, except that the variable `fid` is inserted inside the command. The `fprintf` command then has the form:

```
fprintf(fid,'text %-5.2f additional text',vari
                                       able_name)
```

`fid` is added to the `fprintf` command.

Step *c*:

When the writing of data to the file is complete, the file is closed using the `fclose` command. The `fclose` command has the form:

```
fclose(fid)
```

Additional notes on using the `fprintf` command for saving output to a file:

- The created file is saved in the current directory.
- It is possible to use the `fprintf` command to write to several different files. This is done by first opening the files, assigning a different `fid` to each (e.g. `fid1`, `fid2`, `fid3`, etc.), and then using the `fid` of a specific file in the `fprintf` command to write to that file.

An example of using `fprintf` commands for saving output to two files is shown in the following script file. The program in the file generates two unit conversion tables. One table converts velocity units from miles per hour to kilometers per hour, and the other table converts force units from pounds to newtons. Each conversion table is saved to a different text file (extension .txt).

```
% Script file in which fprintf is used to write output to files.
% Two conversion tables are created and saved to two different files.
% One converts mi/h to km/h, the other converts lb to N.
clear all
Vmph=10:10:100;                     Creating a vector of velocities in mi/h.
Vkmh=Vmph.*1.609;                   Converting mph to km/h.
TBL1=[Vmph; Vkmh];                  Creating a table (matrix) with two rows.
Flb=200:200:2000;                   Creating a vector of forces in lb.
FN=Flb.*4.448;                      Converting lb to N.
TBL2=[Flb; FN];                     Creating a table (matrix) with two rows.
fid1=fopen('VmphtoVkm.txt','w');    Open a .txt file named VmphtoVkm.
fid2=fopen('FlbtoFN.txt','w');      Open a .txt file named FlbtoFN.
fprintf(fid1,'Velocity Conversion Table\n \n');
                    Writing a title and an empty line to the file fid1.
fprintf(fid1,'      mi/h           km/h    \n');
                    Writing two column headings to the file fid1.
fprintf(fid1,'   %8.2f       %8.2f\n',TBL1);
                    Writing the data from the variable TBL1 to the file fid1.
```

```
fprintf(fid2,'Force Conversion Table\n \n');
fprintf(fid2,'     Pounds        Newtons    \n');
fprintf(fid2,'   %8.2f       %8.2f\n',TBL2);
fclose(fid1);
fclose(fid2);
```

Writing the force conversion table (data in variable `TBL2`) to the file fid2.

Closing the files fid1 and fid2.

When the script file above is executed two new .txt files, named VmphtoVkm and FlbtoFN, are created and saved in the current directory. These files can be opened with any application that can read .txt files. Figures 4-3 and 4-4 show how the two files appear when they are opened with Microsoft Word.

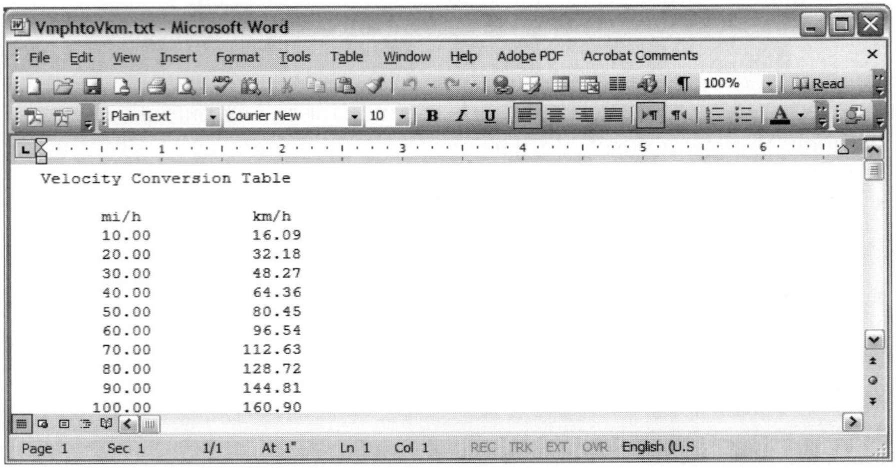

Figure 4-3: The VmphtoVkm.txt file opened in Word.

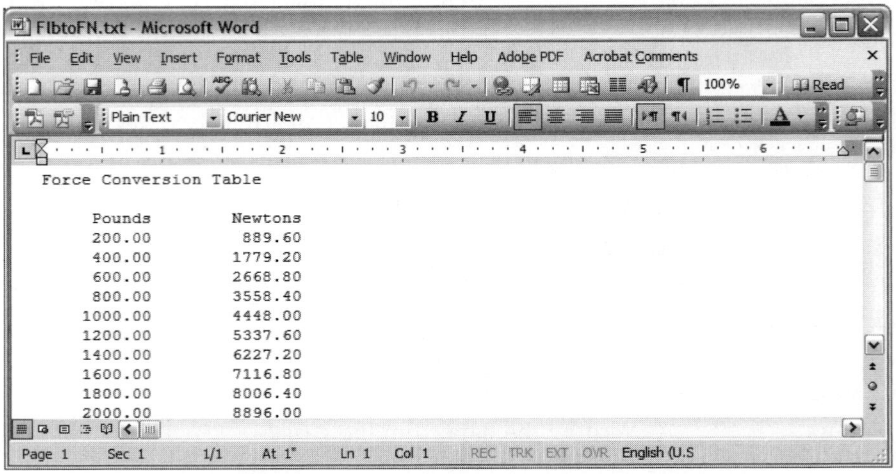

Figure 4-4: The FlbtoFN.txt file opened in Word.

4.4 THE save AND load COMMANDS

The save and load commands are most useful for saving and retrieving data for use in MATLAB. The save command can be used for saving the variables that are currently in the workspace, and the load command is used for retrieving variables that have been previously saved, to the workspace. The workspace can be saved when MATLAB is used in one type of platform (e.g., PC), and retrieved for use in MATLAB in another platform (e.g., Mac). The save and load commands can also be used for exchanging data with applications outside MATLAB. Additional commands that can be used for this purpose are presented in Section 4.5.

4.4.1 The save Command

The save command is used for saving the variables (all or some of them) that are stored in the workspace. The two simplest forms of the save command are:

`save file_name` and `save('file_name')`

When either one of these commands is executed, all of the variables currently in the workspace are saved in a file named file_name.mat that is created in the current directory. In mat files, which are written in a binary format, each variable preserves its name, type, size, and value. These files cannot be read by other applications. The save command can also be used for saving only some of the variables that are in the workspace. For example, to save two variables named var1 and var2 the command is:

`save file_name var1 var2` or

`save('file_name','var1','var2')`

The save command can also be used for saving in ASCII format, which can be read by applications outside MATLAB. Saving in ASCII format is done by adding the argument -ascii in the command (for example, save file_name -ascii). In the ASCII format the variable's name, type, and size are not preserved. The data is saved as characters separated by spaces but without the variable names. For example, the following shows how two variables (a 1×4 vector and a 2×3 matrix) are defined in the Command Window and then saved in ASCII format to a file named DatSavAsci:

```
>> V=[3 16 -4 7.3];                  Create a 1 × 4 vector V.
>> A=[6 -2.1 15.5; -6.1 8 11];       Create a 2 × 3 matrix A.
>> save -ascii DatSavAsci            Save variables to a file named DatSavAsci.
```

Once saved, the file can be opened by any application that can read ASCII files. For example, Figure 4-5 shows the data when the file is opened with Notepad.

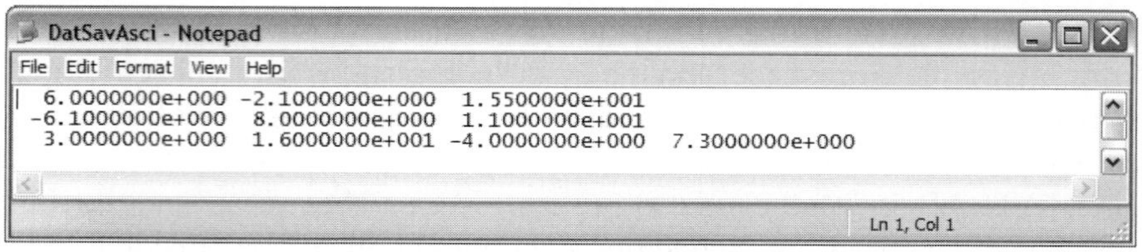

Figure 4-5: Data saved in ASCII format.

Note that the file does not include the names of the variables just the numerical values of the variables (first `A` and then `V`) are listed.

4.4.2 The `load` *Command*

The `load` command can be used for retrieving variables that were saved with the `save` command back to the workspace, and for importing data that was created with other applications and saved in ASCII format or in text (.txt) files. Variables that were saved with the `save` command in .mat files can be retrieved with the command:

```
load file_name
```

or

```
load('file_name')
```

When the command is executed, all the variables in the file (with the name, type, size, and values as were saved) are added (loaded back) to the workspace. If the workspace already has a variable with the same name as a variable that is retrieved with the `load` command, then the variable that is retrieved replaces the existing variable. The `load` command can also be used for retrieving only some of the variables that are in the saved .mat file. For example, to retrieve two variables named `var1` and `var2`, the command is:

```
load file_name var1 var2
```

or

```
load('file_name','var1','var2')
```

The `load` command can also be used to import data that is saved in ASCII or text (.txt) to the workspace. This is possible, however, only if the data in the file is in the form of a variable in MATLAB. Thus, the file can have one number (scalar), a row or a column of numbers (vector), or rows with the same number of numbers in each (matrix). For example, the data shown in Figure 4-5 cannot be loaded with the `load` command (even though it was saved in ASCII format with the `save` command) because the number of elements is not the same in all rows. (Recall that this file was created by saving two different variables.)

When data is loaded from an ASCII or text file into the workspace it has to be assigned to a variable name. Data in ASCII format can be loaded with either of the following two forms of the `load` command:

```
load file_name
```

or

```
VarName=load('file_name')
```

If the data is in a text file, the extension .txt has to be added to the file name. The form of the `load` command is then:

```
load file_name.txt
```

or

```
VarName=load('file_name.txt')
```

In the first form of the command the data is assigned to a variable that has the name of the file. In the second form the data is assigned to a variable named `VarName`.

For example, the data shown in Figure 4-6 (a 3×2 matrix) is typed in Notepad, and then saved as `DataFromText.txt`.

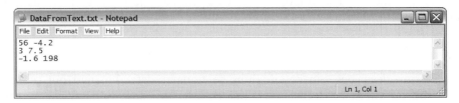

Figure 4-6: Data saved as .txt file.

Next, two forms of the `load` command are used to import the data in the text file to the Workspace of MATLAB. In the first command the data is assigned to a variable named `DfT`. In the second command the data is automatically assigned to a variable named `DataFromText`, which is the name of the text file where the data was saved.

```
>> DfT=load('DataFromText.txt')
DfT =
   56.0000   -4.2000
    3.0000    7.5000
   -1.6000  198.0000
>> load DataFromText.txt
>> DataFromText
DataFromText =
   56.0000   -4.2000
    3.0000    7.5000
   -1.6000  198.0000
```

Load the file `DataFromText` and assign the loaded data to the variable `Dft`.

Use the load command with the file `DataFromText`.

The data is assigned to a variable named `DataFromText`.

Importing data to (or exporting from) other applications can also be done, with MATLAB commands that are presented in the next section.

4.5 IMPORTING AND EXPORTING DATA

MATLAB is often used for analyzing data that was recorded in experiments or generated by other computer programs. This can be done by first importing the data into MATLAB. Similarly, data that is produced by MATLAB sometimes needs to be transferred to other computer applications. There are various types of data (numerical, text, audio, graphics, and images). This section describes only how to import and export numerical data, which is probably the most common type of data that needs to be transferred by new users of MATLAB. For other types of data transfer, look in the Help Window under File I/O.

Importing data can be done either by using commands or by using the Import Wizard. Commands are useful when the format of the data being imported is known. MATLAB has several commands that can be used for importing various types of data. Importing commands can also be included in a script file such that the data is imported when the script is executed. The Import Wizard is useful when the format of the data (or the command that is applicable for importing the data) is not known. The Import Wizard determines the format of the data and automatically imports it.

4.5.1 Commands for Importing and Exporting Data

This section describes—in detail—how to transfer data into and out of Excel spreadsheets. Microsoft Excel is commonly used for storing data, and Excel is compatible with many data recording devices and computer applications. Many people are also capable of importing and exporting various data formats into and from Excel. MATLAB also has commands for transferring data directly to and from formats such as csv and ASCII, and to the spreadsheet program Lotus 123. Details of these and many other commands can be found in the Help Window under File I/O

Importing and exporting data into and from Excel:

Importing data from Excel is done with the `xlsread` command. When the command is executed, the data from the spreadsheet is assigned as an array to a variable. The simplest form of the `xlsread` command is:

```
variable_name=xlsread('filename')
```

- `'filename'` (typed as a string) is the name of the Excel file. The directory of the Excel file must be either the current directory or listed in the search path.
- If the Excel file has more than one sheet, the data will be imported from the first sheet.

When an Excel file has several sheets, the `xlsread` command can be used to import data from a specified sheet. The form of the command is then:

```
variable_name = xlsread('filename','sheet_name')
```

- The name of the sheet is typed as a string.

Another option is to import only a portion of the data that is in the spreadsheet. This is done by typing an additional argument in the command:

```
variable_name = xlsread('filename','sheet_name','range')
```

- The `'range'` (typed as a string) is a rectangular region of the spreadsheet defined by the addresses (in Excel notation) of the cells at opposite corners of the region. For example, `'C2:E5'` is a 4×3 region of rows 2, 3, 4, and 5 and columns *C*, *D*, and *E*.

Exporting data from MATLAB to an Excel spreadsheet is done by using the `xlswrite` command. The simplest form of the command is:

```
xlswrite('filename',variable_name)
```

- `'filename'` (typed as a string) is the name of the Excel file to which the data is exported. The file must be in the current directory. If the file does not exist, a new Excel file with the specified name will be created.
- `variable_name` is the name of the variable in MATLAB with the assigned data that is being exported.
- The arguments `'sheet_name'` and `'range'` can be added to the `xlswrite` command to export to a specified sheet and to a specified range of cells, respectively.

As an example, the data from the Excel spreadsheet shown in Figure 4-7 is imported into MATLAB by using the `xlsread` command.

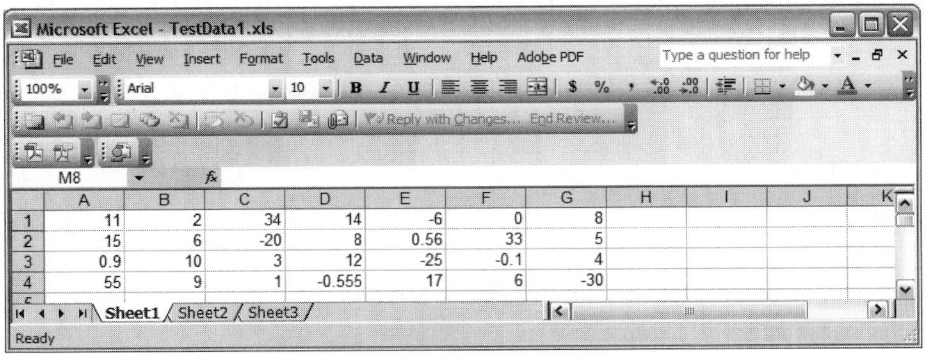

Figure 4-7: Excel spreadsheet with data.

The spreadsheet is saved in a file named TestData1 in a disk in drive A. After the Current Directory is changed to drive A, the data is imported into MATLAB by assigning it to the variable DATA:

```
>> DATA = xlsread('TestData1')
DATA =
   11.0000    2.0000   34.0000   14.0000   -6.0000         0    8.0000
   15.0000    6.0000  -20.0000    8.0000    0.5600   33.0000    5.0000
    0.9000   10.0000    3.0000   12.0000  -25.0000   -0.1000    4.0000
   55.0000    9.0000    1.0000   -0.5550   17.0000    6.0000  -30.0000
```

4.5.2 Using the Import Wizard

Using the Import Wizard is probably the easiest way to import data into MATLAB since the user does not have to know, or to specify, the format of the data. The Import Wizard is activated by selecting **Import Data** in the **File** menu of the Command Window. (It can also be started by typing the command `uiimport`.) The Import Wizard starts by displaying a file selection box that shows all the data files recognized by the Wizard. The user then selects the file that contains the data to be imported, and clicks **Open**. The Import Wizard opens the file and displays a portion of the data in a preview box so that the user can verify that the data is the correct choice. The Import Wizard tries to process the data, and if the wizard is successful, it displays the variables it has created with a portion of the data. The user clicks **next** and the wizard shows the Column Separator that was used. If the variable has the correct data, the user can proceed with the wizard (click **next**); otherwise the user can choose a different Column Separator. In the next window the wizard shows the name and size of the variable to be created in MATLAB. (When the data is all numerical, the variable in MATLAB has the same name as the file from which the data was imported.) When the wizard ends (click **finish**), the data is imported to MATLAB.

As an example, the Import Wizard is used to import numerical ASCII data saved in a .txt file. The data saved with the file name TestData2 is shown in Figure 4-8.

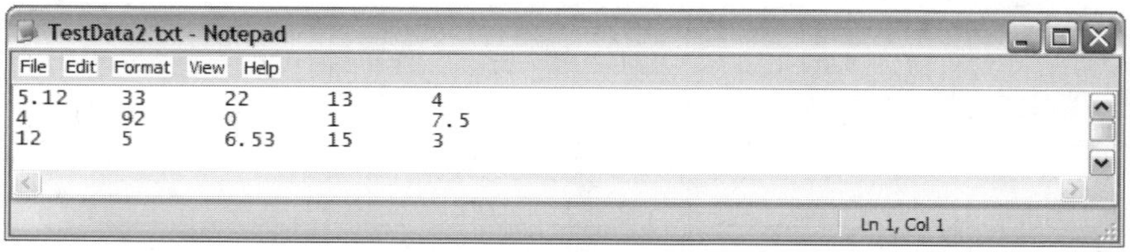

Figure 4-8: Numerical ASCII data.

The display of the Import Wizard during the import process for the TestData2 file is shown in Figures 4-9 and 4-10. Figure 4-10 shows that the name of the variable in MATLAB is `TestData2` and its size is 3×5.

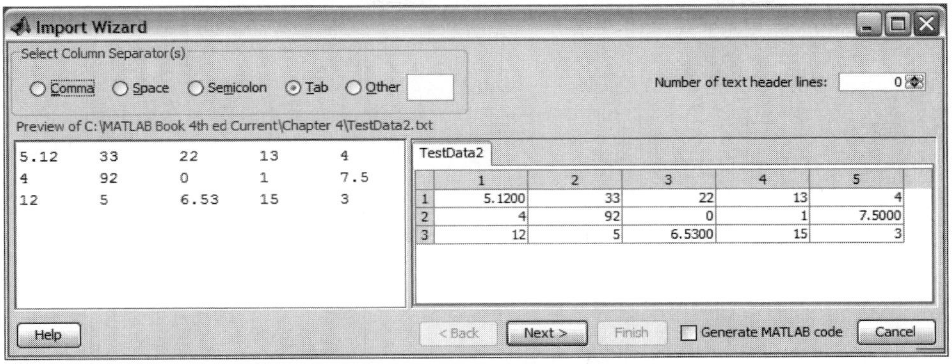

Figure 4-9: Import Wizard, first display.

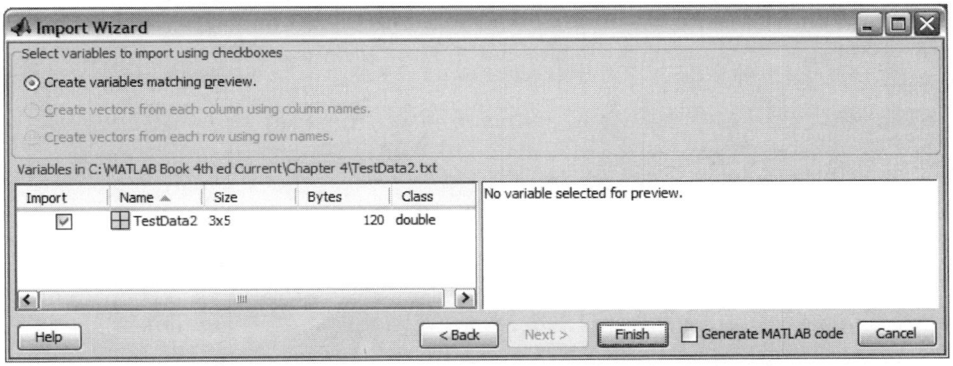

Figure 4-10: Import Wizard, second display.

In the Command Window of MATLAB, the imported data can be displayed by typing the name of the variable.

```
>> TestData2
TestData2 =
    5.1200   33.0000   22.0000   13.0000    4.0000
    4.0000   92.0000         0    1.0000    7.5000
   12.0000    5.0000    6.5300   15.0000    3.0000
```

4.6 EXAMPLES OF MATLAB APPLICATIONS

Sample Problem 4-1: Height and surface area of a silo

A cylindrical silo with radius r has a spherical cap roof with radius R. The height of the cylindrical portion is H. Write a program in a script file that determines the height H for given values of r, R, and the volume V. In addition, the program calculates the surface area of the silo.

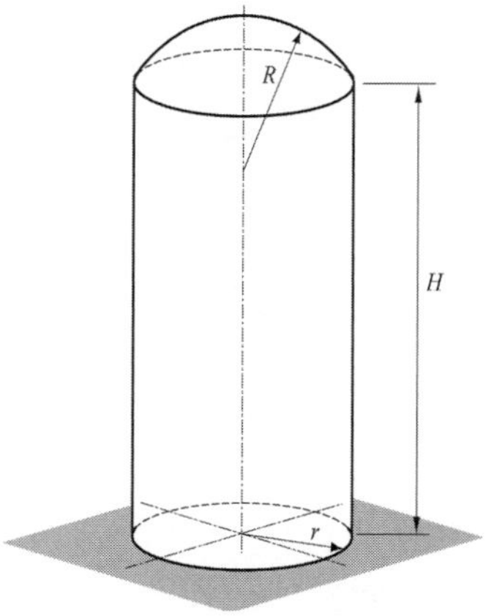

Use the program to calculate the height and surface area of a silo with $r = 30$ ft, $R = 45$ ft, and a volume of 120,000 ft^3. Assign values for r, R, and V in the Command Window.

Solution

The total volume of the silo is obtained by adding the volume of the cylindrical part and the volume of the spherical cap. The volume of the cylinder is given by

$$V_{cyl} = \pi r^2 H$$

and the volume of the spherical cap is given by:

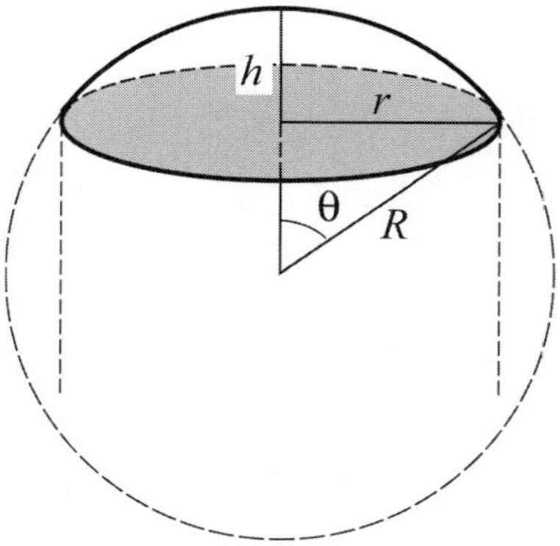

$$V_{cap} = \frac{1}{3}\pi h^2(3R - h)$$

where $h = R - R\cos\theta = R(1 - \cos\theta)$,

and θ is calculated from $\sin\theta = \frac{r}{R}$.

Using the equations above, the height, H, of the cylindrical part can be expressed by

$$H = \frac{V - V_{cap}}{\pi r^2}$$

The surface area of the silo is obtained by adding the surface areas of the cylindrical part and the spherical cap.

$$S = S_{cyl} + S_{cap} = 2\pi r H + 2\pi R h$$

A program in a script file that solves the problem is presented below:

```
theta=asin(r/R);                 Calculating θ.
h=R*(1-cos(theta));              Calculating h.
Vcap=pi*h^2*(3*R-h)/3;           Calculating the volume of the cap.
```

```
H=(V-Vcap)/(pi*r^2);                                  Calculating H.
S=2*pi*(r*H + R*h);                      Calculating the surface area S.
fprintf('The height H is: %f ft.',H)
fprintf('\nThe surface area of the silo is: %f square ft.',S)
```

The Command Window where the script file, named silo, was executed is:

```
>> r=30; R=45; V=200000;                Assigning values to r, R, and V.
>> silo                               Running the script file named silo.
The height H is: 64.727400 ft.
The surface area of the silo is: 15440.777753 square ft.
```

Sample Problem 4-2: Centroid of a composite area

Write a program in a script file that calculates the coordinates of the centroid of a composite area. (A composite area can easily be divided into sections whose centroids are known.) The user needs to divide the area into sections and know the coordinates of the centroid (two numbers) and the area of each section (one number). When the script file is executed, it asks the user to enter the three numbers as a row in a matrix. The user enters as many rows as there are sections. A section that represents a hole is taken to have a negative area. For output, the program displays the coordinates of the centroid of the composite area. Use the program to calculate the centroid of the area shown in the figure.

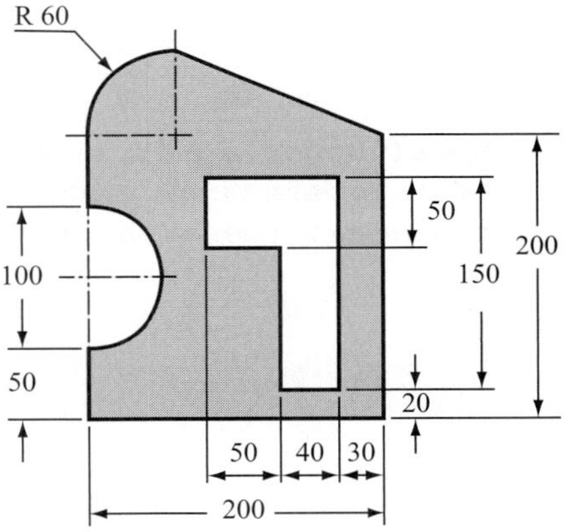

Dimensions in mm

Solution

The area is divided into six sections as shown in the following figure. The total area is calculated by adding the three sections on the left and subtracting the three sections on the right. The location and coordinates of the centroid of each section are marked in the figure, as well as the area of each section.

The coordinates $\bar{X}$ and $\bar{Y}$ of the centroid of the total area are given by

$\bar{X} = \frac{\Sigma A\bar{x}}{\Sigma A}$ and $\bar{Y} = \frac{\Sigma A\bar{y}}{\Sigma A}$, where $\bar{x}$, $\bar{y}$, and A are the coordinates of the centroid and area of each section, respectively.

A script file with a program for calculating the coordinates of the centroid of a composite area is provided below.

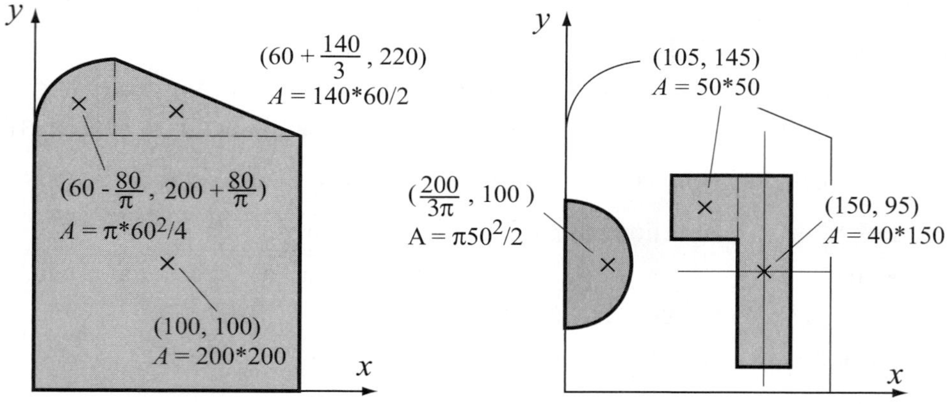

Units: coordinates mm, area mm^2

```
% The program calculates the coordinates of the centroid
% of a composite area.
clear C xs ys As
C=input('Enter a matrix in which each row has three ele-
ments.\nIn each row enter the x and y coordinates of the
centroid and the area of a section.\n');
xs=C(:,1)';          Creating a row vector for the x coordinate of
                     each section (first column of C).
ys=C(:,2)';          Creating a row vector for the y coordinate of
                     each section (second column of C).
As=C(:,3)';          Creating a row vector for the area of each
                     section (third column of C).
A=sum(As);           Calculating the total area.
x=sum(As.*xs)/A;     Calculating the coordinates of the
y=sum(As.*ys)/A;     centroid of the composite area.
fprintf('The coordinates of the centroid are: ( %f, %f )\n',x,y)
```

The script file was saved with the name Centroid. The following shows the Command Window where the script file was executed.

```
>> Centroid
Enter a matrix in which each row has three elements.
In each row enter the x and y coordinates of the centroid
and the area of a section.
```

```
[100 100 200*200
60-80/pi 200+80/pi pi*60^2/4
60+140/3 220 140*60/2
200/(3*pi) 100 -pi*50^2/2
105 145 -50*50
150 95 -40*150]
The coordinates of the centroid are: ( 85.387547 , 131.211809 )
```

Entering the data for matrix C. Each row has three elements: the *x*, *y*, and *A* of a section.

Sample Problem 4-3: Voltage divider

When several resistors are connected in an electrical circuit in series, the voltage across each of them is given by the voltage divider rule:

$$v_n = \frac{R_n}{R_{eq}} v_s$$

where v_n and R_n are the voltage across resistor n and its resistance, respectively, $R_{eq} = \Sigma R_n$ is the equivalent resistance, and v_s is the source voltage. The power dissipated in each resistor is given by:

$$P_n = \frac{R_n}{R_{eq}^2} v_s^2$$

The figure below shows a circuit with seven resistors connected in series.

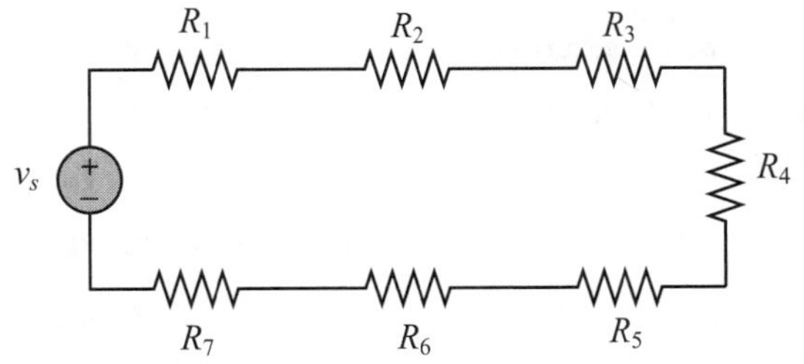

Write a program in a script file that calculates the voltage across each resistor, and the power dissipated in each resistor, in a circuit that has resistors connected in series. When the script file is executed it requests the user to first enter the source voltage and then to enter the resistances of the resistors in a vector. The program displays a table with the resistance listed in the first column, the voltage across the resistor in the second column, and the power dissipated in the resistor in the third column. Following the table, the program displays the current in the circuit and the total power.

Execute the file and enter the following data for v_s and the R's.

$v_s = 24\text{V}$, $R_1 = 20\Omega$, $R_2 = 14\Omega$, $R_3 = 12\Omega$, $R_4 = 18\Omega$, $R_5 = 8\Omega$, $R_6 = 15\Omega$, $R_7 = 10\Omega$.

Solution

A script file that solves the problem is shown below.

```
% The program calculates the voltage across each resistor
% in a circuit that has resistors connected in series.
vs=input('Please enter the source voltage ');
Rn=input('Enter the values of the resistors as elements in a
row vector\n');
Req=sum(Rn);                 Calculate the equivalent resistance.
vn=Rn*vs/Req;                Apply the voltage divider rule.
Pn=Rn*vs^2/Req^2;            Calculate the power in each resistor.
i = vs/Req;                  Calculate the current in the circuit.
Ptotal = vs*i;               Calculate the total power in the circuit.
Table = [Rn', vn', Pn'];     Create a variable table with the
                             vectors Rn, vn, and Pn as columns.
disp(' ')
disp(' Resistance Voltage   Power')     Display headings for
disp('    (Ohms)     (Volts)    (Watts)')   the columns.
disp(' ')                    Display an empty line.
disp(Table)                  Display the variable Table.
disp(' ')
fprintf('The current in the circuit is %f Amps.',i)
fprintf('\nThe total power dissipated in the circuit is %f
Watts.',Ptotal)
```

The Command Window where the script file was executed is:

```
>> VoltageDivider                              Name of the script file.
Please enter the source voltage 24  <---       Voltage entered by the user.
Enter the value of the resistors as elements in a row vector

[20  14  12  18  8  15  10] <---               Resistor values entered as a vector.
  Resistance  Voltage    Power
    (Ohms)     (Volts)    (Watts)
   20.0000     4.9485    1.2244
   14.0000     3.4639    0.8571
   12.0000     2.9691    0.7346
   18.0000     4.4536    1.1019
    8.0000     1.9794    0.4897
```

```
15.0000        3.7113     0.9183
10.0000        2.4742     0.6122

The current in the circuit is 0.247423 Amps.
The total power dissipated in the circuit is 5.938144 Watts.
```

4.7 PROBLEMS

Solve the following problems by first writing a program in a script file and then executing the program.

1. The wind chill temperature, T_{wc}, is the air temperature felt on exposed skin due to wind. In U.S. customary units it is calculated by

$$T_{wc} = 35.74 + 0.6215T - 35.75v^{0.16} + 0.4275T\,v^{0.16}$$

where T is the temperature in degrees F and v is the wind speed in mi/h. Write a MATLAB program in a script file that calculates T_{wc}. For input the program asks the user to enter values for T and v. For output the program displays the message: "The wind chill temperature is: XX," where XX is the value of the wind chill temperature rounded to the nearest integer. Execute the program entering $T = 30°\text{F}$ and $v = 42$ mi/h.

2. The monthly payment M of a loan amount P for y years and with interest rate r can be calculated by the formula:

$$M = \frac{P(r/12)}{1-(1+r/12)^{-12y}}$$

Calculate the monthly payment and the total payment for a \$100,000 loan for 10, 11, 12, ... , 29, 30 years with an interest rate of 4.85%. Display the results in a three-column table where the first column is the number of years, the second is the monthly payment, and the third is the total payment.

3. A torus-shaped water tube is designed to have a volume of 8,000 in.^{3}. The volume of the tube, V, and its surface area, S, are given by:

$$V = \frac{1}{4}\pi^2(a+b)(b-a)^2 \quad \text{and} \quad S = \pi^2(b^2-a^2)$$

If $a = Kb$, determine S and a and b for $K = 0.2$, 0.3, 0.4, 0.6, and 0.7. Display the results in a table.

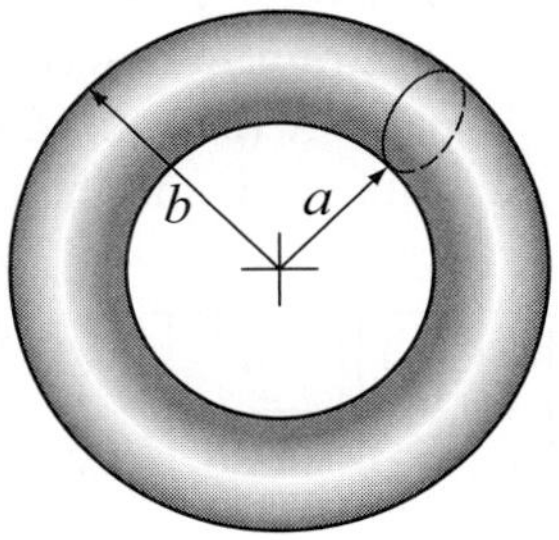

4. An ice cream container shaped as a frustum of a cone with $R_2 = 1.2R_1$ is designed to have a volume of 1,000 cm^3. Determine R_1, R_2, and the surface area, *S*, of the paper for containers with heights *h* of 8, 10, 12, 14, and 16 cm. Display the results in a table.

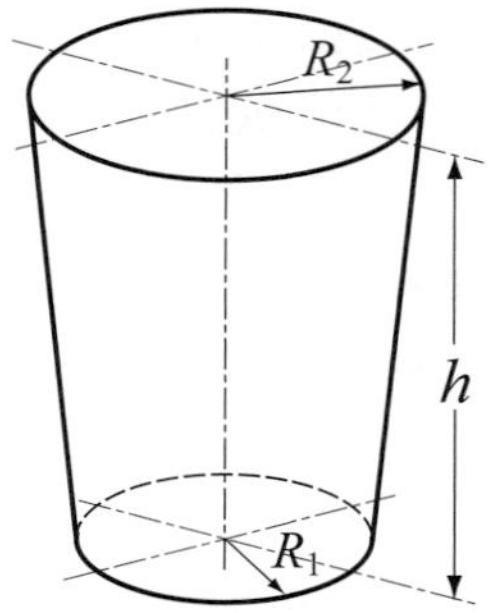

The volume of the container, *V*, and the surface area of the paper are given by:

$$V = \frac{1}{3}\pi h(R_1^2 + R_2^2 + R_1 R_2)$$

$$S = \pi(R_1 + R_2)\sqrt{(R_2 - R_1)^2 + h^2} + \pi(R_1^2 + R_2^2)$$

5. Write a MATLAB program in a script file that calculate the average, standard deviation, and median of a list of grades as well as the number of grades on the list. The program asks the user (`input` command) to enter the grades as elements of a vector. The program then calculates the required quantities using MATLAB's built-in functions `length`, `mean`, `std`, and `median`. The results are displayed in the Command Window in the following format:
"There are XX grades." where XX is the numerical value.
"The average grade is XX." where XX is the numerical value.
"The standard deviation is XX." where XX is the numerical value.
"The median deviation is XX." where XX is the numerical value.
Execute the program and enter the following grades: 81, 65, 61, 78, 94, 80, 65, 76, 77, 95, 82, 49, and 75.

6. The growth of some bacteria populations can be described by

$$N = N_0 e^{kt}$$

where *N* is the number of individuals at time *t*, N_0 is the number at time $t = 0$, and *k* is a constant. Assuming the number of bacteria doubles every hour, determine the number of bacteria every hour for 24 hours starting from an initial single bacterium.

7. A rocket flying straight up measures the angle θ with the horizon at different heights *h*. Write a MATLAB program in a script file that calculates the radius of the earth *R* (assuming the earth is a perfect sphere) at each data point and then determines the average of all the values.

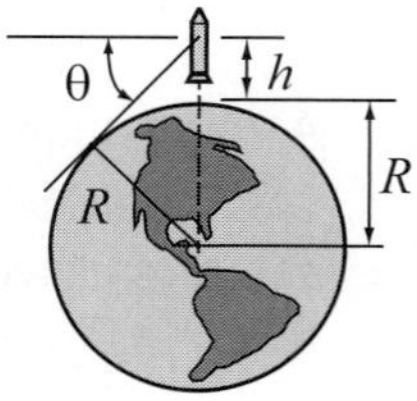

h (km)	4	8	12	16	20	24	28	32	36	40
θ (deg)	2.0	2.9	3.5	4.1	4.5	5.0	5.4	5.7	6.1	6.4

8. A railroad bumper is designed to slow down a rapidly moving railroad car. After a 20,000 kg railroad car traveling at 20 m/s engages the bumper, its displacement x (in meters) and velocity v (in m/s) as a function of time t (in seconds) is given by:

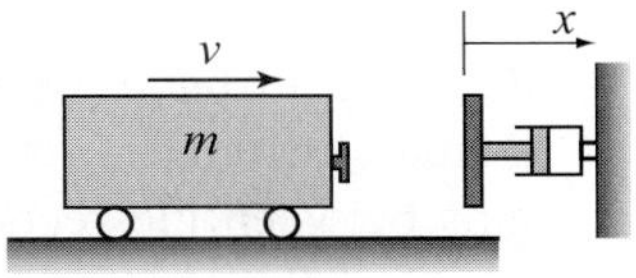

$$x(t) = 4.219(e^{-1.58t} - e^{-6.32t}) \quad \text{and} \quad v(t) = 26.67e^{-6.32t} - 6.67e^{-1.58t}$$

Determine x and v for every two hundredth of a second for the first half second after impact. Display the results in a three-column table in which the first column is time (s), the second is displacement (m), and the third is velocity (m/s).

9. Decay of radioactive materials can be modeled by the equation $A = A_0 e^{kt}$, where A is the amount at time t, A_0 is the amount at $t = 0$, and k is the decay constant ($k \leq 0$). Iodine-132 is a radioisotope that is used in thyroid function tests. Its half-life time is 13.3 hours. Calculate the relative amount of Iodine-132 (A/A_0) in a patient's body 48 hours after receiving a dose. After determining the value of k, define a vector $t = 0, 4, 8, \ldots, 48$ and calculate the corresponding values of A/A_0.

10. The value, B, of a savings account of an amount A that is deposited for n years with a yearly interest rate of r is given by:

$$B = A\left(1 + \frac{r}{100}\right)^n$$

Write a MATLAB program in a script file that calculates the balance B after 10 years for an initial deposit of \$10,000 for yearly interest rates ranging from 2% to 6% with increments of 0.5%. Display the results in a table. The table should have two columns where the first column displays the interest rate and the second displays the corresponding value of B.

11. A rectangular printed page with sides of lengths a and b is designed to have a printed area of 60 in.^{2} and margins of 1.75 in. at the top and bottom and 1.2 in. at both sides. Write a MATLAB program that determine the dimensions of a and b such that the overall area of the page will be as small as possible. In the program define a vector a with values ranging from 5 to 20 with increments of 0.05. Use this vector for calculating the corresponding values of b and the overall area of the page. Then use MATLAB's built-in function `min` to find the dimensions of the smallest page.

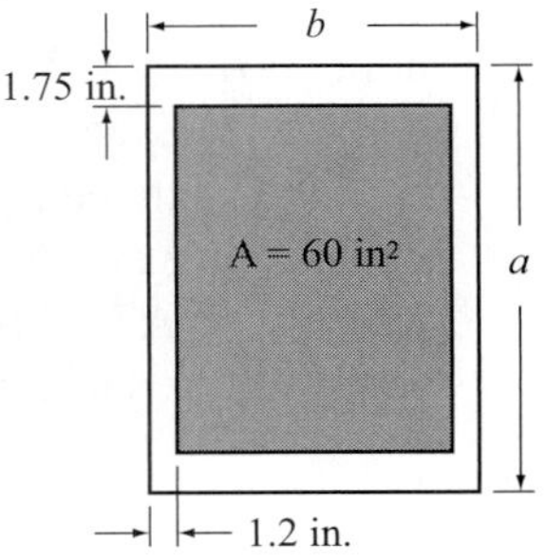

12. A round billboard with radius R = 55 in. is designed to have a rectangular picture placed inside a rectangle with sides a and b. The margins between the rectangle and the picture are 10 in. at the top and bottom and 4 in. at each side. Write a MATLAB program that determines the dimensions a and b such that the overall area of the picture will be as large as possible. In the program define a vector a with values ranging from 5 to 100 with increments of 0.25. Use this vector for calculating the corresponding values of b and the overall area of the picture. Then use MATLAB's built-in function `max` to find the dimensions of the largest rectangle.

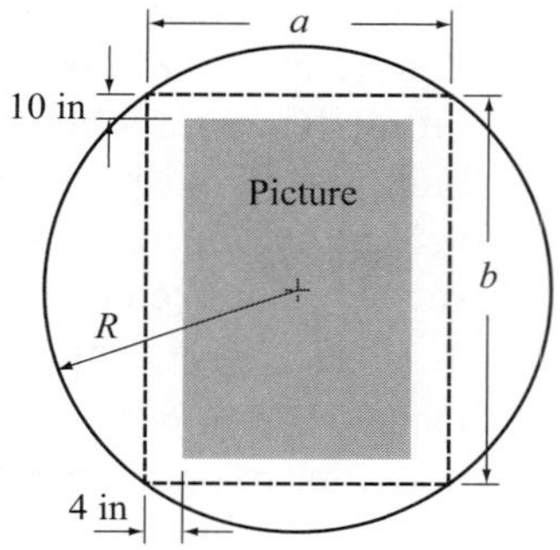

13. The balance of a loan, B, after n monthly payments is given by

$$B = A\left(1 - \frac{r}{1200}\right)^n - \frac{P}{r/1200}\left[\left(1 + \frac{r}{1200}\right)^n - 1\right]$$

where A is the loan amount, P is the amount of a monthly payment, and r is the yearly interest rate entered in % (e.g., 7.5% entered as 7.5). Consider a 5-year, \$20,000 car loan with 6.5% yearly interest that has a monthly payment of \$391.32. Calculate the balance of the loan after every 6 months (i.e., at n = 6, 12, 18, 24, ... , 54, 60). Each time calculate the percent of the loan that is already paid. Display the results in a three-column table, where the first column displays the month, and the second and third columns display the corresponding value of B and percentage of the loan that is already paid, respectively.

14. A large TV screen of height H = 50 ft is placed on the side wall of a tall building. The height from the street to the bottom of the screen is h = 130 ft. The best view of the screen is when θ is maximum. Write a MATLAB program that determines the distance x at which θ is at maximum. Define a vector x with elements ranging from 30 to 300 with spacing of 0.5. Use this vector to calculate the corresponding values of θ. Then use MATLAB's built-in function `min` to find the value of x that corresponds to the largest value of θ.

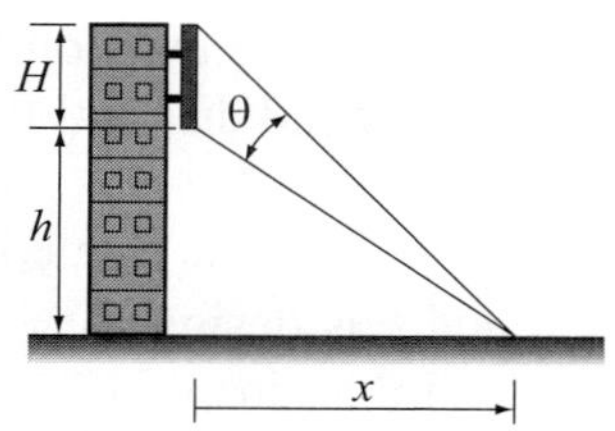

15. A student has a summer job as a lifeguard at the beach. After spotting a swimmer in trouble, he tries to deduce the path by which he can reach the swimmer in the shortest time. The path of shortest distance (path A) is obviously not the best since it maximizes the time spent swimming (he can run faster than he can swim). Path B minimizes the time spent swimming but is probably not the best since it is the longest (reasonable) path. Clearly the optimal path is somewhere in between paths A and B.

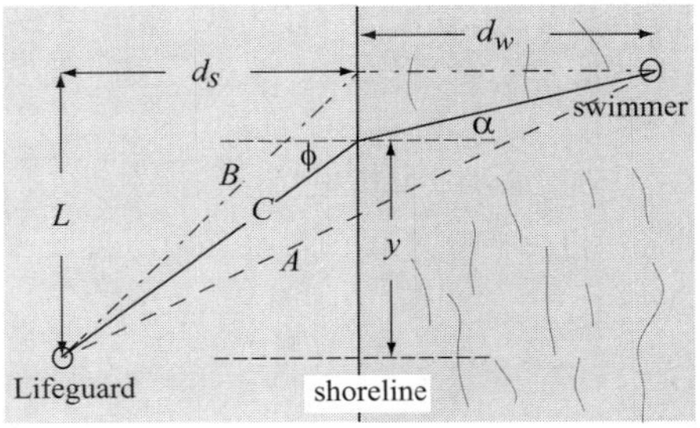

Consider an intermediate path C and determine the time required to reach the swimmer in terms of the running speed $v_{run} = 3$ m/s the swimming speed $v_{swim} = 1$ m/s; the distances $L = 48$ m, $d_s = 30$ m, and $d_w = 42$ m; and the lateral distance y at which the lifeguard enters the water. Create a vector y that ranges between path A and path B ($y = 20, 21, 22, \ldots, 48$ m) and compute a time t for each y. Use MATLAB built-in function `min` to find the minimum time t_{min} and the entry point y for which it occurs. Determine the angles that correspond to the calculated value of y and investigate whether your result satisfies Snell's law of refraction:

$$\frac{\sin\phi}{\sin\alpha} = \frac{v_{run}}{v_{swim}}$$

16. The airplane shown is flying at a constant speed of $v = 50$ m/s in a circular path of radius $\rho = 2000$ m and is being tracked by a radar station positioned a distance $h = 500$ m below the bottom of the plane path (point A). The airplane is at point A at $t = 0$, and the angle α as a function of time is given (in radians) by $\alpha = \frac{v}{\rho}t$. Write a MATLAB program that calculates θ and r as functions of time. The program should first determine the time at which $\alpha = 90°$. Then construct a vector t having 15 elements over the interval $0 \le t \le t_{90°}$, and calculate θ and r at each time. The program should print the values of ρ, h, and v, followed by a 15×3 table where the first column is t, the second is the angle θ in degrees, and the third is the corresponding value of r.

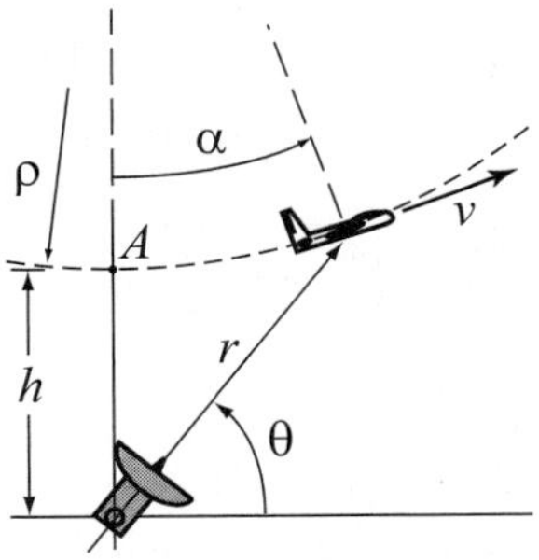

17. Early explorers often estimated altitude by measuring the temperature of boiling water. Use the following two equations to make a table that modern-day hikers could use for the same purpose.

$$p = 29.921(1 - 6.8753 \times 10^{-6}h), \qquad T_b = 49.161 \ln p + 44.932$$

where p is atmospheric pressure in inches of mercury, T_b is boiling temperature in °F, and h is altitude in feet. The table should have two columns, the first altitude and the second boiling temperature. The altitude should range between –500 ft and 10,000 ft at increments of 500 ft.

18. The variation of vapor pressure p (in units of mm Hg) of benzene with temperature in the range of $0 \le T \le 42°\text{C}$ can be modeled with the equation (Handbook of Chemistry and Physics, CRC Press)

$$\log_{10} p = b - \frac{0.05223a}{T}$$

where $a = 34172$ and $b = 7.9622$ are material constants and T is absolute temperature (K). Write a program in a script file that calculates the pressure for various temperatures. The program should create a vector of temperatures from $T = 0°\text{C}$ to $T = 42°$°C with increments of 2 degrees, and display a two-column table p and T, where the first column temperatures in °C, and the second column the corresponding pressures in mm Hg.

19. For many gases the temperature dependence of the heat capacity C_p of can be described in terms of a cubic equation:

$$C_p = a + bT + cT^2 + dT^3$$

The following table gives the coefficients of the cubic equation for four gases. C_p is in joules/(g mol)(°C) and T is in °C.

Gas	a	b	c	d
SO_2	38.91	3.904×10^{-2}	-3.105×10^{-5}	8.606×10^{-9}
SO_3	48.50	9.188×10^{-2}	-8.540×10^{-5}	32.40×10^{-9}
O_2	29.10	1.158×10^{-2}	-0.6076×10^{-5}	1.311×10^{-9}
N_2	29.00	0.2199×10^{-2}	-0.5723×10^{-5}	-2.871×10^{-9}

Calculate the heat capacity for each gas at temperatures ranging between 200 and 400 °C at 20 °C increments. To present the results, create an 11×5 matrix where the first column is the temperature, and the second through fifth columns are the heat capacities of SO_2, SO_3, O_2, and N_2, respectively.

20. The heat capacity of an ideal mixture of four gases $C_{p_{mixture}}$ can be expressed in terms of the heat capacity of the components by the mixture equation

$$C_{p_{mixture}} = x_1 C_{p1} + x_2 C_{p2} + x_3 C_{p3} + x_4 C_{p4}$$

where x_1, x_2, x_3, and x_4 are the fractions of the components, and C_{p1}, C_{p2}, C_{p3}, and C_{p4} are the corresponding heat capacities. A mixture of unknown quantities of the four gases SO_2, SO_3, O_2, and N_2 is given. To determine the fractions of the components, the following values of the heat capacity of the mixture were measured at three temperatures:

Temperature °C	25	150	300
$C_{p_{mixture}}$ joules/(g mol)(°C)	39.82	44.72	49.10

Use the equation and data in the Problem 19 to determine the heat capacity of each of the four components at the three temperatures. Then use the mixture equation to write three equations for the mixture at the three temperatures. The fourth equation is $x_1 + x_2 + x_3 + x_4 = 1$. Determine x_1, x_2, x_3, and x_4 by solving the linear system of equations.

21. When several resistors are connected in an electrical circuit in parallel, the current through each of them is given by $i_n = \frac{v_s}{R_n}$ where i_n and R_n are the current through resistor n and its resistance, respectively, and v_s is the source voltage. The equivalent resistance, R_{eq}, can be determined from the equation

$$\frac{1}{R_{eq}} = \frac{1}{R_1} + \frac{1}{R_2} + \ldots + \frac{1}{R_n}$$

The source current is given by $i_s = v_s / R_{eq}$, and the power, P_n, dissipated in each resistor is given by $P_n = v_s i_n$.

Write a program in a script file that calculates the current through each resistor and the power dissipated in a circuit that has resistors connected in parallel. When the script file runs, it asks the user first to enter the source voltage and then to enter the resistors' resistance in a vector. The program displays a table with the resistance shown in the first column, the current through the resistor in the second column, and the power dissipated in the resistor in the third column. Following the table, the program displays the source current and the total power. Use the script file to solve the following circuit.

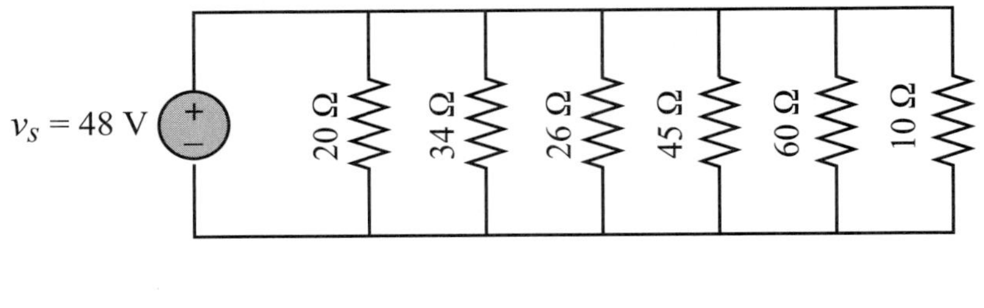

22. A truss is a structure made of members joined at their ends. For the truss shown in the figure, the forces in the nine members are determined by solving the following system of nine equations.

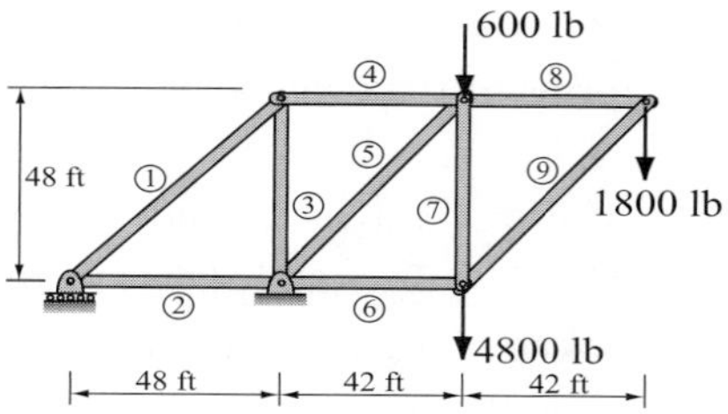

$$-\cos(45°)F_1 + F_4 = 0$$
$$-F_3 - \sin(45°)F_1 = 0$$
$$-F_2 + \sin(45°)F_5 + F_6 = 0$$
$$-\cos(48.81°)F_5 - F_4 + F_8 = 0,\quad -\sin(48.81°)F_5 - F_7 = 600$$
$$-\sin(48.81°)F_9 = 1800,\quad -F_8 - \cos(48.81°)F_9 = 0,$$
$$F_7 + \sin(48.81°)F_9 = 4800,\quad \cos(48.81°)F_9 - F_6 = 0$$

Write the equations in matrix form and use MATLAB to determine the forces in the members. A positive force means tensile force and a negative force means compressive force. Display the results in a table.

23. A truss is a structure made of members joined at their ends. For the truss shown in the figure, the forces in the 11 members are determined by solving the following system of 11 equations.

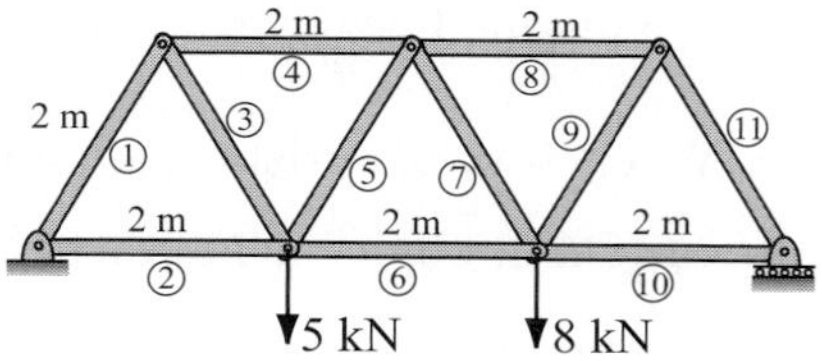

$$\frac{1}{2}F_1 + F_2 = 0,\quad \frac{\sqrt{3}}{2}F_1 = -6$$
$$-\frac{1}{2}F_1 + \frac{1}{2}F_3 + F_4 = 0,\quad -\frac{\sqrt{3}}{2}F_1 - \frac{\sqrt{3}}{2}F_3 = 0,\quad -F_2 - \frac{1}{2}F_3 + \frac{1}{2}F_5 + F_6 = 0$$
$$\frac{\sqrt{3}}{2}F_3 + \frac{\sqrt{3}}{2}F_5 = 5,\quad -F_4 - \frac{1}{2}F_5 + \frac{1}{2}F_7 + F_8 = 0,\quad -\frac{\sqrt{3}}{2}F_5 - \frac{\sqrt{3}}{2}F_7 = 0$$
$$-F_6 - \frac{1}{2}F_7 + \frac{1}{2}F_9 + F_{10} = 0,\quad \frac{\sqrt{3}}{2}F_7 + \frac{\sqrt{3}}{2}F_9 = 8,\quad -F_8 - \frac{1}{2}F_9 + \frac{1}{2}F_{11} = 0$$

Write the equations in matrix form and use MATLAB to determine the forces in the members. A positive force means tensile force and a negative force means compressive force. Display the results in a table.

24. The graph of the function $f(x) = ax^4 + bx^3 + cx^2 + dx + e$ passes through the points (–4, –7.6), (–2, –17.2), (0.2, 9.2), (1, –1.6), and (4, –36.4). Determine the constants a, b, c, d, and e. (Write a system of five equations with five unknowns and use MATLAB to solve the equations.)

25. The surface of many airfoils can be described with an equation of the form

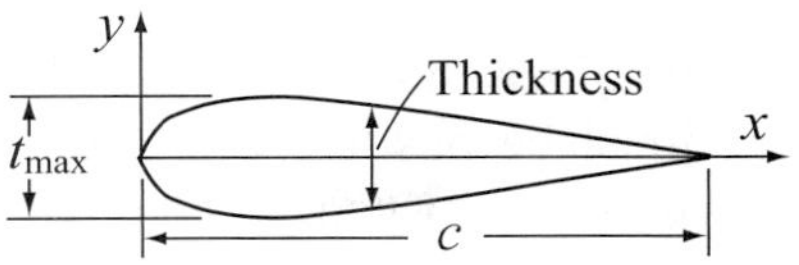

$$y = \mp \frac{tc}{0.2}[a_0\sqrt{x/c} + a_1(x/c) +$$
$$+ a_2(x/c)^2 + a_3(x/c)^3 + a_4(x/c)^4]$$

where t is the maximum thickness as a fraction of the chord length c (e.g., $t_{max} = ct$). Given that $c = 1$ m and $t = 0.2$ m, the following values for y have been measured for a particular airfoil:

x (m)	0.15	0.35	0.5	0.7	0.85
y (m)	0.08909	0.09914	0.08823	0.06107	0.03421

Determine the constants a_0, a_1, a_2, a_3, and a_4. (Write a system of five equations and five unknowns and use MATLAB to solve the equations.)

26. During a golf match, a certain number of points are awarded for each eagle and a different number for each birdie. No points are awarded for par, and a certain number of points are deducted for each bogey and a different number deducted for each double bogey (or worse). The newspaper report of an important match neglected to mention what these point values were, but did provide the following table of the results:

Golfer	Eagles	Birdies	Pars	Bogeys	Doubles	Points
A	1	2	10	1	1	5
B	2	3	11	0	1	12
C	1	4	10	1	10	11
D	1	3	10	12	0	8

From the information in the table write four equations in terms of four unknowns. Solve the equations for the unknown points awarded for eagles and birdies and points deducted for bogeys and double bogeys.

27. The dissolution of copper sulfide in aqueous nitric acid is described by the following chemical equation:

$$aCuS + bNO_3^- + cH^+ \rightarrow dCu^{2+} + eSO_4^{2-} + fNO + gH_2O$$

where the coefficients a, b, c, d, e, f, and g are the numbers of the various molecule participating in the reaction and are unknown. The unknown coefficients are determined by balancing each atom on left and right and then balancing the ionic charge. The resulting equations are:

$a = d$, $a = e$, $b = f$, $3b = 4e + f + g$, $c = 2g$, $-b + c = 2d - 2e$

There are seven unknowns and only six equations. A solution can still be obtained, however, by taking advantage of the fact that all the coefficients must be positive integers. Add a seventh equation by guessing $a = 1$ and solve the system of equations. The solution is valid if all the coefficients are positive integers. If this is not the case, take $a = 2$ and repeat the solution. Continue the process until all the coefficients in the solution are positive integers.

28. The wind chill temperature, T_{wc}, is the air temperature felt on exposed skin due to wind. In U.S. customary units it is calculated by:

$$T_{wc} = 35.74 + 0.6215T - 35.75v^{0.16} + 0.4275T\,v^{0.16}$$

where T is the temperature in degrees F, and v is the wind speed in mi/h. Write a MATLAB program in a script file that displays the following chart of wind chill temperature for given air temperature and wind speed in the Command Window:

	Temperature (F)								
	40	30	20	10	0	-10	-20	-30	-40
Speed (mi/h)									
10	34	21	9	-4	-16	-28	-41	-53	-66
20	30	17	4	-9	-22	-35	-48	-61	-74
30	28	15	1	-12	-26	-39	-53	-67	-80
40	27	13	-1	-15	-29	-43	-57	-71	-84
50	26	12	-3	-17	-31	-45	-60	-74	-88
60	25	10	-4	-19	-33	-48	-62	-76	-91

29. The stress intensity factor due to the crack shown depends upon a geometrical parameter C_I given by:

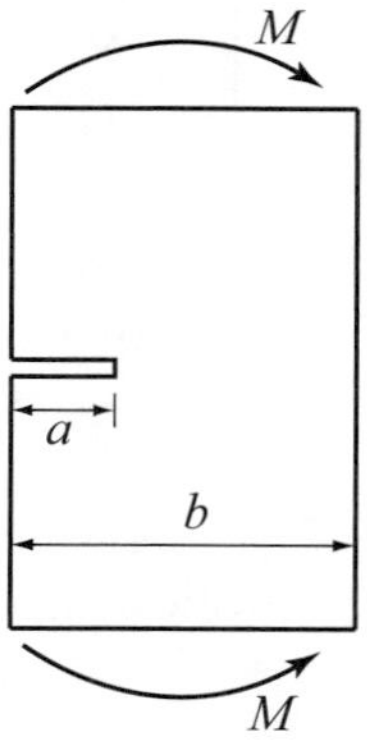

$$C_I = \sqrt{\frac{2}{\pi\alpha}\tan\frac{\pi\alpha}{2}}\left[\frac{0.923 + 0.199\left(1 - \sin\frac{\pi\alpha}{2}\right)^4}{\cos\frac{\pi\alpha}{2}}\right]$$

where $\alpha = \frac{a}{b}$. Calculate C_I for α between 0.05 and 0.95 at 0.05 increments, and display the results in a two-column table with the first column showing α and the second C_I.

Chapter 5
Two-Dimensional Plots

Plots are a very useful tool for presenting information. This is true in any field, but especially in science and engineering, where MATLAB is mostly used. MATLAB has many commands that can be used for creating different types of plots. These include standard plots with linear axes, plots with logarithmic and semi-logarithmic axes, bar and stairs plots, polar plots, three-dimensional contour surface and mesh plots, and many more. The plots can be formatted to have a desired appearance. The line type (solid, dashed, etc.), color, and thickness can be prescribed, line markers and grid lines can be added, as can titles and text comments. Several graphs can be created in the same plot, and several plots can be placed on the same page. When a plot contains several graphs and/or data points, a legend can be added to the plot as well.

This chapter describes how MATLAB can be used to create and format many types of two-dimensional plots. Three-dimensional plots are addressed separately in Chapter 9. An example of a simple two-dimensional plot that was created with MATLAB is shown in Figure 5-1. The figure contains two curves that show the variation of light intensity with distance. One curve is constructed from data points measured in an experiment, and the other curve shows the variation of light as predicted by a theoretical model. The axes in the figure are both linear, and different types of lines (one solid and one dashed) are used for the curves. The theoretical curve is shown with a solid line, while the experimental points are connected with a dashed line. Each data point is marked with a circular marker. The dashed line that connects the experimental points is actually red when the plot is displayed in the Figure Window. As shown, the plot in Figure 5-1 is formatted to have a title, axis titles, a legend, markers, and a boxed text label.

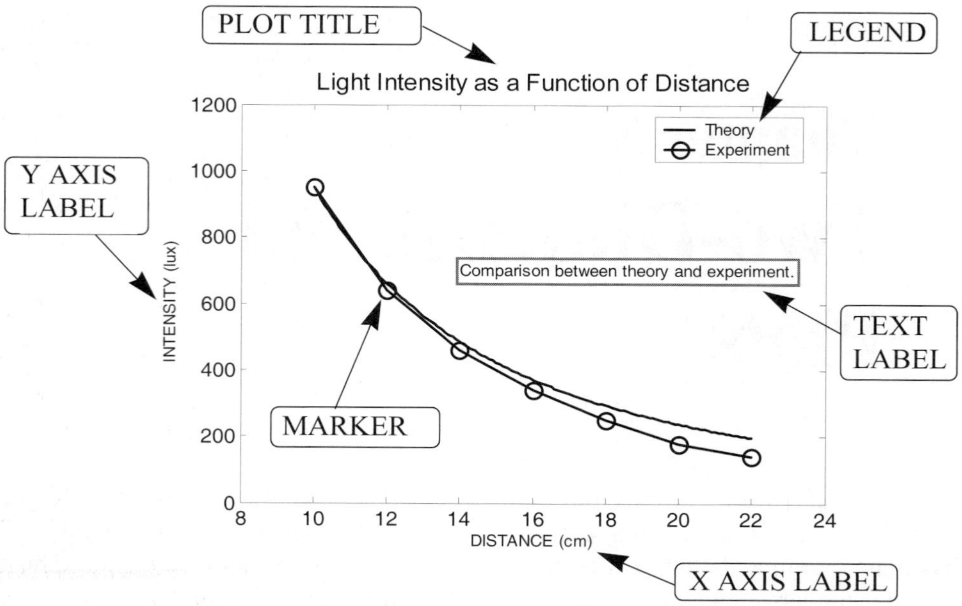

Figure 5-1: Example of a formatted two-dimensional plot.

5.1 THE `plot` *COMMAND*

The `plot` command is used to create two-dimensional plots. The simplest form of the command is:

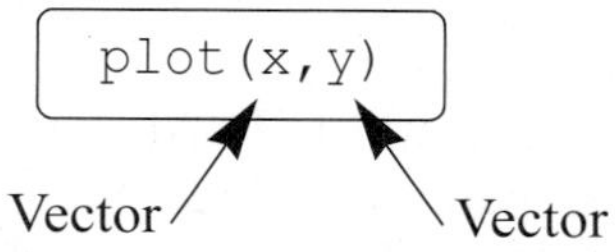

The arguments `x` and `y` are each a vector (one-dimensional array). The two vectors *must* have the same number of elements. When the `plot` command is executed, a figure is created in the Figure Window. If not already open, the Figure Window opens automatically when the command is executed. The figure has a single curve with the `x` values on the abscissa (horizontal axis) and the `y` values on the ordinate (vertical axis). The curve is constructed of straight-line segments that connect the points whose coordinates are defined by the elements of the vectors `x` and `y`. Each of the vectors, of course, can have any name. The vector that is typed first in the `plot` command is used for the horizontal axis, and the vector that is typed second is used for the vertical axis.

The figure that is created has axes with a linear scale and default range. For example, if a vector x has the elements 1, 2, 3, 5, 7, 7.5, 8, 10, and a vector y has the elements 2, 6.5, 7, 7, 5.5, 4, 6, 8, a simple plot of y versus x can be created by typing the following in the Command Window:

```
>> x=[1   2   3   5   7   7.5   8   10];
>> y=[2   6.5   7   7   5.5   4   6   8];
>> plot(x,y)
```

Once the plot command is executed, the Figure Window opens and the plot is displayed, as shown in Figure 5-2.

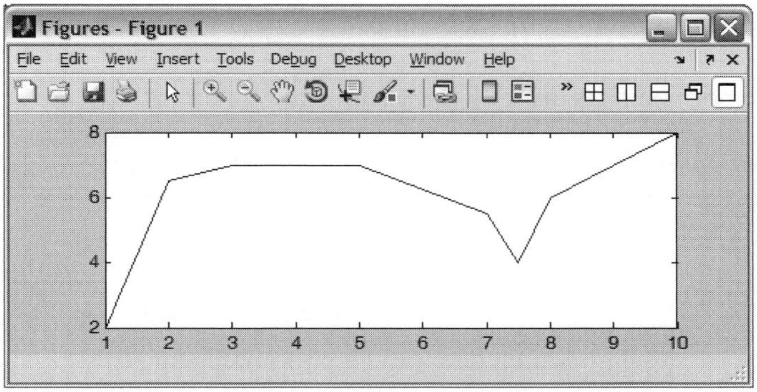

Figure 5-2: The Figure Window with a simple plot.

The plot appears on the screen in blue, which is the default line color.

The plot command has additional, optional arguments that can be used to specify the color and style of the line and the color and type of markers, if any are desired. With these options the command has the form:

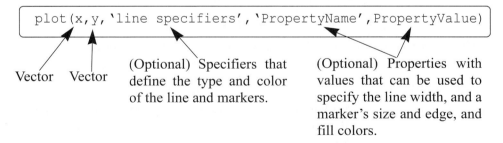

```
plot(x,y,'line specifiers','PropertyName',PropertyValue)
```

Vector Vector

(Optional) Specifiers that define the type and color of the line and markers.

(Optional) Properties with values that can be used to specify the line width, and a marker's size and edge, and fill colors.

Line Specifiers:

Line specifiers are optional and can be used to define the style and color of the line and the type of markers (if markers are desired). The line style specifiers are:

Line Style	Specifier
solid (default)	-
dashed	--

Line Style	Specifier
dotted	:
dash-dot	-.

The line color specifiers are:

Line Color	Specifier
red	`r`
green	`g`
blue	`b`
cyan	`c`

Line Color	Specifier
magenta	`m`
yellow	`y`
black	`k`
white	`w`

The marker type specifiers are:

Marker Type	Specifier		Marker Type	Specifier
plus sign	`+`		square	`s`
circle	`o`		diamond	`d`
asterisk	`*`		five-pointed star	`p`
point	`.`		six-pointed star	`h`
cross	`x`		triangle (pointed left)	`<`
triangle (pointed up)	`^`		triangle (pointed right)	`>`
triangle (pointed down)	`v`			

Notes about using the specifiers:

- The specifiers are typed inside the `plot` command as strings.
- Within the string the specifiers can be typed in any order.
- The specifiers are optional. This means that none, one, two, or all three types can be included in a command.

Some examples:

`plot(x,y)`	A blue solid line connects the points with no markers (default).
`plot(x,y,'r')`	A red solid line connects the points.
`plot(x,y,'--y')`	A yellow dashed line connects the points.
`plot(x,y,'*')`	The points are marked with * (no line between the points).
`plot(x,y,'g:d')`	A green dotted line connects the points that are marked with diamond markers.

`Property Name` and `Property Value`:

Properties are optional and can be used to specify the thickness of the line, the size of the marker, and the colors of the marker's edge line and fill. The Property Name is typed as a string, followed by a comma and a value for the property, all inside the `plot` command.

Four properties and their possible values are:

Property name	Description	Possible property values
`LineWidth` (or `linewidth`)	Specifies the width of the line.	A number in units of points (default 0.5).
`MarkerSize` (or `markersize`)	Specifies the size of the marker.	A number in units of points.
`MarkerEdgeColor` (or `markeredgecolor`)	Specifies the color of the marker, or the color of the edge line for filled markers.	Color specifiers from the table above, typed as a string.
`MarkerFaceColor` (or `markerfacecolor`)	Specifies the color of the filling for filled markers.	Color specifiers from the table above, typed as a string.

For example, the command

```
plot(x,y,'-mo','LineWidth',2,'markersize',12,
      'MarkerEdgeColor','g','markerfacecolor','y')
```

creates a plot that connects the points with a magenta solid line and circles as markers at the points. The line width is 2 points and the size of the circle markers is 12 points. The markers have a green edge line and yellow filling.

A note about line specifiers and properties:

The three line specifiers, which indicate the style and color of the line, and the type of the marker can also be assigned with a `PropertyName` argument followed by a `PropertyValue` argument. The Property Names for the line specifiers are:

Specifier	Property Name	Possible property values
Line style	`linestyle` (or `LineStyle`)	Line style specifier from the table above, typed as a string.
Line color	`color` (or `Color`)	Color specifier from the table above, typed as a string.
Marker	`marker` (or `Marker`)	Marker specifier from the table above, typed as a string.

As with any command, the `plot` command can be typed in the Command Window, or it can be included in a script file. It also can be used in a function file (explained in Chapter 7). It should also be remembered that before the `plot` command can be executed the vectors `x` and `y` must have assigned elements. This can be done, as was explained in Chapter 2, by entering values directly, by using commands, or as the result of mathematical operations. The next two subsections show examples of creating simple plots.

5.1.1 Plot of Given Data

In this case given data is used to create vectors that are then used in the `plot` command. The following table contains sales data of a company from 1988 to 1994.

Year	1988	1989	1990	1991	1992	1993	1994
Sales (millions)	8	12	20	22	18	24	27

To plot this data, the list of years is assigned to one vector (named `yr`), and the corresponding sales data is assigned to a second vector (named `sle`). The Command Window where the vectors are created and the `plot` command is used is shown below:

```
>> yr=[1988:1:1994];
>> sle=[8  12  20  22  18  24  27];
>> plot(yr,sle,'--r*','linewidth',2,'markersize',12)
>>
```

Line Specifiers: dashed red line and asterisk marker.

Property Name and Property Value: the line width is 2 points and the marker size is 12 points.

Once the `plot` command is executed, the Figure Window with the plot, as shown in Figure 5-3, opens. The plot appears on the screen in red.

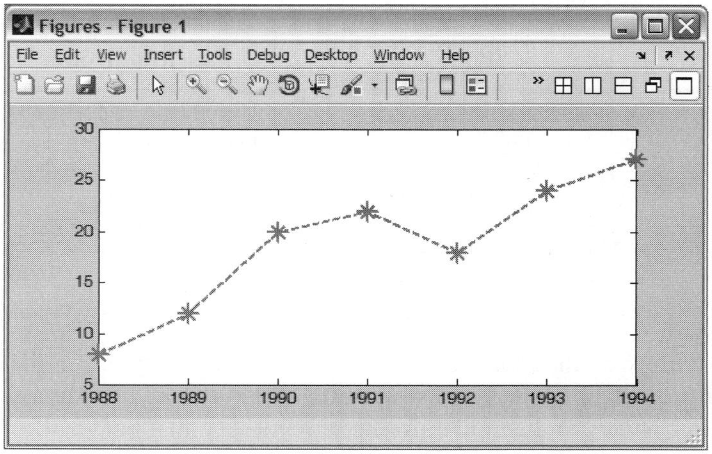

Figure 5-3: The Figure Window with a plot of the sales data.

5.1.2 Plot of a Function

In many situations there is a need to plot a given function. This can be done in MATLAB by using the `plot` or the `fplot` command. The use of the `plot` command is explained below. The `fplot` command is explained in detail in the next section.

In order to plot a function $y = f(x)$ with the `plot` command, the user needs to first create a vector of values of x for the domain over which the function will be plotted. Then a vector y is created with the corresponding values of $f(x)$ by using element-by-element calculations (see Chapter 3). Once the two vectors are defined, they can be used in the `plot` command.

As an example, the `plot` command is used to plot the function $y = 3.5^{-0.5x}\cos(6x)$ for $-2 \le x \le 4$. A program that plots this function is shown in the following script file.

```
%  A script file that creates a plot of
%  the function: 3.5.^(-0.5*x).*cos(6x)
x=[-2:0.01:4];
y=3.5.^(-0.5*x).*cos(6*x);
plot(x,y)
```

Create vector x with the domain of the function.

Create vector y with the function value at each x.

Plot y as a function of x.

Once the script file is executed, the plot is created in the Figure Window, as shown in Figure 5-4. Since the plot is made up of segments of straight lines that connect the points, to obtain an accurate plot of a function, the spacing between the elements of the vector x must be appropriate. Smaller spacing is needed for a

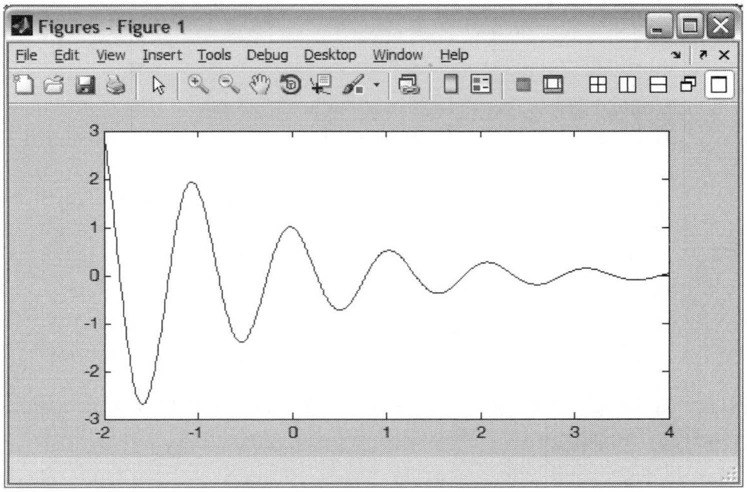

Figure 5-4: The Figure Window with a plot of the function $y = 3.5^{-0.5x}\cos(6x)$.

function that changes rapidly. In the last example a small spacing of 0.01 produced the plot that is shown in Figure 5-4. However, if the same function in the same domain is plotted with much larger spacing—for example, 0.3—the plot that is obtained, shown in Figure 5-5, gives a distorted picture of the function. Note

```
x=[-2:0.3:4];
y=3.5.^(-0.5*x).*cos(6*x);
plot(x,y)
```

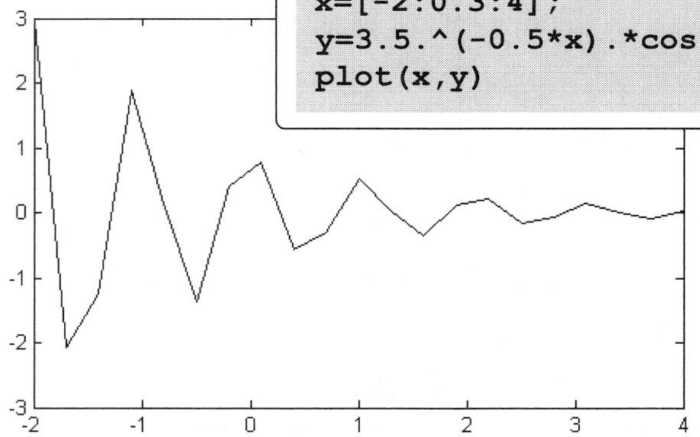

Figure 5-5: A plot of the function $y = 3.5^{-0.5x}\cos(6x)$ with large spacing.

also that in Figure 5-4 the plot is shown with the Figure Window, while in Figure 5-5 only the plot is shown. The plot can be copied from the Figure Window (in the **Edit** menu, select **Copy Figure**) and then pasted into other applications.

5.2 THE `fplot` *COMMAND*

The `fplot` command plots a function with the form $y = f(x)$ between specified limits. The command has the form:

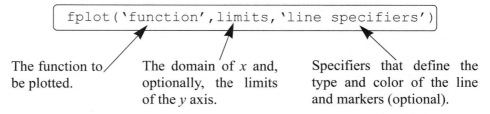

```
fplot('function',limits,'line specifiers')
```

The function to be plotted.

The domain of x and, optionally, the limits of the y axis.

Specifiers that define the type and color of the line and markers (optional).

`'function'`: The function can be typed directly as a string inside the command. For example, if the function that is being plotted is $f(x) = 8x^2 + 5\cos(x)$, it is typed as: `'8*x^2+5*cos(x)'`. The functions can include MATLAB built-in functions and functions that are created by the user (covered in Chapter 6).

- The function to be plotted can be typed as a function of any letter. For example, the function in the previous paragraph can be typed as `'8*z^2+5*cos(z)'` or `'8*t^2+5*cos(t)'`.

- The function cannot include previously defined variables. For example, in the function above it is not possible to assign 8 to a variable, and then use the variable when the function is typed in the fplot command.

limits: The limits argument is a vector with two elements that specify the domain of x [xmin,xmax], or a vector with four elements that specifies the domain of x and the limits of the y-axis [xmin,xmax,ymin,ymax].

line specifiers: The line specifiers are the same as in the plot command. For example, a plot of the function $y = x^2 + 4\sin(2x) - 1$ for $-3 \le x \le 3$ can be created with the fplot command by typing:

```
>> fplot('x^2+4*sin(2*x)-1',[-3 3])
```

in the Command Window. The figure that is obtained in the Figure Window is shown in Figure 5-6.

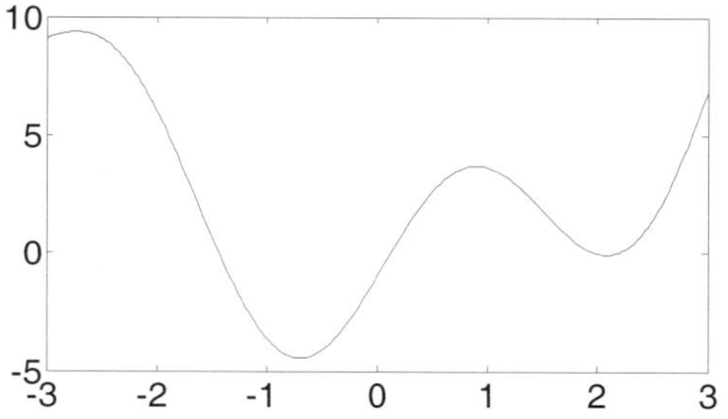

Figure 5-6: A plot of the function $y = x^2 + 4\sin(2x) - 1$**.**

5.3 *PLOTTING MULTIPLE GRAPHS IN THE SAME PLOT*

In many situations there is a need to make several graphs in the same plot. This is shown, for example, in Figure 5-1 where two graphs are plotted in the same figure. There are three methods to plot multiple graphs in one figure. One is by using the plot command, the second is by using the hold on and hold off commands, and the third is by using the line command.

5.3.1 Using the plot *Command*

Two or more graphs can be created in the same plot by typing pairs of vectors inside the plot command. The command

```
plot(x,y,u,v,t,h)
```

creates three graphs—y vs. x, v vs. u, and h vs. t—all in the same plot. The vectors of each pair must be of the same length. MATLAB automatically plots the graphs in different colors so that they can be identified. It is also possible to add line specifiers following each pair. For example the command

```
plot(x,y,'-b',u,v,'--r',t,h,'g:')
```

plots `y` vs. `x` with a solid blue line, `v` vs.`u` with a dashed red line, and `h` vs. `t` with a dotted green line.

Sample Problem 5-1: Plotting a function and its derivatives

Plot the function $y = 3x^3 - 26x + 10$, and its first and second derivatives, for $-2 \le x \le 4$, all in the same plot.

Solution

The first derivative of the function is: $y' = 9x^2 - 26$.

The second derivative of the function is: $y'' = 18x$.

A script file that creates a vector x and calculates the values of y, y', and y'' is:

```
x=[-2:0.01:4];
y=3*x.^3-26*x+6;
yd=9*x.^2-26;
ydd=18*x;
plot(x,y,'-b',x,yd,'--r',x,ydd,':k')
```

- `x=[-2:0.01:4];` — Create vector `x` with the domain of the function.
- `y=3*x.^3-26*x+6;` — Create vector `y` with the function value at each `x`.
- `yd=9*x.^2-26;` — Create vector `yd` with values of the first derivative.
- `ydd=18*x;` — Create vector `ydd` with values of the second derivative.
- `plot(x,y,'-b',x,yd,'--r',x,ydd,':k')` — Create three graphs, `y` vs. `x`, `yd` vs. `x`, and `ydd` vs. `x`, in the same figure.

The plot that is created is shown in Figure 5-7.

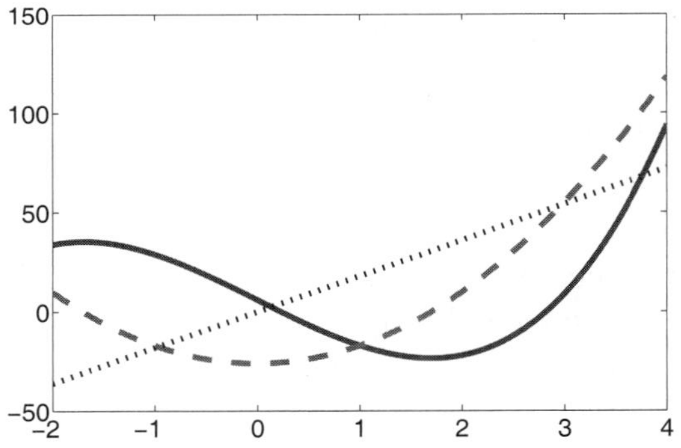

Figure 5-7: A plot of the function $y = 3x^3 - 26x + 10$ and its first and second derivatives.

5.3.2 *Using the* `hold on` *and* `hold off` *Commands*

To plot several graphs using the `hold on` and `hold off` commands, one graph is plotted first with the `plot` command. Then the `hold on` command is typed. This keeps the Figure Window with the first plot open, including the axis properties and formatting (see Section 5.4) if any was done. Additional graphs can be added with `plot` commands that are typed next. Each `plot` command creates a graph that is added to that figure. The `hold off` command stops this process. It returns MATLAB to the default mode, in which the `plot` command erases the previous plot and resets the axis properties.

As an example, a solution of Sample Problem 5-1 using the `hold on` and `hold off` commands is shown in the following script file:

```
x=[-2:0.01:4];
y=3*x.^3-26*x+6;
yd=9*x.^2-26;
ydd=18*x;
plot(x,y,'-b')                      The first graph is created.
hold on
plot(x,yd,'--r')                    Two more graphs are added to the figure.
plot(x,ydd,':k')
hold off
```

5.3.3 *Using the* `line` *Command*

With the `line` command additional graphs (lines) can be added to a plot that already exists. The form of the line command is:

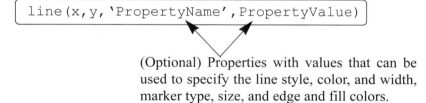

```
line(x,y,'PropertyName',PropertyValue)
```

(Optional) Properties with values that can be used to specify the line style, color, and width, marker type, size, and edge and fill colors.

The format of the `line` command is almost the same as the `plot` command (see Section 5.1). The `line` command does not have the line specifiers, but the line style, color, and marker can be specified with the Property Name and property value features. The properties are optional and if none are entered MATLAB uses default properties and values. For example, the command:

```
line(x,y,'linestyle','--','color','r','marker','o')
```

will add a dashed red line with circular markers to a plot that already exists.

The major difference between the `plot` and `line` commands is that the `plot` command starts a new plot every time it is executed, while the `line` command adds lines to a plot that already exists. To make a plot that has several graphs, a plot command is typed first and then line commands are typed for additional graphs. (If a line command is entered before a plot command an error message is displayed.)

The solution to Sample Problem 5-1, which is the plot in Figure 5-7, can be obtained by using the plot and line commands as shown in the following script file:

```
x=[-2:0.01:4];
y=3*x.^3-26*x+6;
yd=9*x.^2-26;
ydd=18*x;
plot(x,y,'LineStyle','-','color','b')
line(x,yd,'LineStyle','--','color','r')
line(x,ydd,'linestyle',':','color','k')
```

5.4 FORMATTING A PLOT

The plot and fplot commands create bare plots. Usually, however, a figure that contains a plot needs to be formatted to have a specific look and to display information in addition to the graph itself. This can include specifying axis labels, plot title, legend, grid, range of custom axis, and text labels.

Plots can be formatted by using MATLAB commands that follow the plot or fplot command, or interactively by using the plot editor in the Figure Window. The first method is useful when a plot command is a part of a computer program (script file). When the formatting commands are included in the program, a formatted plot is created every time the program is executed. On the other hand, formatting that is done in the Figure Window with the plot editor after a plot has been created holds only for that specific plot, and will have to be repeated the next time the plot is created.

5.4.1 Formatting a Plot Using Commands

The formatting commands are entered after the plot or the fplot command. The various formatting commands are:

The xlabel and ylabel commands:

Labels can be placed next to the axes with the xlabel and ylabel command which have the form:

```
xlabel('text as string')
ylabel('text as string')
```

The title command:

A title can be added to the plot with the command:

```
title('text as string')
```

The text is placed at the top of the figure as a title.

The `text` command:

A text label can be placed in the plot with the `text` or `gtext` commands:

```
text(x,y,'text as string')
gtext('text as string')
```

The `text` command places the text in the figure such that the first character is positioned at the point with the coordinates `x`, `y` (according to the axes of the figure). The `gtext` command places the text at a position specified by the user. When the command is executed, the Figure Window opens and the user specifies the position with the mouse.

The `legend` command:

The `legend` command places a legend on the plot. The legend shows a sample of the line type of each graph that is plotted, and places a label, specified by the user, beside the line sample. The form of the command is:

```
legend('string1','string2', ..... ,pos)
```

The strings are the labels that are placed next to the line sample. Their order corresponds to the order in which the graphs were created. The `pos` is an optional number that specifies where in the figure the legend is to be placed. The options are:

`pos = -1`	Places the legend outside the axes boundaries on the right side.
`pos = 0`	Places the legend inside the axes boundaries in a location that interferes the least with the graphs.
`pos = 1`	Places the legend at the upper-right corner of the plot (default).
`pos = 2`	Places the legend at the upper-left corner of the plot.
`pos = 3`	Places the legend at the lower-left corner of the plot.
`pos = 4`	Places the legend at the lower-right corner of the plot.

Formatting the text within the `xlabel`, `ylabel`, `title`, `text` and `legend` commands:

The text in the string that is included in the command and is displayed when the command is executed can be formatted. The formatting can be used to define the font, size, position (superscript, subscript), style (italic, bold, etc.), and color of the characters, the color of the background, and many other details of the display. Some of the more common formatting possibilities are described below. A complete explanation of all the formatting features can be found in the Help Window under Text and Text Properties. The formatting can be done either by adding modifiers inside the string, or by adding to the command optional `PropertyName` and `PropertyValue` arguments following the string.

The modifiers are characters that are inserted within the string. Some of the modifiers that can be added are:

Modifier	Effect
`\bf`	bold font
`\it`	italic style
`\rm`	normal font

Modifier	Effect
`\fontname{fontname}`	specified font is used
`\fontsize{fontsize}`	specified font size is used

These modifiers affect the text from the point at which they are inserted until the end of the string. It is also possible to have the modifiers applied to only a section of the string by typing the modifier and the text to be affected inside braces { }.

Subscript and superscript:

A single character can be displayed as a subscript or a superscript by typing _ (the underscore character) or ^ in front of the character, respectively. Several consecutive characters can be displayed as a subscript or a superscript by typing the characters inside braces { } following the _ or the ^.

Greek characters:

Greek characters can be included in the text by typing `\name of the letter` within the string. To display a lowercase Greek letter the name of the letter should be typed in all lowercase English characters, To display a capital Greek letter the name of the letter should start with a capital letter. Some examples are:

Characters in the string	Greek letter
`\alpha`	α
`\beta`	β
`\gamma`	γ
`\theta`	θ
`\pi`	π
`\sigma`	σ

Characters in the string	Greek letter
`\Phi`	Φ
`\Delta`	Δ
`\Gamma`	Γ
`\Lambda`	Λ
`\Omega`	Ω
`\Sigma`	Σ

Formatting of the text that is displayed by the `xlabel`, `ylabel`, `title`, and `text` commands can also be done by adding optional `PropertyName` and `PropertyValue` arguments following the string inside the command. With this option the `text` command, for example, has the form:

```
text(x,y,'text as string',PropertyName,PropertyValue)
```

In the other three commands the `PropertyName` and `PropertyValue` arguments are added in the same way. The `PropertyName` is typed as a string, and the `PropertyValue` is typed as a number if the property value is a number and as a string if the property value is a word or a letter character. Some of the Property Names and corresponding possible Property Values are:

Property name	Description	Possible property values
`Rotation`	Specifies the orientation of the text.	Scalar (degrees) Default: 0
`FontAngle`	Specifies italic or normal style characters.	`normal`, `italic` Default: `normal`
`FontName`	Specifies the font for the text.	Font name that is available in the system.
`FontSize`	Specifies the size of the font.	Scalar (points) Default: 10
`FontWeight`	Specifies the weight of the characters.	`light`, `normal`, `bold` Default: `normal`
`Color`	Specifies the color of the text.	Color specifiers (see Section 5.1).
`Background-Color`	Specifies the background color (rectangular area).	Color specifiers (see Section 5.1).
`EdgeColor`	Specifies the color of the edge of a rectangular box around the text.	Color specifiers (see Section 5.1). Default: none.
`LineWidth`	Specifies the width of the edge of a rectangular box around the text.	Scalar (points) Default: 0.5

The `axis` command:

When the `plot(x,y)` command is executed, MATLAB creates axes with limits that are based on the minimum and maximum values of the elements of `x` and `y`. The `axis` command can be used to change the range and the appearance of the axes. In many situations a graph looks better if the range of the axes extend beyond the range of the data. The following are some of the possible forms of the `axis` command:

`axis([xmin,xmax,ymin,ymax])` Sets the limits of both the x and y axes (`xmin`, `xmax`, `ymin`, and `ymax` are numbers).

`axis equal` Sets the same scale for both axes.

`axis square` Sets the axes region to be square.

`axis tight` Sets the axis limits to the range of the data.

The `grid` command:

`grid on` Adds grid lines to the plot.

`grid off` Removes grid lines from the plot.

An example of formatting a plot by using commands is given in the following script file which was used to generate the formatted plot in Figure 5-1.

```
x=[10:0.1:22];
y=95000./x.^2;
xd=[10:2:22];
yd=[950  640  460  340  250  180  140];
plot(x,y,'-','LineWidth',1.0)
xlabel('DISTANCE (cm)')
ylabel('INTENSITY (lux)')
title('\fontname{Arial}Light Intensity as a Function of Distance','FontSize',14)
axis([8 24 0 1200])
text(14,700,'Comparison between theory and experiment.','EdgeColor','r','LineWidth',2)
hold on
plot(xd,yd,'ro--','linewidth',1.0,'markersize',10)
legend('Theory','Experiment',0)
hold off
```

Formatting text inside the `title` command.

Formatting text inside the `text` command.

5.4.2 *Formatting a Plot Using the Plot Editor*

A plot can be formatted interactively in the Figure Window by clicking on the plot and/or using the menus. Figure 5-8 shows the Figure Window with the plot of Figure 5-1. The Plot Editor can be used to introduce new formatting items or to modify formatting that was initially introduced with the formatting commands.

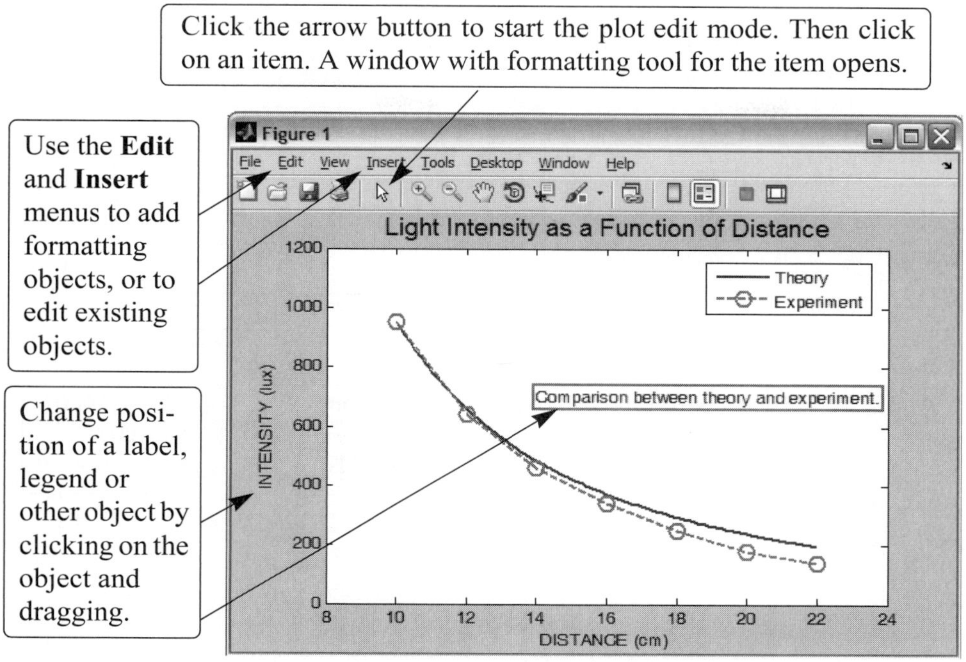
Click the arrow button to start the plot edit mode. Then click on an item. A window with formatting tool for the item opens.
Use the **Edit** and **Insert** menus to add formatting objects, or to edit existing objects.
Change position of a label, legend or other object by clicking on the object and dragging.
Figure 1
File Edit View Insert Tools Desktop Window Help
Light Intensity as a Function of Distance
Theory
Experiment
Comparison between theory and experiment.
INTENSITY (lux)
DISTANCE (cm)
0
200
400
600
800
1000
1200
8
10
12
14
16
18
20
22
24

Figure 5-8: Formatting a plot using the Plot Editor.

5.5 *PLOTS WITH LOGARITHMIC AXES*

Many science and engineering applications require plots in which one or both axes have a logarithmic (log) scale. Log scales provide means for presenting data over a wide range of values. It also provides a tool for identifying characteristics of data and possible forms of mathematical relationships that can be appropriate for modeling the data (see Section 8.2.2).

MATLAB commands for making plots with log axes are:

`semilogy(x,y)`	Plots `y` versus `x` with a log (base 10) scale for the y axis and linear scale for the x axis.
`semilogx(x,y)`	Plots `y` versus `x` with a log (base 10) scale for the x axis and linear scale for the y axis.
`loglog(x,y)`	Plots `y` versus `x` with a log (base 10) scale for both axes.

Line specifiers and Property Name and Property Value arguments can be added to the commands (optional) just as in the `plot` command. As an example, Figure 5-9 shows a plot of the function $y = 2^{(-0.2x + 10)}$ for $0.1 \le x \le 60$. The figure shows four plots of the same function: one with linear axes, one with a log scale for the y axis, one with a log scale for the x axis, and one with a log scale on both axes.

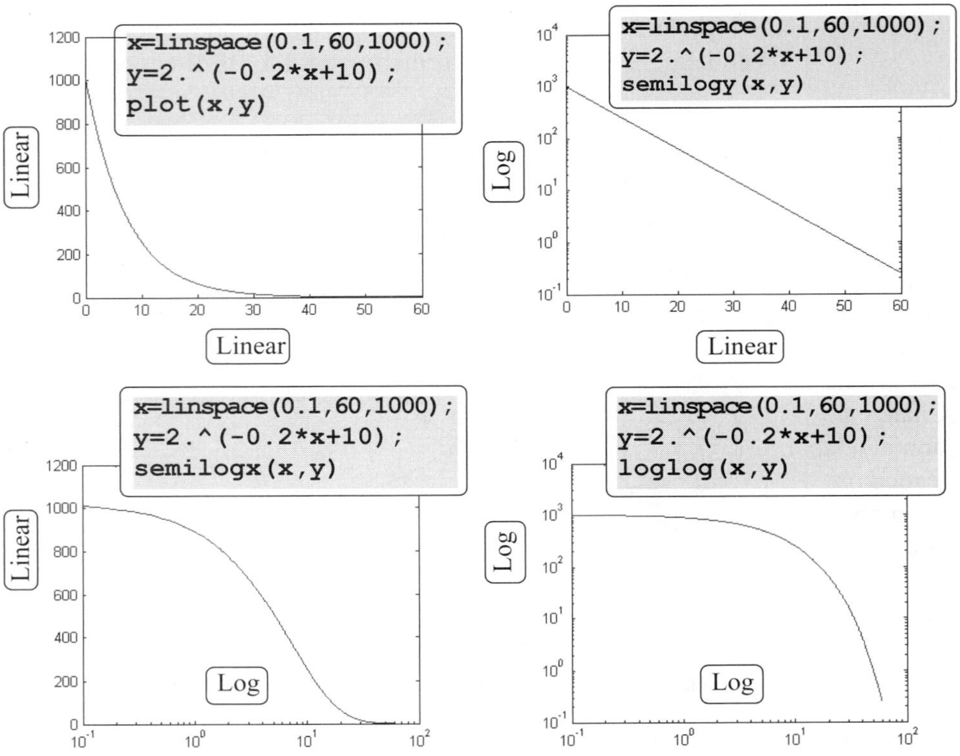

Figure 5-9: Plots of $y = 2^{(-0.2x + 10)}$ with linear, semilog, and log-log scales.

Notes for plots with logarithmic axes:

- The number zero cannot be plotted on a log scale (since a log of zero is not defined).
- Negative numbers cannot be plotted on log scales (since a log of a negative number is not defined).

5.6 PLOTS WITH ERROR BARS

Experimental data that is measured and then displayed in plots frequently contains error and scatter. Even data that is generated by computational models includes error or uncertainty that depends on the accuracy of the input parameters and the assumptions in the mathematical models that are used. One method of plotting data that displays the error, or uncertainty, is by using error bars. An error bar is typically a short vertical line that is attached to a data point in a plot. It shows the magnitude of the error that is associated with the value that is displayed by the data point. For example, Figure 5-10 shows a plot with error bars for the experimental data from Figure 5-1.

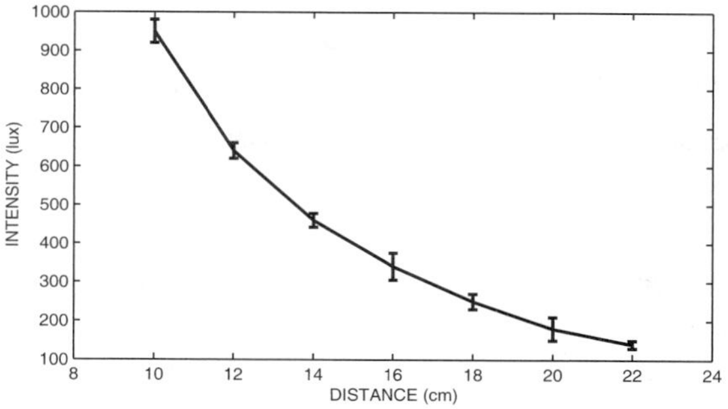

Figure 5-10: A plot with error bars.

Plots with error bars can be done in MATLAB with the `errorbar` command. Two forms of the command, one for making plots with symmetric error bars (with respect to the value of the data point) and the other for nonsymmetric error bars at each point, are presented. When the error is symmetric, the error bar extends the same length above and below the data point and the command has the form:

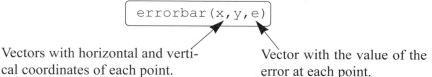

```
errorbar(x,y,e)
```

Vectors with horizontal and vertical coordinates of each point.

Vector with the value of the error at each point.

- The lengths of the three vectors `x`, `y`, and `e` must be the same.
- The length of the error bar is twice the value of `e`. At each point the error bar extends from `y(i)-e(i)` to `y(i)+e(i)`.

The plot in Figure 5-10, which has symmetric error bars, was done by executing the following code:

```
xd=[10:2:22];
yd=[950 640 460 340 250 180 140];
ydErr=[30 20 18 35 20 30 10]
errorbar(xd,yd,ydErr)
xlabel('DISTANCE (cm)')
ylabel('INTENSITY (lux)')
```

The command for making a plot with error bars that are not symmetric is:

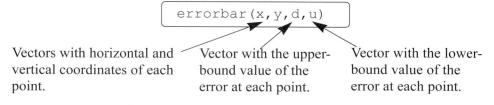

```
errorbar(x,y,d,u)
```

Vectors with horizontal and vertical coordinates of each point.

Vector with the upper-bound value of the error at each point.

Vector with the lower-bound value of the error at each point.

- The lengths of the four vectors `x`, `y`, `d`, and `u` must be the same.
- At each point the error bar extends from `y(i)-d(i)` to `y(i)+u(i)`.

5.7 PLOTS WITH SPECIAL GRAPHICS

All the plots that have been presented so far in this chapter are line plots in which the data points are connected by lines. In many situations plots with different graphics or geometry can present data more effectively. MATLAB has many options for creating a wide variety of plots. These include bar, stairs, stem, and pie plots and many more. Following are some of the special graphics plots that can be created with MATLAB. A complete list of the plotting functions that MATLAB offers and information on how to use them can be found in the Help Window. In this window first choose “Functions by Category,” then select “Graphics” and then select “Basic Plots and Graphs” or “Specialized Plotting.”

Bar (vertical and horizontal), stairs, and stem plots are presented in the following charts using the sales data from Section 5.1.1.

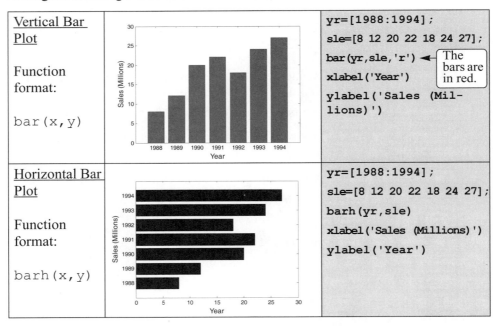

Vertical Bar Plot
 	
Function format:
 	
`bar(x,y)`
	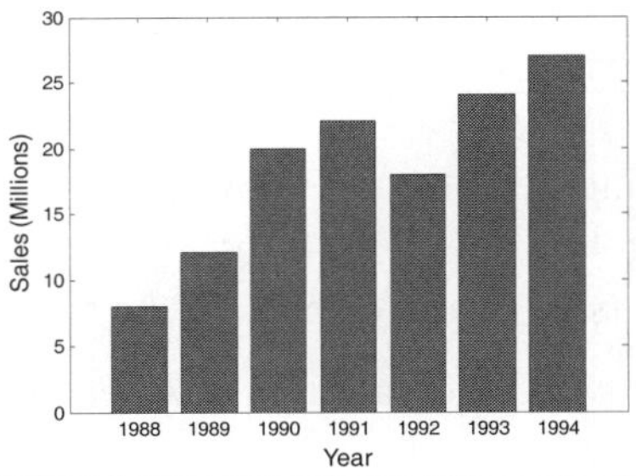	`yr=[1988:1994];`
		`sle=[8 12 20 22 18 24 27];`
		`bar(yr,sle,'r')` ← The bars are in red.
		`xlabel('Year')`
		`ylabel('Sales (Mil-`
		`lions)')`

Horizontal Bar Plot
 	
Function format:
 	
`barh(x,y)`
	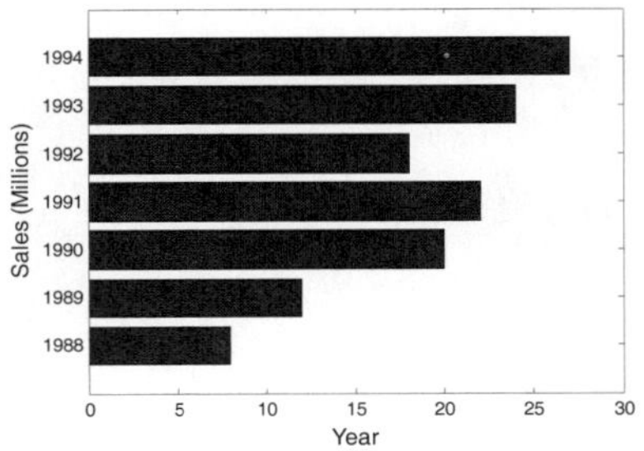	`yr=[1988:1994];`
		`sle=[8 12 20 22 18 24 27];`
		`barh(yr,sle)`
		`xlabel('Sales (Millions)')`
		`ylabel('Year')`

<u>Stairs Plot</u>
 	
Function format:
 	
`stairs(x,y)`
	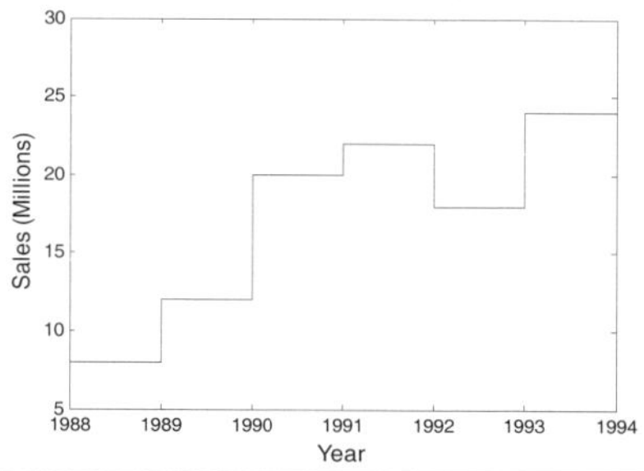	`yr=[1988:1994];`
		`sle=[8 12 20 22 18 24 27];`
		`stairs(yr,sle)`

<u>Stem Plot</u>
 	
Function Format
 	
`stem(x,y)`
	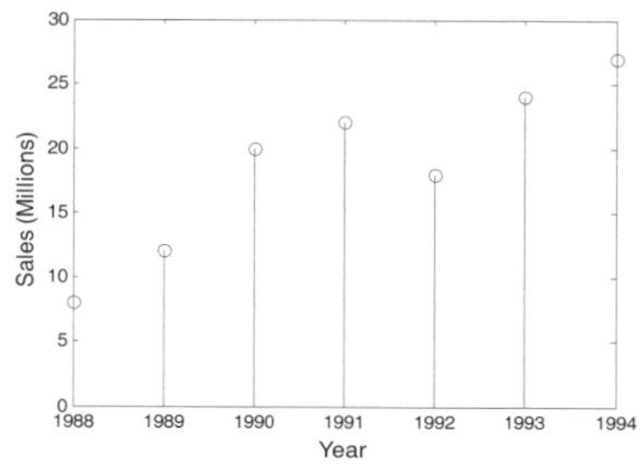	`yr=[1988:1994];`
		`sle=[8 12 20 22 18 24 27];`
		`stem(yr,sle)`

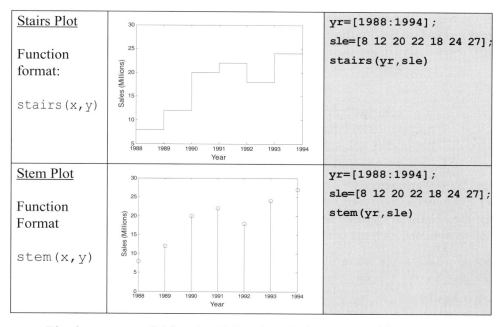

Pie charts are useful for visualizing the relative sizes of different but related quantities. For example, the table below shows the grades that were assigned to a class. The data is used to create the pie chart that follows.

Grade	A	B	C	D	E
Number of students	11	18	26	9	5

<u>Pie Plot</u>
 	
Function format:
 	
`pie(x)`
	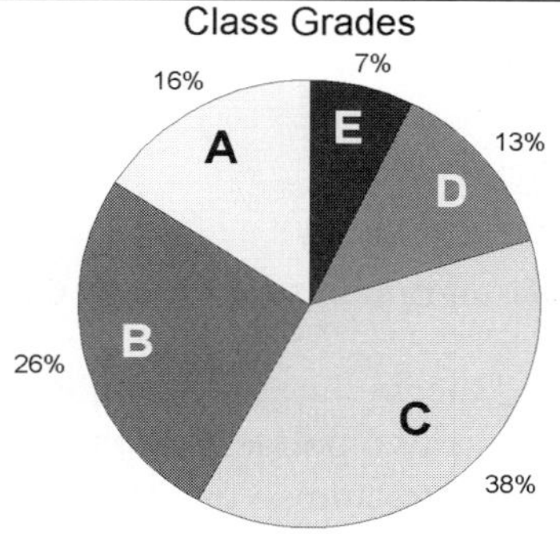	`grd=[11 18 26 9 5];`
		`pie(grd)`
		`title('Class Grades')`
		
		MATLAB draws the sections in different colors. The letters (grades) were added using the Plot Editor.

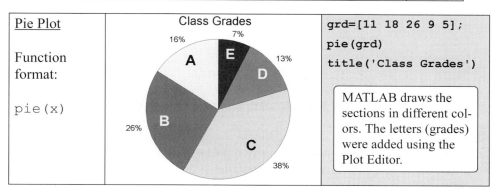

5.8 *HISTOGRAMS*

Histograms are plots that show the distribution of data. The overall range of a given set of data points is divided into subranges (bins), and the histogram shows how many data points are in each bin. The histogram is a vertical bar plot in which the width of each bar is equal to the range of the corresponding bin and the height of the bar corresponds to the number of data points in the bin. Histograms are created in MATLAB with the `hist` command. The simplest form of the command is:

```
hist(y)
```

`y` is a vector with the data points. MATLAB divides the range of the data points into 10 equally spaced subranges (bins) and then plots the number of data points in each bin.

For example, the following data points are the daily maximum temperature (in °F) in Washington, DC, during the month of April 2002: 58 73 73 53 50 48 56 73 73 66 69 63 74 82 84 91 93 89 91 80 59 69 56 64 63 66 64 74 63 69 (data from the U.S. National Oceanic and Atmospheric Administration). A histogram of this data is obtained with the commands:

```
>> y=[58 73 73 53 50 48 56 73 73 66 69 63 74 82 84 91 93 89
91 80 59 69 56 64 63 66 64 74 63 69];
>> hist(y)
```

The plot that is generated is shown in Figure 5-11 (the axis titles were added using the Plot Editor). The smallest value in the data set is 48 and the largest is 93,

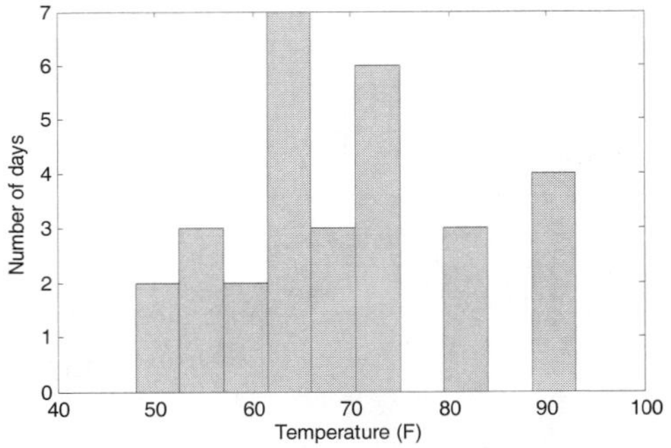

Figure 5-11: Histogram of temperature data.

which means that the range is 45 and the width of each bin is 4.5. The range of the first bin is from 48 to 52.5 and contains two points. The range of the second bin is from 52.5 to 57 and contains three points, and so on. Two of the bins (75 to 79.5 and 84 to 88.5) do not contain any points.

Since the division of the data range into 10 equally spaced bins might not be the division that is preferred by the user, the number of bins can be defined to be different than 10. This can be done either by specifying the number of bins, or by specifying the center point of each bin as shown in the following two forms of the hist command:

`hist(y,nbins)` or `hist(y,x)`

nbins is a scalar that defines the number of bins. MATLAB divides the range in equally spaced subranges.

x is a vector that specifies the location of the center of each bin (the distance between the centers does not have to be the same for all the bins). The edges of the bins are at the middle point between the centers.

In the example above the user might prefer to divide the temperature range into three bins. This can be done with the command:

```
>> hist(y,3)
```

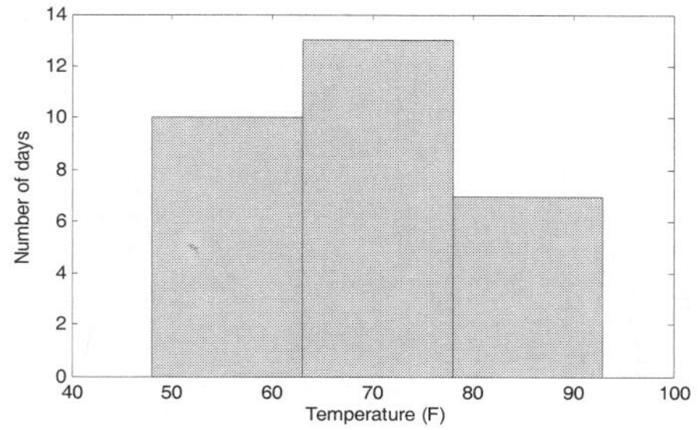

As shown in the top graph, the histogram that is generated has three equally spaced bins.

The number and width of the bins can also be specified by a vector x whose elements define the centers of the bins. For example, shown in the lower graph is a histogram that displays the temperature data from above in six bins with an equal width of 10 degrees. The elements of the vector x for this plot are 45, 55, 65, 75, 85, and 95. The plot was obtained with the following commands:

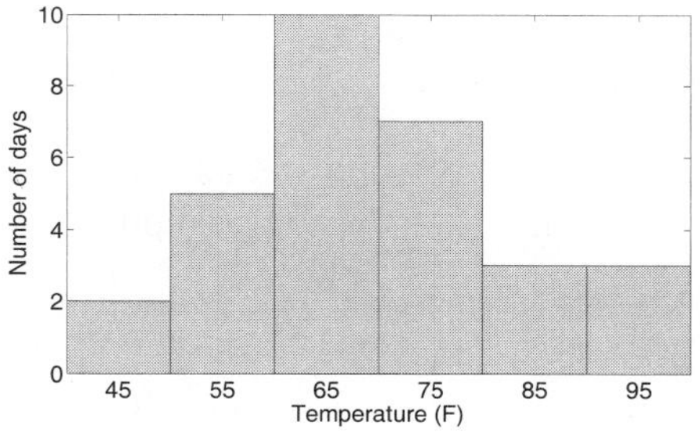


```
>> x=[45:10:95]
x =
    45    55    65    75    85    95
>> hist(y,x)
```

The hist command can be used with options that provide numerical output in addition to plotting a histogram. An output of the number of data points in each bin can be obtained with one of the following commands:

`n=hist(y)` `n=hist(y,nbins)` `n=hist(y,x)`

The output n is a vector. The number of elements in n is equal to the number of bins, and the value of each element of n is the number of data points (frequency count) in the corresponding bin. For example, the histogram in Figure 5-11 can also be created with the following command:

```
>> n = hist(y)
n =
   2   3   2   7   3   6   0   3   0   4
```

The vector n shows how many elements are in each bin.

The vector n shows that the first bin has two data points, the second bin has three data points, and so on.

An additional, optional numerical output is the location of the bins. This output can be obtained with one of the following commands:

```
[n xout]=hist(y)
```

```
[n xout]=hist(y,nbins)
```

xout is a vector in which the value of each element is the location of the center of the corresponding bin. For example, for the histogram in Figure 5-11:

```
>> [n xout]=hist(y)
n =
   2   3   2   7   3   6   0   3   0   4
xout =
   50.2500   54.7500   59.2500   63.7500   68.2500   72.7500
77.2500   81.7500   86.2500   90.7500
```

The vector xout shows that the center of the first bin is at 50.25, the center of the second bin is at 54.75, and so on.

5.9 Polar Plots

Polar coordinates, in which the position of a point in a plane is defined by the angle θ and the radius (distance) to the point, are frequently used in the solution of science and engineering problems. The polar command is used to plot functions in polar coordinates. The command has the form:

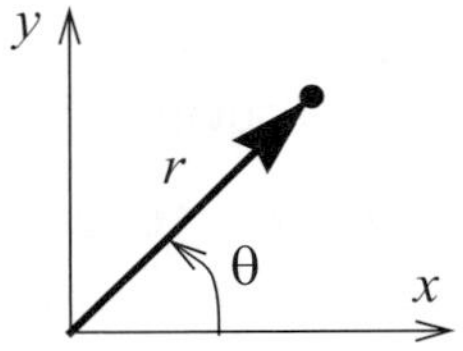

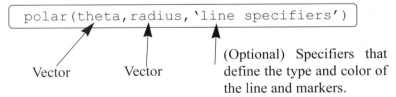

```
polar(theta,radius,'line specifiers')
```

theta: Vector

radius: Vector

'line specifiers': (Optional) Specifiers that define the type and color of the line and markers.

where theta and radius are vectors whose elements define the coordinates of the points to be plotted. The polar command plots the points and draws the polar grid. The line specifiers are the same as in the plot command. To plot a function $r = f(\theta)$ in a certain domain, a vector for values of θ is created first, and then a vector r with the corresponding values of $f(\theta)$ is created using element-by-element calculations. The two vectors are then used in the `polar` command.

For example, a plot of the function $r = 3\cos^2(0.5\theta) + \theta$ for $0 \le \theta \le 2\pi$ is shown below.

```
t=linspace(0,2*pi,200);
r=3*cos(0.5*t).^2+t;
polar(t,r)
```

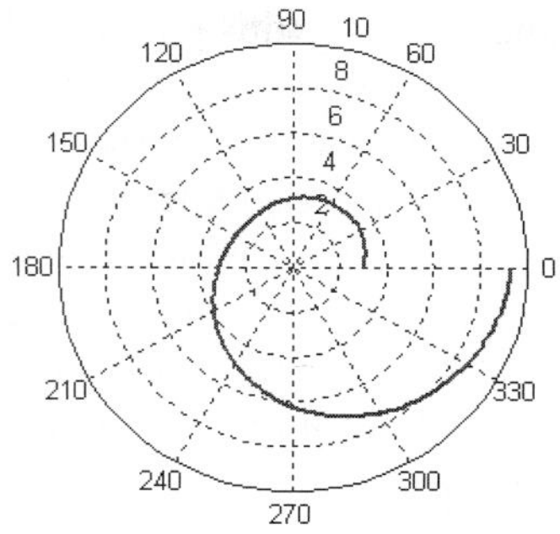

5.10 PUTTING MULTIPLE PLOTS ON THE SAME PAGE

Multiple plots can be created on the same page with the `subplot` command, which has the form:

```
subplot(m,n,p)
```

The command divides the Figure Window (and the page when printed) into $m \times n$ rectangular subplots. The subplots are arranged like elements in an $m \times n$ matrix where each element is a subplot. The subplots are numbered from 1 through $m \cdot n$. The upper left subplot is numbered 1 and the lower right subplot is numbered $m \cdot n$. The numbers increase from left to right within a row, from the first row to the last. The command `subplot(m,n,p)` makes the subplot `p` current. This means that the next plot command (and any formatting commands) will create a plot (with the corresponding format) in this subplot. For example, the command `subplot(3,2,1)` creates six areas arranged in three rows and two columns as shown, and makes the upper left subplot current. An example of using the `subplot` command is shown in the solution of Sample Problem 5-2.

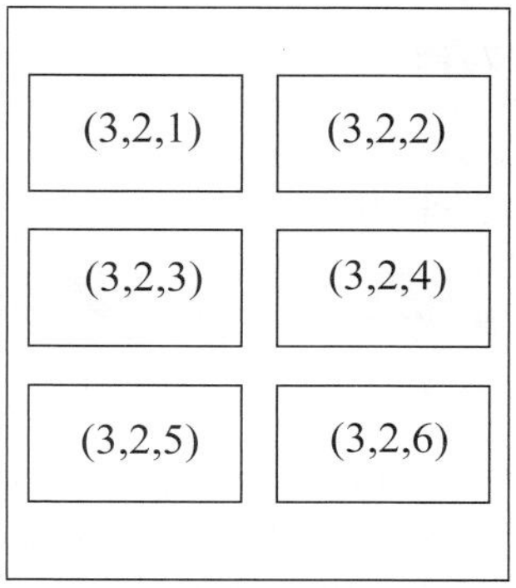

5.11 MULTIPLE FIGURE WINDOWS

When `plot` or any other command that generates a plot is executed, the Figure Window opens (if not already open) and displays the plot. MATLAB labels the Figure Window as Figure 1 (see the top left corner of the Figure Window that is displayed in Figure 5-4). If the Figure Window is already open when the `plot` or any other command that generates a plot is executed, a new plot is displayed in the

Figure Window (replacing the existing plot). Commands that format plots are applied to the plot in the Figure Window that is open.

It is possible, however, to open additional Figure Windows and have several of them open (with plots) at the same time. This is done by typing the command `figure`. Every time the command `figure` is entered, MATLAB opens a new Figure Window. If a command that creates a plot is entered after a `figure` command, MATLAB generates and displays the new plot in the last Figure Window that was opened, which is called the active or current window. MATLAB labels the new Figure Windows successively; i.e., Figure 2, Figure 3, and so on. For example, after the following three commands are entered, the two Figure Windows that are shown in Figure 5-12 are displayed.

```
>> fplot('x*cos(x)',[0,10])                Plot displayed in Figure 1 window.
>> figure                                  Figure 2 window opens.
>> fplot('exp(-0.2*x)*cos(x)',[0,10])      Plot displayed in Figure 2 window.
```

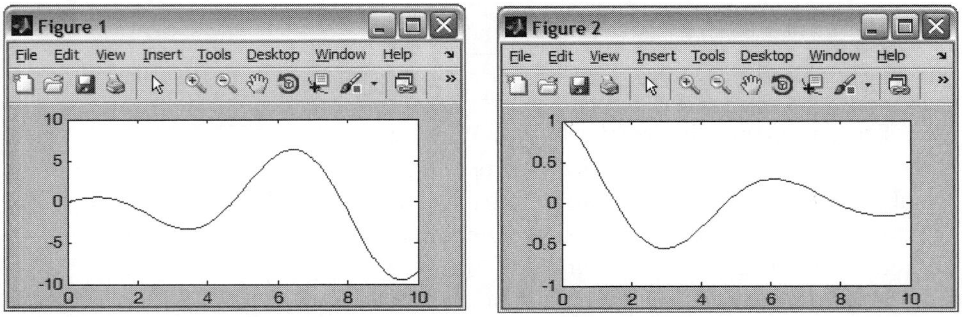

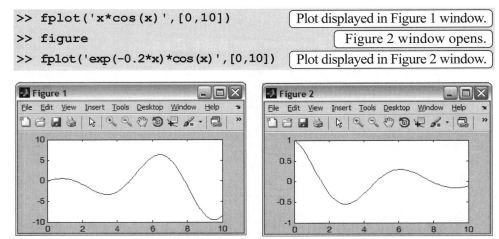

Figure 5-12: Two open Figure Windows.

The `figure` command can also have an input argument that is a number (integer), of the form `figure(n)`. The number corresponds to the number of the corresponding Figure Window. When the command is executed, window number `n` becomes the active Figure Window (if a Figure Window with this number does not exist, a new window with this number opens). When commands that create new plots are executed, the plots that they generate are displayed in the active Figure Window. In the same way, commands that format plots are applied to the plot in the active window. The `figure(n)` command provides means for having a program in a script file that opens and makes plots in a few defined Figure Windows. (If several `figure` commands are used in a program instead, new Figure Windows will open every time the script file is executed.)

Figure Windows can be closed with the `close` command. Several forms of the command are:

`close` closes the active Figure Window.

`close(n)` closes the *n*th Figure Window.

`close all` closes all Figure Windows that are open.

5.12 *EXAMPLES OF MATLAB APPLICATIONS*

Sample Problem 5-2: Piston-crank mechanism

The piston-rod-crank mechanism is used in many engineering applications. In the mechanism shown in the following figure, the crank is rotating at a constant speed of 500 rpm.

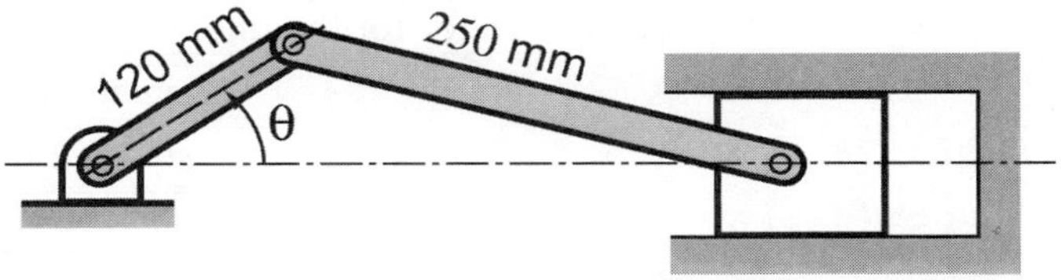

Calculate and plot the position, velocity, and acceleration of the piston for one revolution of the crank. Make the three plots on the same page. Set $\theta = 0°$ when $t = 0$.

Solution

The crank is rotating with a constant angular velocity $\dot{\theta}$. This means that if we set $\theta = 0°$ when $t = 0$, then at time t the angle θ is given by $\theta = \dot{\theta}t$, and that $\ddot{\theta} = 0$ at all times.

The distances d_1 and h are given by:

$d_1 = r\cos\theta$ and $h = r\sin\theta$

With h known, the distance d_2 can be calculated using the Pythagorean Theorem:

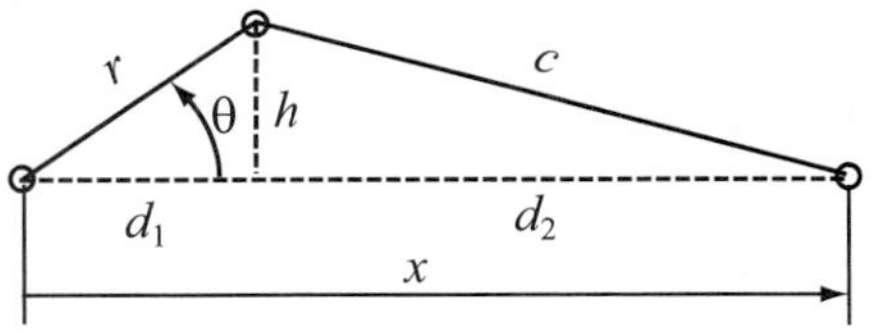

$$d_2 = (c^2 - h^2)^{1/2} = (c^2 - r^2\sin^2\theta)^{1/2}$$

The position x of the piston is then given by:

$$x = d_1 + d_2 = r\cos\theta + (c^2 - r^2\sin^2\theta)^{1/2}$$

The derivative of x with respect to time gives the velocity of the piston:

$$\dot{x} = -r\dot{\theta}\sin\theta - \frac{r^2\dot{\theta}\sin 2\theta}{2(c^2 - r^2\sin^2\theta)^{1/2}}$$

The second derivative of x with respect to time gives the acceleration of the piston:

$$\ddot{x} = -r\dot{\theta}^2\cos\theta - \frac{4r^2\dot{\theta}^2\cos 2\theta(c^2 - r^2\sin^2\theta) + (r^2\dot{\theta}\sin 2\theta)^2}{4(c^2 - r^2\sin^2\theta)^{3/2}}$$

In the equation above, $\ddot{\theta}$ was taken to be zero.

A MATLAB program (script file) that calculates and plots the position, velocity, and acceleration of the piston for one revolution of the crank is shown below.

```
THDrpm=500; r=0.12; c=0.25;                 Define θ̇, r, and c.
THD=THDrpm*2*pi/60;                          Change the units of θ̇ from rpm to rad/s.
tf=2*pi/THD;                                 Calculate the time for one revolution of the crank.
t=linspace(0,tf,200);                        Create a vector for the time with 200 elements.
TH=THD*t;                                    Calculate θ for each t.
d2s=c^2-r^2*sin(TH).^2;                      Calculate d2 squared for each θ.
x=r*cos(TH)+sqrt(d2s);                       Calculate x for each θ.
xd=-r*THD*sin(TH)-(r^2*THD*sin(2*TH))./(2*sqrt(d2s));
xdd=-r*THD^2*cos(TH)-(4*r^2*THD^2*cos(2*TH).*d2s+
(r^2*sin(2*TH)*THD).^2)./(4*d2s.^(3/2));
subplot(3,1,1)                               Calculate ẋ and ẍ for each θ.
plot(t,x)                                    Plot x vs. t.
grid                                         Format the first plot.
xlabel('Time (s)')
ylabel('Position (m)')
subplot(3,1,2)
plot(t,xd)                                   Plot ẋ vs. t.
grid                                         Format the second plot.
xlabel('Time (s)')
ylabel('Velocity (m/s)')
subplot(3,1,3)
plot(t,xdd)                                  Plot ẍ vs. t.
grid                                         Format the third plot.
xlabel('Time (s)')
ylabel('Acceleration (m/s^2)')
```

When the script file runs it generates the three plots on the same page as shown in Figure 5-13. The figure nicely shows that the velocity of the piston is zero at the end points of the travel range where the piston changes the direction of the motion. The acceleration is maximum (directed to the left) when the piston is at the right end.

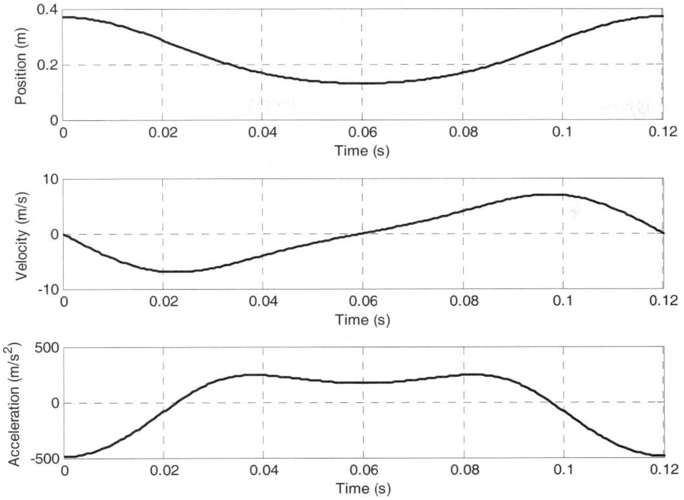

Figure 5-13: Position, velocity, and acceleration of the piston vs. time.

Sample Problem 5-3: Electric Dipole

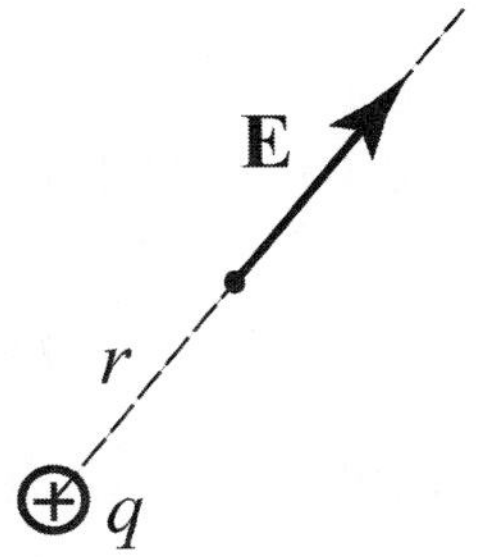

The electric field at a point due to a charge is a vector **E** with magnitude E given by Coulomb's law:

$$E = \frac{1}{4\pi\varepsilon_0}\frac{q}{r^2}$$

where $\varepsilon_0 = 8.8541878 \times 10^{-12}\frac{C^2}{N \cdot m^2}$ is the permittivity constant, q is the magnitude of the charge, and r is the distance between the charge and the point. The direction of **E** is along the line that connects the charge with the point. **E** points outward from q if q is positive, and toward q if q is negative. An electric dipole is created when a positive charge and a negative charge of equal magnitude are placed some distance apart. The electric field, **E**, at any point is obtained by superposition of the electric field of each charge.

An electric dipole with $q = 12 \times 10^{-9}$ C is created, as shown in the figure. Determine and plot the magnitude of the electric field along the x axis from $x = -5$ cm to $x = 5$ cm.

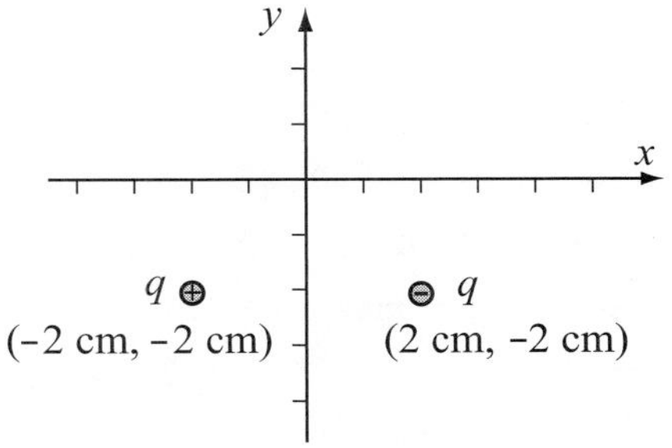

Solution

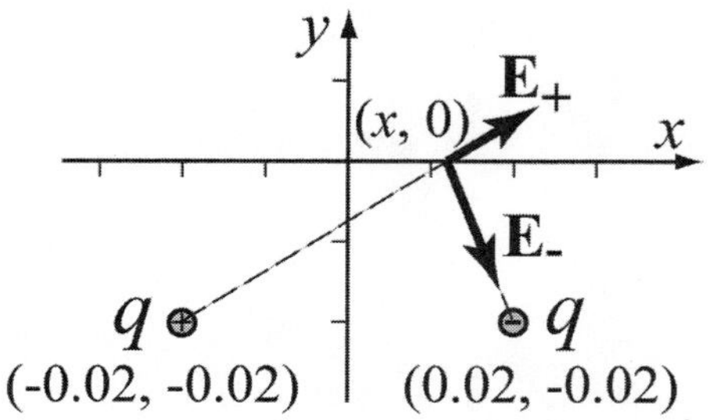

The electric field **E** at any point $(x, 0)$ along the x axis is obtained by adding the electric field vectors due to each of the charges.

$$\mathbf{E} = \mathbf{E}_- + \mathbf{E}_+$$

The magnitude of the electric field is the length of the vector **E**.

The problem is solved by following these steps:

Step 1: Create a vector x for points along the x axis.

Step 2: Calculate the distance (and distance squared) from each charge to the points on the x axis.

$$r_{minus} = \sqrt{(0.02 - x)^2 + 0.02^2} \qquad r_{plus} = \sqrt{(x + 0.02x)^2 + 0.02^2}$$

Step 3: Write unit vectors in the direction from each charge to the points on the x axis.

$$\mathbf{E}_{minusUV} = \frac{1}{r_{minus}}((0.02 - x)\mathbf{i} - 0.02\mathbf{j})$$

$$\mathbf{E}_{plusUV} = \frac{1}{r_{plus}}((x + 0.02)\mathbf{i} + 0.02\mathbf{j})$$

Step 4: Calculate the magnitude of the vector $\mathbf{E}_-$ and $\mathbf{E}_+$ at each point by using Coulomb's law.

$$E_{minusMAG} = \frac{1}{4\pi\varepsilon_0}\frac{q}{r_{minus}^2} \qquad E_{plusMAG} = \frac{1}{4\pi\varepsilon_0}\frac{q}{r_{plus}^2}$$

Step 5: Create the vectors $\mathbf{E}_-$ and $\mathbf{E}_+$ by multiplying the unit vectors by the magnitudes.

Step 6: Create the vector **E** by adding the vectors $\mathbf{E}_-$ and $\mathbf{E}_+$.

Step 7: Calculate E, the magnitude (length) of **E**.

Step 8: Plot E as a function of x.

A program in a script file that solves the problem is:

```
q=12e-9;
epsilon0=8.8541878e-12;
x=[-0.05:0.001:0.05]';
rminusS=(0.02-x).^2+0.02^2;
rminus=sqrt(rminusS);
rplusS=(x+0.02).^2+0.02^2;
rplus=sqrt(rplusS);
```

Create a column vector x.

Step 2. Each variable is a column vector.

```
EminusUV=[((0.02-x)./rminus), (-0.02./rminus)];
EplusUV=[((x+0.02)./rplus), (0.02./rplus)];
EminusMAG=(q/(4*pi*epsilon0))./rminusS;
EplusMAG=(q/(4*pi*epsilon0))./rplusS;
Eminus=[EminusMAG.*EminusUV(:,1), EminusMAG.*EminusUV(:,2)];
Eplus=[EplusMAG.*EplusUV(:,1), EplusMAG.*EplusUV(:,2)];
E=Eminus+Eplus;
EMAG=sqrt(E(:,1).^2+E(:,2).^2);
plot(x,EMAG,'k','linewidth',1)
xlabel('Position along the x-axis (m)','FontSize',12)
ylabel('Magnitude of the electric field (N/C)','FontSize',12)
title('ELECTRIC FIELD DUE TO AN ELECTRIC DIPOLE','FontSize',12)
```

Steps 3 & 4. Each variable is a two column matrix. Each row is the vector for the corresponding x.

Step 5.

Step 6.

Step 7.

When this script file is executed in the Command Window the following figure is created in the Figure Window:

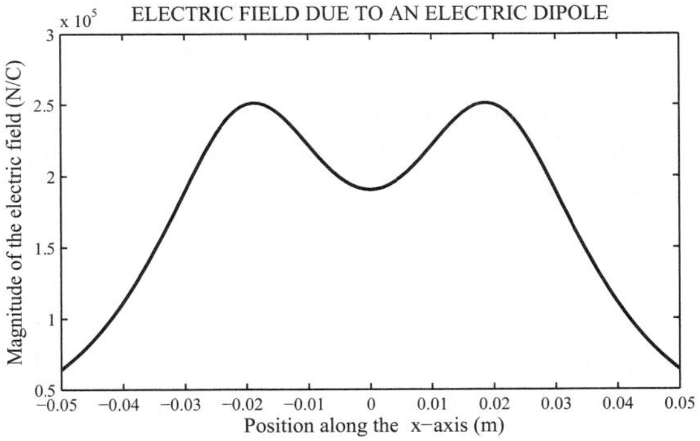

5.13 PROBLEMS

1. Plot the function $f(t) = \dfrac{(x+5)^2}{4+3x^2}$ for $-3 \le x \le 5$.

2. Plot the function $f(t) = \dfrac{5\sin(x)}{x+e^{-0.75x}} - \dfrac{3x}{5}$ for $-5 \le x \le 10$.

3. Make two separate plots of the function $f(x) = (x+1)(x-2)(2x-0.25)-e^x$, one plot for $0 \le x \le 3$ and one for $-3 \le x \le 6$.

4. Use the `fplot` command to plot the function $f(x) = \sqrt{|\cos(3x)|} + \sin^2(4x)$ in the domain $-2 \le x \le 2$.

5. Use the `fplot` command to plot the function

   $f(x) = e^{2\sin(0.4x)}5\cos(4x)$ in the domain $-20 \le x \le 30$.

6. A parametric equation is given by

   $$x = 1.5\sin(5t), \quad y = 1.5\cos(3t)$$

   Plot the function for $0 \le t \le 2\pi$. Format the plot such that the both axes will range from –2 to 2.

7. Plot the function $f(x) = \dfrac{x^2 + 3x + 3}{0.8(x + 1)}$ for $-4 \le x \le 3$. Notice that the function has a vertical asymptote at $x = -1$. Plot the function by creating two vectors for the domain of x. The first vector (name it $x1$) includes elements from –4 to –1.1, and the second vector (name it $x2$) includes elements from –0.9 to 3. For each x vector create a y vector (mane them $y1$ and $y2$) with the corresponding values of y according to the function. To plot the function make two curves in the same plot ($y1$ vs. $x1$, and $y2$ vs. $x2$).

8. A parametric equation is given by

   $$x = \frac{3t}{1 + t^3}, \quad y = \frac{3t^2}{1 + t^3}$$

   (Note that the denominator approaches 0 when t approaches –1) Plot the function (the plot is called the Folium of Descartes) by plotting two curves in the same plot—one for $-30 \le t \le -1.6$ and the other for $-0.6 \le t \le 40$.

9. Plot the function $f(x) = \dfrac{x^2 - 4x - 7}{x^2 - x - 6}$ for $-6 \le x \le 6$. Notice that the function has two vertical asymptotes. Plot the function by dividing the domain of x into three parts: one from –6 to near the left asymptote, one between the two asymptotes, and one from near the right asymptote to 6. Set the range of the y axis from –20 to 20.

10. A cycloid is a curve (shown in the figure) traced by a point on a circle that rolls along a line. The parametric equation of a cycloid is given by

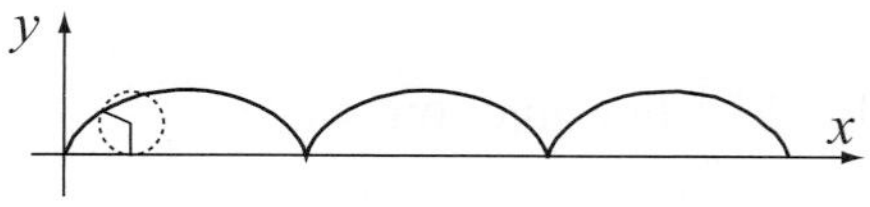

    $$x = r(t - \sin t) \text{ and } y = r(t - \cos t)$$

    Plot a cycloid with $r = 1.5$ and $0 \le t \le 4\pi$.

11. Plot the function $f(x) = \cos x \sin(2x)$ and its derivative, both on the same plot, for $\pi \le x \le \pi$. Plot the function with a solid line, and the derivative with a dashed line. Add a legend and label the axes.

12. The Gateway Arch in St. Louis is shaped according to the equation

$$y = 693.8 - 68.8\cosh\left(\frac{x}{99.7}\right) \text{ ft}$$

Make a plot of the arch.

13. An electrical circuit that includes a voltage source v_S with an internal resistance r_S and a load resistance R_L is shown in the figure. The power P dissipated in the load is given by

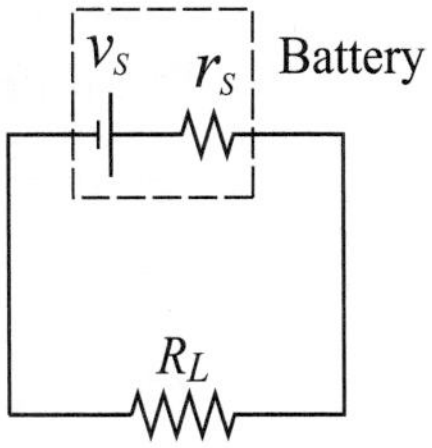

$$P = \frac{v_S^2 R_L}{(R_L + r_S)^2}$$

Plot the power P as a function of R_L for $1 \le R_L \le 10\,\Omega$, given that $v_S = 12$ V and $r_S = 2.5\,\Omega$.

14. Two ship, A and B, travel at a speed of $v_A = 27$ mi/h and $v_B = 14$ mi/h, respectively. The directions they are moving and their location at 8 A.M. are shown in the figure. Plot the distance between the ships as a function of time for the next 4 hours. The horizontal axis should show the actual time of day starting at 8 A.M., while the vertical axis should show the distance. Label the axes.

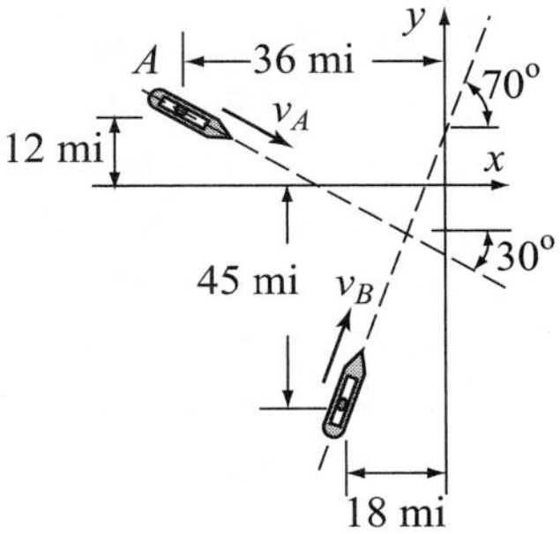

15. The plasma concentration C_P of orally delivered drugs is a function of the rate of absorption, K_{ab}, and the rate of elimination, K_{el}:

$$C_P = A\frac{K_{ab}}{K_{ab} - K_{el}}(e^{-K_{el}t} - e^{-K_{ab}t})$$

where A is a constant (associated with the specific drag) and t is time. Consider a case where $A = 140$ mg/L, $K_{ab} = 1.6$ h^{-1}, and $K_{ab} = 0.45$ h^{-1}. Make a plot that displays C_P vs. time for $0 \le t \le 10$.

16. The position as a function of time of a squirrel running on a grass field is given in polar coordinates by:

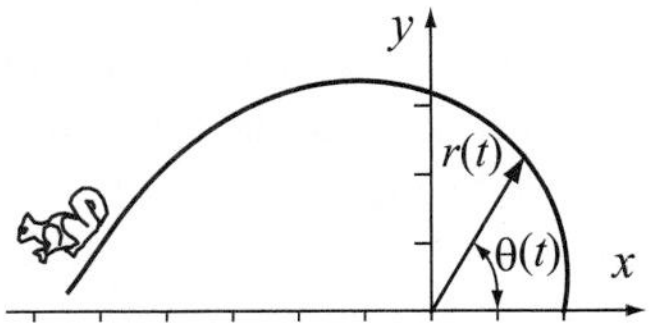

$$r(t) = 20 + 30(1 - e^{-0.1t}) \text{ m}$$
$$\theta(t) = \pi(1 - e^{-0.2t})$$

(*a*) Plot the trajectory (position) of the squirrel for $0 \le t \le 20$ s.

(*b*) Create a (second) plot for the speed of the squirrel, given by $v = r\frac{d\theta}{dt}$, as a function of time for $0 \le t \le 20$ s.

17. In astronomy, the relationship between the relative luminosity L/L_{Sun} (brightness relative to the sun), the relative radius R/R_{Sun}, and the relative temperature T/T_{Sun} of a star is modeled by:

$$\frac{L}{L_{Sun}} = \left(\frac{R}{R_{Sun}}\right)^2\left(\frac{T}{T_{Sun}}\right)^4$$

The HR (Hertzsprung-Russell) diagram is a plot of L/L_{Sun} versus the temperature. The following data is given:

	Sun	Spica	Regulus	Alioth	Barnard's Star	Epsilon Indi	Beta Crucis
Temp (K)	5,840	22,400	13,260	9,400	3,130	4,280	28,200
L/L_{Sun}	1	13,400	150	108	0.0004	0.15	34,000
R/R_{Sun}	1	7.8	3.5	3.7	0.18	0.76	8

To compare the data with the model, use MATLAB to plot an HR diagram. The diagram should have two sets of points. One uses the values of L/L_{Sun} from the table (use asterisk markers), and the other uses values of L/L_{Sun} that are calculated from the equation by using R/R_{Sun} from the table (use circle markers). In the HR diagram both axes are logarithmic. In addition, the values of temperature on the horizontal axis are decreasing from left to right. This is done with the command `set(gca,'XDir','reverse')`. Label the axes and use a legend.

18. The position x as a function of time of a particle that moves along a straight line is given by

$$x(t) = 0.41t^4 - 10.8t^3 + 64t^2 - 8.2t + 4.4 \text{ ft}$$

The velocity $v(t)$ of the particle is determined by the derivative of $x(t)$ with respect to t, and the acceleration $a(t)$ is determined by the derivative of $v(t)$ with respect to t.

Derive the expressions for the velocity and acceleration of the particle, and make plots of the position, velocity, and acceleration as functions of time for $0 \le t \le 8$ s. Use the `subplot` command to make the three plots on the same page with the plot of the position on the top, the velocity in the middle, and the acceleration at the bottom. Label the axes appropriately with the correct units.

19. In a typical tension test a dog bone shaped specimen is pulled in a machine. During the test, the force F needed to pull the specimen and the length L of a gauge section are measured. This data is used for plotting a stress-strain diagram of the material. Two definitions, engineering and true, exist for stress and strain. The engineering stress σ_e and strain ε_e are defined by

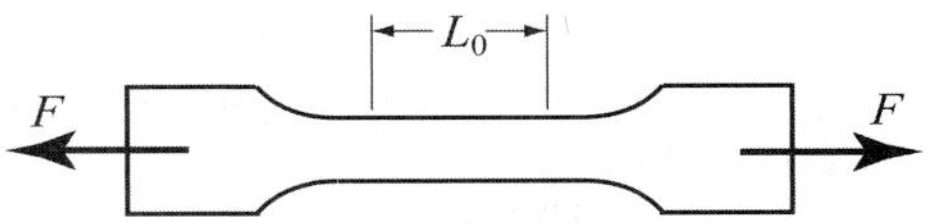

$\sigma_e = \frac{F}{A_0}$ and $\varepsilon_e = \frac{L - L_0}{L_0}$, where L_0 and A_0 are the initial gauge length and the initial cross-sectional area of the specimen, respectively. The true stress σ_t and strain ε_t are defined by $\sigma_t = \frac{F}{A_0}\frac{L}{L_0}$ and $\varepsilon_t = \ln\frac{L}{L_0}$.

The following are measurements of force and gauge length from a tension test with an aluminum specimen. The specimen has a round cross section with radius 6.4 mm (before the test). The initial gauge length is $L_0 = 25$ mm. Use the data to calculate and generate the engineering and true stress-strain curves, both on the same plot. Label the axes and label the curves.
Units: When the force is measured in newtons (N), and the area is calculated in m^2, the unit of the stress is pascals (Pa).

F (N)	0	13,345	26,689	40,479	42,703	43,592	44,482	44,927
L (mm)	25	25.037	25.073	25.113	25.122	25.125	25.132	25.144
F (N)	45,372	46,276	47,908	49,035	50,265	53,213	56,161	
L (mm)	25.164	25.208	25.409	25.646	26.084	27.398	29.150	

20. The area of the aortic valve, A_V in cm^2, can be estimated by the equation (Hakki Formula)

$$A_V = \frac{Q}{\sqrt{PG}}$$

where Q is the cardiac output in L/min, and PG is the difference between the left ventricular systolic pressure and the aortic systolic pressure (in mm Hg). Make one plot with two curves of A_V versus PG, for $2 \le PG \le 60$ mm Hg—one curve for $Q = 4$ L/min and the other for $Q = 5$ L/min. Label the axes and use a legend.

21. A series *RLC* circuit with an AC voltage source is shown. The amplitude of the current, I, in this circuit is given by

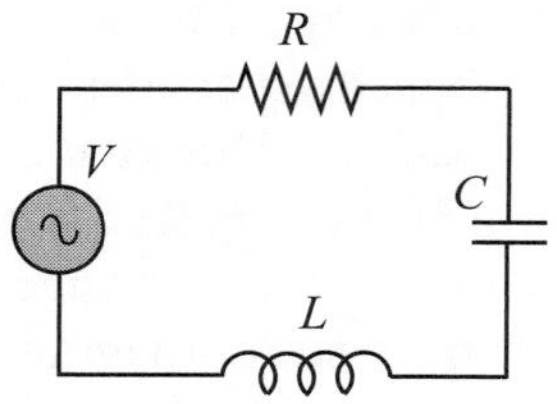

$$I = \frac{v_m}{\sqrt{R^2 + (\omega_d L - 1/(\omega_d C))^2}}$$

where $\omega_d = 2\pi f_d$ in which f_d is the driving frequency; R and C are the resistance of the resistor and capacitance of the capacitor, respectively; and v_m is the amplitude of V. For the circuit in the figure $R = 80\ \Omega$, $C = 18 \times 10^{-6}$ F, $L = 260 \times 10^{-3}$ H, and $v_m = 10$ V.

Make a plot of I as a function of f_d for $10 \le f \le 10000$ Hz. Use a linear scale for I and a log scale for f_d.

22. The speed distribution, $N(v)$, of gas molecules can be modeled by Maxwell's speed distribution law:

$$N(v) = 4\pi\left(\frac{m}{2\pi kT}\right)^{3/2} v^2 e^{\frac{-mv^2}{2kT}}$$

where m (kg) is the mass of each molecule, v (m/s) is the speed, T (K) is the temperature, and $k = 1.38 \times 10^{-23}$ J/K is Boltzmann's constant. Make a plot of $N(v)$ versus v for $0 \le v \le 1200$ m/s for oxygen molecules ($m = 5.3 \times 10^{-26}$ kg). Make two graphs in the same plot, one for $T = 80$ K and the other for $T = 300$ K. Label the axes and display a legend.

23. A resistor, $R = 4\ \Omega$, and an inductor, $L = 1.3$ H, are connected in a circuit to a voltage source as shown in Figure (*a*) (an *RL* circuit). When the voltage

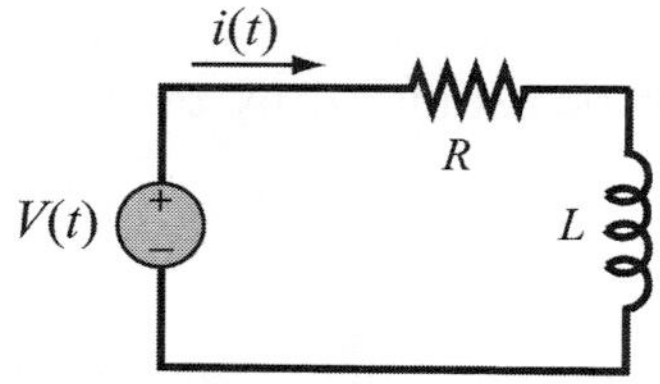

(*a*)

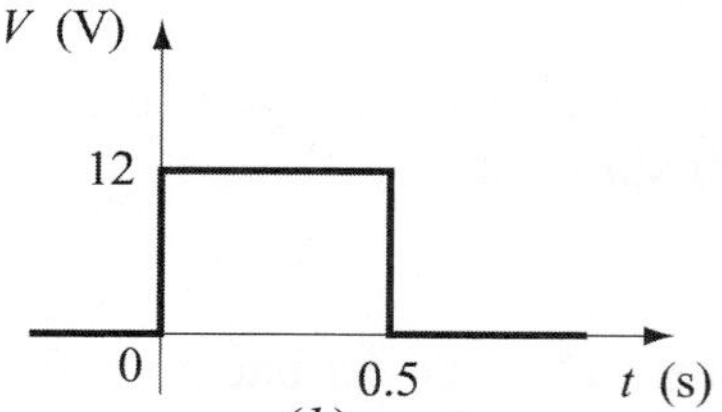

(*b*)

source applies a rectangular voltage pulse with an amplitude of $V = 12$ V and a duration of 0.5 s, as shown in Figure (*b*), the current $i(t)$ in the circuit as a function of time is given by:

$$i(t) = \frac{V}{R}(1 - e^{(-Rt)/L}) \quad \text{for} \quad 0 \le t \le 0.5 \text{ s}$$

$$i(t) = e^{-(Rt)/L}\frac{V}{R}(e^{(0.5R)/L} - 1) \quad \text{for} \quad 0.5 \le t \text{ s}$$

Make a plot of the current as a function of time for $0 \le t \le 2$ s.

24. The shape of a symmetrical four digit NACA airfoil is described by the equation

$$y = \pm\frac{tc}{0.2}\left[0.2969\sqrt{\frac{x}{c}} - 0.1260\frac{x}{c} - 0.3516\left(\frac{x}{c}\right)^2 + 0.2843\left(\frac{x}{c}\right)^3 - 0.1015\left(\frac{x}{c}\right)^4\right]$$

where c is the cord length and t is the maximum thickness as a fraction of the cord length (tc = maximum thickness). Symmetrical four digit NACA airfoils are designated NACA 00XX, where XX is $100t$ (i.e., NACA 0012 has $t = 0.12$). Plot the shape of a NACA 0020 airfoil with a cord length of 1.5 m.

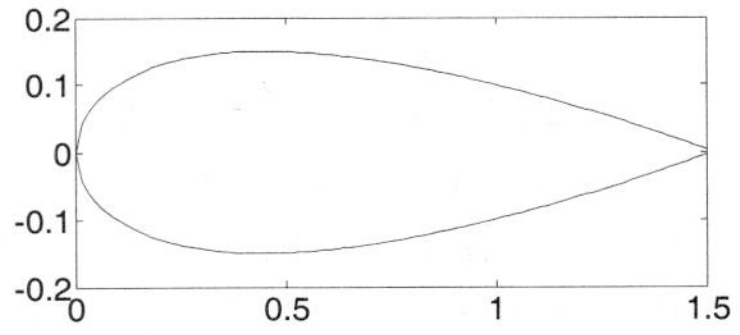

25. The dynamic storage modulus G' and loss modulus G'' are measures of a material mechanical response to harmonic loading. For many biological materials these moduli can be described by Fung's model:

$$G'(\omega) = G_\infty\left\{1 + \frac{c}{2}\ln\left[\frac{1 + (\omega\tau_2)^2}{1 + (\omega\tau_1)^2}\right]\right\} \quad \text{and} \quad G''(\omega) = cG_\infty[\tan^{-1}(\omega\tau_2) - \tan^{-1}(\omega\tau_1)]$$

where ω is the frequency of the harmonic loading, and G_∞, c, τ_1, and τ_2 are material constants. Plot G' and G'' versus ω (two separate plots on the same page) for $G_\infty = 5$ ksi, $c = 0.05$, $\tau_1 = 0.05$ s, and $\tau_2 = 500$ s. Let ω vary between 0.0001 and 1000 s^{-1}. Use a log scale for the ω axis.

26. The vibrations of the body of a helicopter due to the periodic force applied by the rotation of the rotor can be modeled by a frictionless spring-mass-damper system subjected to an external periodic force. The position $x(t)$ of the mass is given by the equation:

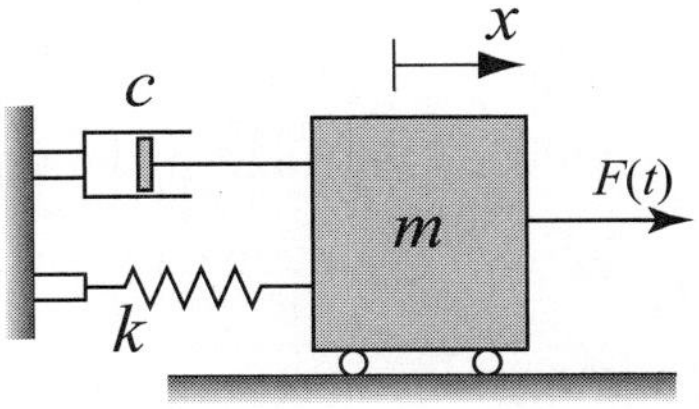

$$x(t) = \frac{2f_0}{\omega_n^2 - \omega^2}\sin\left(\frac{\omega_n - \omega}{2}t\right)\sin\left(\frac{\omega_n - \omega}{2}t\right)$$

where $F(t) = F_0\sin\omega t$, and $f_0 = F_0/m$, ω is the frequency of the applied force, and ω_n is the natural frequency of the helicopter. When the value of ω is close to the value of ω_n, the vibration consists of fast oscillation with slowly changing amplitude called beat. Use $F_0/m = 12$ N/kg, $\omega_n = 10$ rad/s, and $\omega = 12$ rad/s to plot $x(t)$ as a function of t for $0 \le t \le 10$ s.

27. Consider the diode circuit shown in the figure. The current i_D and the voltage v_D can be determined from the solution of the following system of equations:

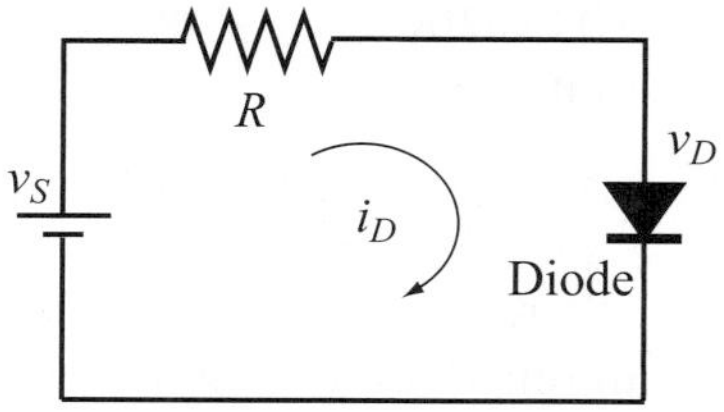

$$i_D = I_0\left(e^{\frac{qv_D}{kT}} - 1\right), \quad i_D = \frac{v_S - v_D}{R}$$

The system can be solved numerically or graphically. The graphical solution is found by plotting i_D as a function of v_D from both equations. The solution is the intersection of the two curves. Make the plots and estimate the solution for the case where $I_0 = 10^{-14}$ A, $v_S = 1.5$ V, $R = 1200\ \Omega$, and $\frac{kT}{q} = 30$ mV.

28. The ideal gas equation states that $\frac{PV}{RT} = n$, where P is the pressure, V is the volume, T is the temperature, $R = 0.08206$ (L atm)/(mol K) is the gas constant, and n is the number of moles. For one mole ($n = 1$) the quantity $\frac{PV}{RT}$ is a constant equal to 1 at all pressures. Real gases, especially at high pressures, deviate from this behavior. Their response can be modeled with the van der Waals equation

$$P = \frac{nRT}{V - nb} - \frac{n^2 a}{V^2}$$

where a and b are material constants. Consider 1 mole ($n = 1$) of nitrogen gas at $T = 300$ K. (For nitrogen gas $a = 1.39$ (L^2 atm)/mol^2, and $b = 0.0391$ L/mol.) Use the van der Waals equation to calculate P as a function of V for $0.08 \le V \le 6$ L, using increments of 0.02 L. At each value of V calculate the value of $\frac{PV}{RT}$ and make a plot of $\frac{PV}{RT}$ versus P. Does the response of nitrogen agree with the ideal gas equation?

29. When monochromatic light passes through a narrow slit it produces on a screen a diffraction pattern consisting of bright and dark fringes. The intensity of the bright fringes, I, as a function of θ can be calculated by

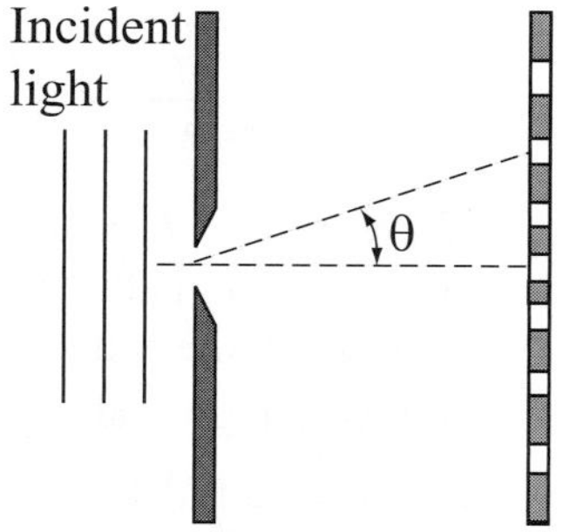

$$I = I_{max}\left(\frac{\sin\alpha}{\alpha}\right)^2, \text{ where } \alpha = \frac{\pi a}{\lambda}\sin\theta$$

where λ is the light wave length and a is the width of the slit. Plot the relative intensity I/I_{max} as a function of θ for $-20° \le \theta \le 20°$. Make one plot that contains three graphs for the cases $a = 10\lambda$, $a = 5\lambda$, and $a = \lambda$. Label the axes, and display a legend.

30. In order to supply fluid to point *D*, a new pipe *CD* with diameter of d_2 is connected to an existing pipe with diameter of d_1 at point *C* between points *A* and *B*. The resistance, *R*, to fluid flow along the path *ACD* is given by

$$R = \frac{L_1 - L_2 \cot\theta}{r_1^4}K + \frac{L_2}{r_2^4 \sin\theta}K$$

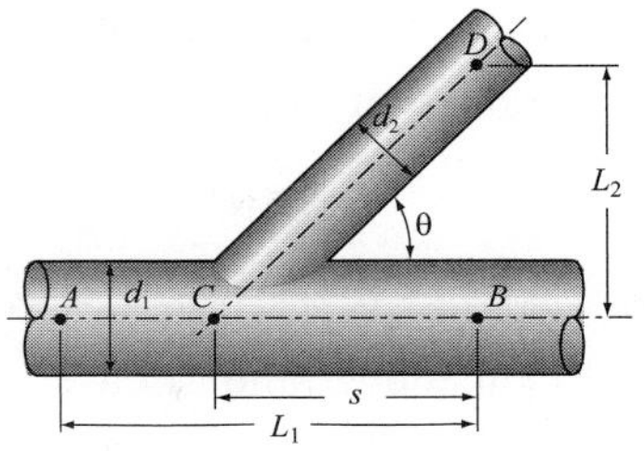

where *K* is a constant. Determine the location of point *C* (the distance *s*) that minimizes the flow resistance *R*. Define a vector θ with elements ranging from $30°$ to $85°$ with spacing of $0.5°$. Calculate R/K for each value of θ, and make a plot of R/K versus θ. Use MATLAB's built-in function `min` to find the minimum value of R/K and the corresponding θ, and then calculate the value of s. Use $d_1 = 1.75$ in., $d_2 = 1.5$ in., $L_1 = 50$ ft, $L_2 = 40$ ft.

31. A simply supported beam is subjected to a constant distributed load *w* over half of its length and a moment *M*, as shown in the figure. The deflection *y*, as a function of *x*, is given by the equations

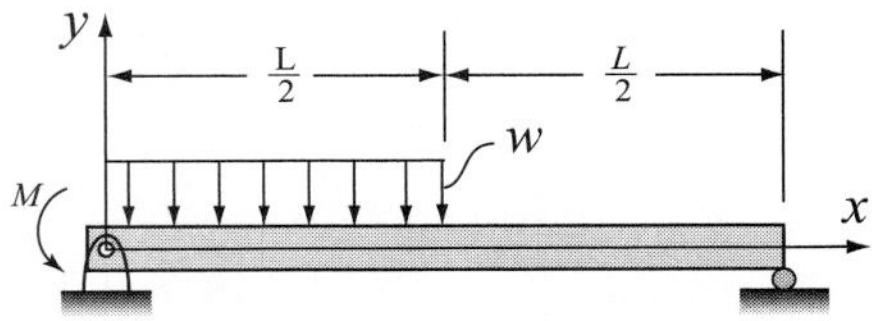

$$y = \frac{-wx}{384EI}(16x^3 - 24Lx^2 + 9L^2) + \frac{Mx}{6EIL}(x^2 - 3Lx + 2L^2) \text{ for } 0 \le x \le \frac{1}{2}L$$

$$y = \frac{-wL}{384EI}(8x^3 - 24Lx^2 + 17L^2x - L^3) + \frac{Mx}{6EIL}(x^2 - 3Lx + 2L^2) \text{ for } \frac{1}{2}L \le x \le L$$

where *E* is the elastic modulus, *I* is the moment of inertia, and *L* is the length of the beam. For the beam shown in the figure $L = 20$ m, $E = 200 \times 10^9$ Pa (steel), $I = 348 \times 10^{-6}$ m^4, $w = 5.4 \times 10^3$ N/m, and $M = 200 \times 10^3$ N m. Make a plot of the deflection of the beam *y* as a function of *x*.

32 The ideal gas law relates the pressure *P*, volume *V*, and temperature *T* of an ideal gas:

$$PV = nRT$$

where *n* is the number of moles and $R = 8.3145$ J/(K mol). Plots of pressure versus volume at constant temperature are called isotherms. Plot the isotherms for one mole of an ideal gas for volume ranging from 1 to 10 m^3, at temperatures of $T = 100$, 200, 300, and 400 K (four curves in one plot). Label the axes and display a legend. The units for pressure are Pa.

33. The voltage difference v_{AB} between points A and B of the Wheatstone bridge circuit is given by:

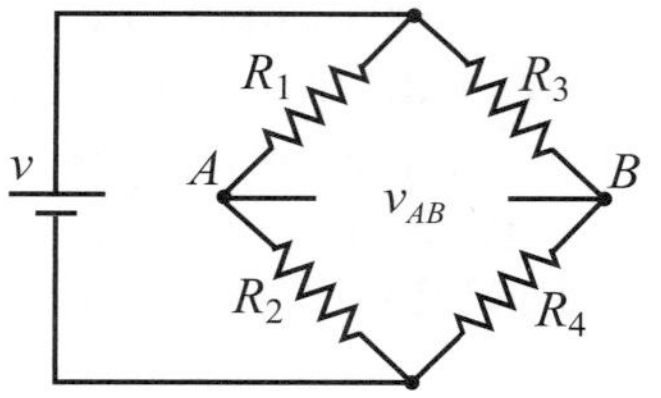

$$v_{AB} = v\left(\frac{R_2}{R_1 + R_2} - \frac{R_4}{R_3 + R_4}\right)$$

Consider the case where $v = 12$ V, $R_3 = R_4 = 250\ \Omega$, and make the following plots:

(*a*) v_{AB} versus R_1 for $0 \le R_1 \le 500\ \Omega$, given $R_2 = 120\ \Omega$.

(*b*) v_{AB} versus R_2 for $0 \le R_2 \le 500\ \Omega$, given $R_1 = 120\ \Omega$.

Plot both plots on a single page (two plots in a column).

34. The resonant frequency f (in Hz) for the circuit shown is given by:

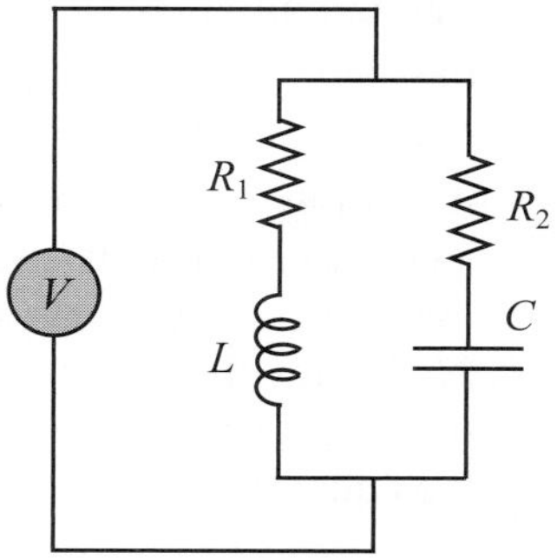

$$f = \frac{1}{2\pi}\sqrt{LC\frac{R_1^2 C - L}{R_2^2 C - L}}$$

Given $L = 0.2$ H, $C = 2 \times 10^{-6}$ F, make the following plots:

(*a*) f versus R_2 for $500 \le R_2 \le 2000\ \Omega$, given $R_1 = 1500\ \Omega$.

(*b*) f versus R_1 for $500 \le R_1 \le 2000\ \Omega$, given $R_2 = 1500\ \Omega$.

Plot both plots on a single page (two plots in a column).

35. The taylor series for sin(x) is:

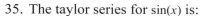

$$x - \frac{x^3}{3!} + \frac{x^5}{5!} - \frac{x^7}{7!} + \frac{x^9}{9!} - \frac{x^{11}}{11!} + \ldots$$

Plot the figure on the right, which shows, for $-2\pi \le x \le 2\pi$, the graph of the function sin(x) and graphs of the Taylor series expansion of sin(x) with one, two, and five terms. Label the axes and display a legend.

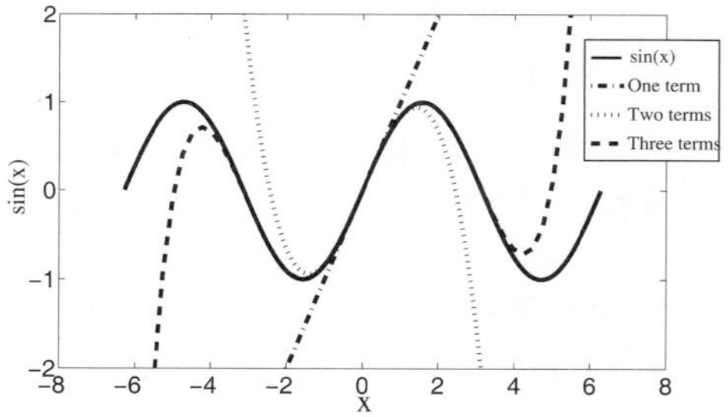

Chapter 6
Programming in MATLAB

A computer program is a sequence of computer commands. In a simple program the commands are executed one after the other in the order they are typed. In this book, for example, all the programs that have been presented so far in script files are simple programs. Many situations, however, require more sophisticated programs in which commands are not necessarily executed in the order they are typed, or different commands (or groups of commands) are executed when the program runs with different input variables. For example, a computer program that calculates the cost of mailing a package uses different mathematical expressions to calculate the cost depending on the weight and size of the package, the content (books are less expensive to mail), and the type of service (airmail, ground, etc.). In other situations there might be a need to repeat a sequence of commands several times within a program. For example, programs that solve equations numerically repeat a sequence of calculations until the error in the answer is smaller than some measure.

MATLAB provides several tools that can be used to control the flow of a program. Conditional statements (Section 6.2) and the `switch` structure (Section 6.3) make it possible to skip commands or to execute specific groups of commands in different situations. `For` loops and `while` loops (Section 6.4) make it possible to repeat a sequence of commands several times.

It is obvious that changing the flow of a program requires some kind of decision-making process within the program. The computer must decide whether to execute the next command or to skip one or more commands and continue at a different line in the program. The program makes these decisions by comparing values of variables. This is done by using relational and logical operators, which are explained in Section 6.1.

It should also be noted that user-defined functions (introduced in Chapter 7) can be used in programming. A user-defined function can be used as a subprogram. When the main program reaches the command line that has the user-defined function, it provides input to the function and "waits" for the results. The user-defined function carries out the calculations and transfers the results back to the main program, which then continues to the next command.

6.1 RELATIONAL AND LOGICAL OPERATORS

A relational operator compares two numbers by determining whether a comparison statement (e.g., $5 < 8$) is true or false. If the statement is true, it is assigned a value of 1. If the statement is false, it is assigned a value of 0. A logical operator examines true/false statements and produces a result that is true (1) or false (0) according to the specific operator. For example, the logical AND operator gives 1 only if both statements are true. Relational and logical operators can be used in mathematical expressions and, as will be shown in this chapter, are frequently used in combination with other commands, to make decisions that control the flow of a computer program.

Relational operators:

Relational operators in MATLAB are:

Relational operator	Description
<	Less than
>	Greater than
<=	Less than or equal to
>=	Greater than or equal to
==	Equal to
~=	Not Equal to

Note that the "equal to" relational operator consists of two = signs (with no space between them), since one = sign is the assignment operator. In other relational operators that consist of two characters there also is no space between the characters (<=, >=, ~=).

- Relational operators are used as arithmetic operators within a mathematical expression. The result can be used in other mathematical operations, in addressing arrays, and together with other MATLAB commands (e.g., `if`) to control the flow of a program.
- When two numbers are compared, the result is 1 (logical true) if the comparison, according to the relational operator, is true, and 0 (logical false) if the comparison is false.
- If two scalars are compared, the result is a scalar 1 or 0. If two arrays are compared (only arrays of the same size can be compared), the comparison is done *element-by-element*, and the result is a logical array of the same size with 1s and 0s according to the outcome of the comparison at each address.

- If a scalar is compared with an array, the scalar is compared with every element of the array, and the result is a logical array with 1s and 0s according to the outcome of the comparison of each element.

Some examples are:

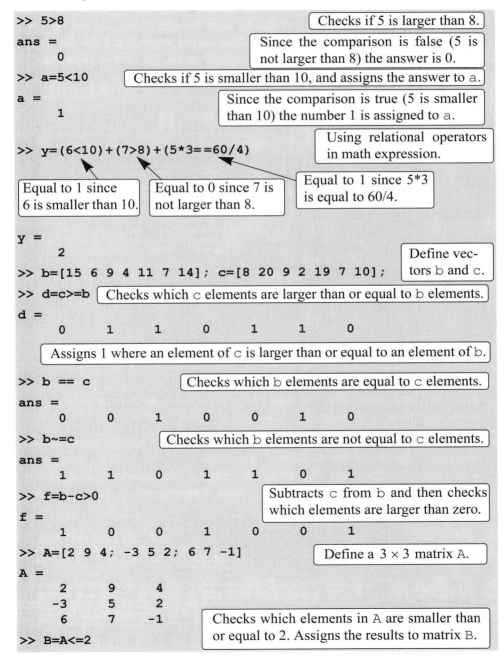

```
>> 5>8
```
Checks if 5 is larger than 8.

```
ans =
     0
```
Since the comparison is false (5 is not larger than 8) the answer is 0.

```
>> a=5<10
```
Checks if 5 is smaller than 10, and assigns the answer to `a`.

```
a =
     1
```
Since the comparison is true (5 is smaller than 10) the number 1 is assigned to `a`.

```
>> y=(6<10)+(7>8)+(5*3==60/4)
```
Using relational operators in math expression.

(6<10): Equal to 1 since 6 is smaller than 10.

(7>8): Equal to 0 since 7 is not larger than 8.

(5*3==60/4): Equal to 1 since 5*3 is equal to 60/4.

```
y =
     2
>> b=[15 6 9 4 11 7 14]; c=[8 20 9 2 19 7 10];
```
Define vectors `b` and `c`.

```
>> d=c>=b
```
Checks which `c` elements are larger than or equal to `b` elements.

```
d =
     0     1     1     0     1     1     0
```
Assigns 1 where an element of `c` is larger than or equal to an element of `b`.

```
>> b == c
```
Checks which `b` elements are equal to `c` elements.

```
ans =
     0     0     1     0     0     1     0
>> b~=c
```
Checks which `b` elements are not equal to `c` elements.

```
ans =
     1     1     0     1     1     0     1
>> f=b-c>0
```
Subtracts `c` from `b` and then checks which elements are larger than zero.

```
f =
     1     0     0     1     0     0     1
>> A=[2 9 4; -3 5 2; 6 7 -1]
```
Define a 3×3 matrix `A`.

```
A =
     2     9     4
    -3     5     2
     6     7    -1
>> B=A<=2
```
Checks which elements in `A` are smaller than or equal to 2. Assigns the results to matrix `B`.

```
B =
     1     0     0
     1     0     1
     0     0     1
```

- The results of a relational operation with vectors, which are vectors with 0s and 1s, are called logical vectors and can be used for addressing vectors. When a logical vector is used for addressing another vector, it extracts from that vector the elements in the positions where the logical vector has 1s. For example:

```
>> r = [8 12 9 4 23 19 10]          Define a vector r.
r =
     8    12     9     4    23    19    10
>> s=r<=10      Checks which r elements are smaller than or equal to 10.
s =
     1     0     1     1     0     0     1
                       A logical vector s with 1s at positions where
                       elements of r are smaller than or equal to 10.
>> t=r(s)       Use s for addresses in vector r to create vector t.
t =                            Vector t consists of elements of
     8     9     4    10       r in positions where s has 1s.
>> w=r(r<=10)          The same procedure can be done in one step.
w =
     8     9     4    10
```

- Numerical vectors and arrays with the numbers 0s and 1s are not the same as logical vectors and arrays with 0s and 1s. Numerical vectors and arrays can not be used for addressing. Logical vectors and arrays, however, can be used in arithmetic operations. The first time a logical vector or an array is used in arithmetic operations it is changed to a numerical vector or array.

- Order of precedence: In a mathematical expression that includes relational and arithmetic operations, the arithmetic operations (+, –, *, /, \) have precedence over relational operations. The relational operators themselves have equal precedence and are evaluated from left to right. Parentheses can be used to alter the order of precedence. Examples are:

```
>> 3+4<16/2                         + and / are executed first.
ans =                       The answer is 1 since 7 < 8 is true.
     1
>> 3+(4<16)/2    4 < 16 is executed first, and is equal to 1, since it is true.
ans =                          3.5 is obtained from 3 + 1/2.
    3.5000
```

Logical operators:

Logical operators in MATLAB are:

Logical operator	Name	Description
& Example: A&B	AND	Operates on two operands (A and B). If both are true, the result is true (1); otherwise the result is false (0).
| Example: A|B	OR	Operates on two operands (A and B). If either one, or both, are true, the result is true (1); otherwise (both are false) the result is false (0).
~ Example: ~A	NOT	Operates on one operand (A). Gives the opposite of the operand; true (1) if the operand is false, and false (0) if the operand is true.

- Logical operators have numbers as operands. A nonzero number is true, and a zero number is false.
- Logical operators (like relational operators) are used as arithmetic operators within a mathematical expression. The result can be used in other mathematical operations, in addressing arrays, and together with other MATLAB commands (e.g., `if`) to control the flow of a program.
- Logical operators (like relational operators) can be used with scalars and arrays.
- The logical operations AND and OR can have both operands as scalars, arrays, or one array and one scalar. If both are scalars, the result is a scalar 0 or 1. If both are arrays, they must be of the same size and the logical operation is done *element-by-element*. The result is an array of the same size with 1s and 0s according to the outcome of the operation at each position. If one operand is a scalar and the other is an array, the logical operation is done between the scalar and each of the elements in the array and the outcome is an array of the same size with 1s and 0s.
- The logical operation NOT has one operand. When it is used with a scalar the outcome is a scalar 0 or 1. When it is used with an array, the outcome is an array of the same size with 1s in positions where the array has nonzero numbers and 0s in positions where the array has 0s.

Following are some examples:

```
>> 3&7
```

3 AND 7.

```
ans =
     1
>> a=5|0
a =
     1
>> ~25
ans =
     0
>> t=25*((12&0)+(~0)+(0|5))
t =
    50
>> x=[9 3 0 11 0 15]; y=[2 0 13 -11 0 4];
>> x&y
ans =
     1     0     0     1     0     1
>> z=x|y
z =
     1     1     1     1     0     1
>> ~(x+y)
ans =
     0     0     0     1     1     0
```

3 and 7 are both true (nonzero), so the outcome is 1.

5 OR 0 (assign to variable a).

1 is assigned to a since at least one number is true (nonzero).

NOT 25.

The outcome is 0 since 25 is true (nonzero) and the opposite is false.

Using logical operators in a math expression.

Define two vectors x and y.

The outcome is a vector with 1 in every position where both x and y are true (nonzero elements), and 0s otherwise.

The outcome is a vector with 1 in every position where either or both x and y are true (nonzero elements), and 0s otherwise.

The outcome is a vector with 0 in every position where the vector x + y is true (nonzero elements), and 1 in every position where x + y is false (zero elements).

Order of precedence:

Arithmetic, relational, and logical operators can be combined in mathematical expressions. When an expression has such a combination, the result depends on the order in which the operations are carried out. The following is the order used by MATLAB:

Precedence	Operation
1 (highest)	Parentheses (if nested parentheses exist, inner ones have precedence)
2	Exponentiation
3	Logical NOT (~)
4	Multiplication, division
5	Addition, subtraction
6	Relational operators (>, <, >=, <=, ==, ~=)
7	Logical AND (&)
8 (lowest)	Logical OR (|)

If two or more operations have the same precedence, the expression is executed in order from left to right.

It should be pointed out here that the order shown above is the one used since MATLAB 6. Previous versions of MATLAB used a slightly different order (& did not have precedence over |), so the user must be careful. Compatibility problems between different versions of MATLAB can be avoided by using parentheses even when they are not required.

The following are examples of expressions that include arithmetic, relational, and logical operators:

```
>> x=-2; y=5;
```

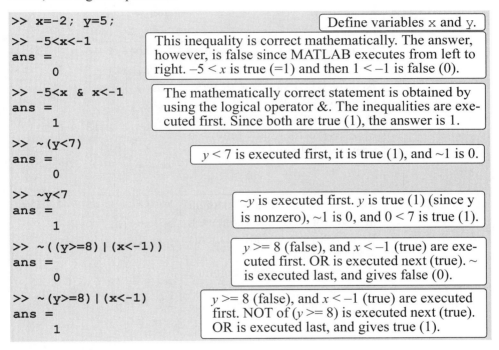

Define variables x and y.

```
>> -5<x<-1
ans =
     0
```

This inequality is correct mathematically. The answer, however, is false since MATLAB executes from left to right. $-5 < x$ is true (=1) and then $1 < -1$ is false (0).

```
>> -5<x & x<-1
ans =
     1
```

The mathematically correct statement is obtained by using the logical operator &. The inequalities are executed first. Since both are true (1), the answer is 1.

```
>> ~(y<7)
ans =
     0
```

$y < 7$ is executed first, it is true (1), and ~1 is 0.

```
>> ~y<7
ans =
     1
```

~*y* is executed first. *y* is true (1) (since y is nonzero), ~1 is 0, and $0 < 7$ is true (1).

```
>> ~((y>=8)|(x<-1))
ans =
     0
```

$y >= 8$ (false), and $x < -1$ (true) are executed first. OR is executed next (true). ~ is executed last, and gives false (0).

```
>> ~(y>=8)|(x<-1)
ans =
     1
```

$y >= 8$ (false), and $x < -1$ (true) are executed first. NOT of ($y >= 8$) is executed next (true). OR is executed last, and gives true (1).

Built-in logical functions:

MATLAB has built-in functions that are equivalent to the logical operators. These functions are:

`and(A,B)`	equivalent to `A&B`
`or(A,B)`	equivalent to `A|B`
`not(A)`	equivalent to `~A`

In addition, MATLAB has other logical built-in functions, some of which are described in the following table:

Function	Description	Example

`xor(a,b)`	Exclusive or. Returns true (1) if one operand is true and the other is false.	`>> xor(7,0)`
		`ans =`
		`     1`
		`>> xor(7,-5)`
		`ans =`
		`     0`

`all(A)`	Returns 1 (true) if all elements in a vector `A` are true (nonzero). Returns 0 (false) if one or more elements are false (zero).
	If `A` is a matrix, treats columns of `A` as vectors, and returns a vector with 1s and 0s.
		`>> A=[6 2 15 9 7 11];`
		`>> all(A)`
		`ans =`
		`     1`
		`>> B=[6 2 15 9 0 11];`
		`>> all(B)`
		`ans =`
		`     0`

`any(A)`	Returns 1 (true) if any element in a vector `A` is true (nonzero). Returns 0 (false) if all elements are false (zero).
	If `A` is a matrix, treats columns of `A` as vectors, and returns a vector with 1s and 0s.
		`>> A=[6 0 15 0 0 11];`
		`>> any(A)`
		`ans =`
		`     1`
		`>> B = [0 0 0 0 0 0];`
		`>> any(B)`
		`ans =`
		`     0`

`find(A)`
 	
`find(A>d)`
	If `A` is a vector, returns the indices of the nonzero elements.
	
	If `A` is a vector, returns the address of the elements that are larger than `d` (any relational operator can be used).
		`>> A=[0 9 4 3 7 0 0 1 8];`
		`>> find(A)`
		`ans =`
		`     2     3     4     5     8     9`
		`>> find(A>4)`
		`ans =`
		`     2     5     9`

The operations of the four logical operators, `and`, `or`, `xor`, and `not` can be summarized in a truth table:

INPUT		OUTPUT				
A	B	AND A&B	OR A|B	XOR (A,B)	NOT ~A	NOT ~B
false	false	false	false	false	true	true
false	true	false	true	true	true	false
true	false	false	true	true	false	true
true	true	true	true	false	false	false

Sample Problem 6-1: Analysis of temperature data

The following were the daily maximum temperatures (in °F) in Washington, DC, during the month of April 2002: 58 73 73 53 50 48 56 73 73 66 69 63 74 82 84 91 93 89 91 80 59 69 56 64 63 66 64 74 63 69 (data from the U.S. National Oceanic and Atmospheric Administration). Use relational and logical operations to determine the following:

(*a*) The number of days the temperature was above 75°.

(*b*) The number of days the temperature was between 65° and 80°.

(*c*) The days of the month when the temperature was between 50° and 60°.

Solution

In the script file below the temperatures are entered in a vector. Relational and logical expressions are then used to analyze the data.

```
T=[58 73 73 53 50 48 56 73 73 66 69 63 74 82 84 ...
   91 93 89 91 80 59 69 56 64 63 66 64 74 63 69];
Tabove75=T>=75;
NdaysTabove75=sum(Tabove75)
Tbetween65and80=(T>=65)&(T<=80);
NdaysTbetween65and80=sum(Tbetween65and80)
datesTbetween50and60=find((T>=50)&(T<=60))
```

A vector with 1s at addresses where `T` >= 75.

Add all the 1s in the vector `Tabove75`.

A vector with 1s at addresses where `T` >= 65 and `T` <= 80.

Add all the 1s in the vector `Tbetween65and80`.

The function `find` returns the address of the elements in `T` that have values between 50 and 60.

The script file (saved as Exp6_1) is executed in the Command Window:

```
>> Exp6_1
NdaysTabove75 =
     7
NdaysTbetween65and80 =
    12
datesTbetween50and60 =
     1     4     5     7    21    23
```

For 7 days the temp was above 75.

For 12 days the temp was between 65 and 80.

Dates of the month with temp between 50 and 60.

6.2 CONDITIONAL STATEMENTS

A conditional statement is a command that allows MATLAB to make a decision of whether to execute a group of commands that follow the conditional statement, or to skip these commands. In a conditional statement a conditional expression is stated. If the expression is true, a group of commands that follow the statement are executed. If the expression is false, the computer skips the group. The basic form of a conditional statement is:

`if` conditional expression consisting of relational and/or logical operators.

Examples:

```
if a < b
if c >= 5
if a == b
if a ~= 0
if (d<h)&(x>7)
if (x~=13)|(y<0)
```

All the variables must have assigned values.

- Conditional statements can be a part of a program written in a script file or a user-defined function (Chapter 7).
- As shown below, for every `if` statement there is an `end` statement.

The `if` statement is commonly used in three structures, `if-end`, `if-else-end`, and `if-elseif-else-end`, which are described next.

6.2.1 The `if-end` Structure

The `if-end` conditional statement is shown schematically in Figure 6-1. The figure shows how the commands are typed in the program, and a flowchart that symbolically shows the flow, or the sequence, in which the commands are executed. As the program executes, it reaches the `if` statement. If the conditional expres-

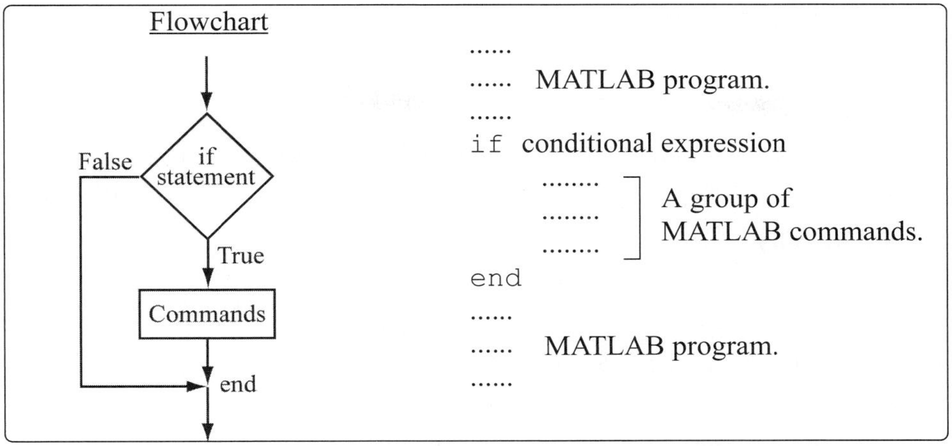

Figure 6-1: The structure of the `if-end` conditional statement.

sion in the `if` statement is true (1), the program continues to execute the commands that follow the `if` statement all the way down to the `end` statement. If the conditional expression is false (0), the program skips the group of commands between the `if` and the `end`, and continues with the commands that follow the `end`.

The words `if` and `end` appear on the screen in blue, and the commands between the `if` statement and the `end` statement are automatically indented (they don't have to be), which makes the program easier to read. An example where the `if-end` statement is used in a script file is shown in Sample Problem 6-2.

Sample Problem 6-2: Calculating worker's pay

A worker is paid according to his hourly wage up to 40 hours, and 50% more for overtime. Write a program in a script file that calculates the pay to a worker. The program asks the user to enter the number of hours and the hourly wage. The program then displays the pay.

Solution

The program in a script file is shown below. The program first calculates the pay by multiplying the number of hours by the hourly wage. Then an `if` statement checks whether the number of hours is greater than 40. If so, the next line is executed and the extra pay for the hours above 40 is added. If not, the program skips to the `end`.

```
t=input('Please enter the number of hours worked  ');
h=input('Please enter the hourly wage in $  ');
Pay=t*h;
if t>40
```

```
    Pay=Pay+(t-40)*0.5*h;
end
fprintf('The worker''s pay is  $ %5.2f',Pay)
```

Application of the program (in the Command Window) for two cases is shown below (the file was saved as Workerpay):

```
>> Workerpay
Please enter the number of hours worked  35
Please enter the hourly wage in $  8
The worker's pay is  $ 280.00
>> Workerpay
Please enter the number of hours worked  50
Please enter the hourly wage in $  10
The worker's pay is  $ 550.00
```

6.2.2 *The* `if-else-end` *Structure*

The `if-else-end` structure provides a means for choosing one group of commands, out of a possible two groups, for execution. The `if-else-end` structure is shown in Figure 6-2. The figure shows how the commands are typed in the program, and a flowchart that illustrates the flow, or the sequence, in which the

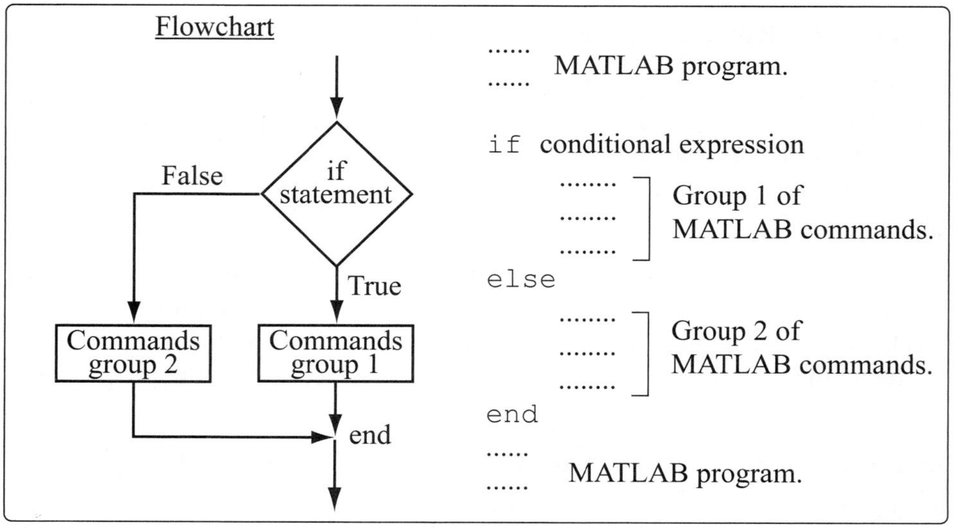

Figure 6-2: The structure of the `if-else-end` **conditional statement.**

commands are executed. The first line is an `if` statement with a conditional expression. If the conditional expression is true, the program executes group 1 of commands between the `if` and the `else` statements and then skips to the `end`. If the conditional expression is false, the program skips to the `else` and then executes group 2 of commands between the `else` and the `end`.

6.2.3 The `if-elseif-else-end` *Structure*

The `if-elseif-else-end` structure is shown in Figure 6-3. The figure shows how the commands are typed in the program, and gives a flowchart that illustrates the flow, or the sequence, in which the commands are executed. This structure includes two conditional statements (`if` and `elseif`) that make it possible to select one out of three groups of commands for execution. The first line is an `if` statement with a conditional expression. If the conditional expression is true, the program executes group 1 of commands between the `if` and the `elseif` statements and then skips to the `end`. If the conditional expression in the `if` statement is false, the program skips to the `elseif` statement. If the conditional expression in the `elseif` statement is true, the program executes group 2 of commands between the `elseif` and the `else` and then skips to the `end`. If the conditional expression in the `elseif` statement is false, the program skips to the `else` and executes group 3 of commands between the `else` and the `end`.

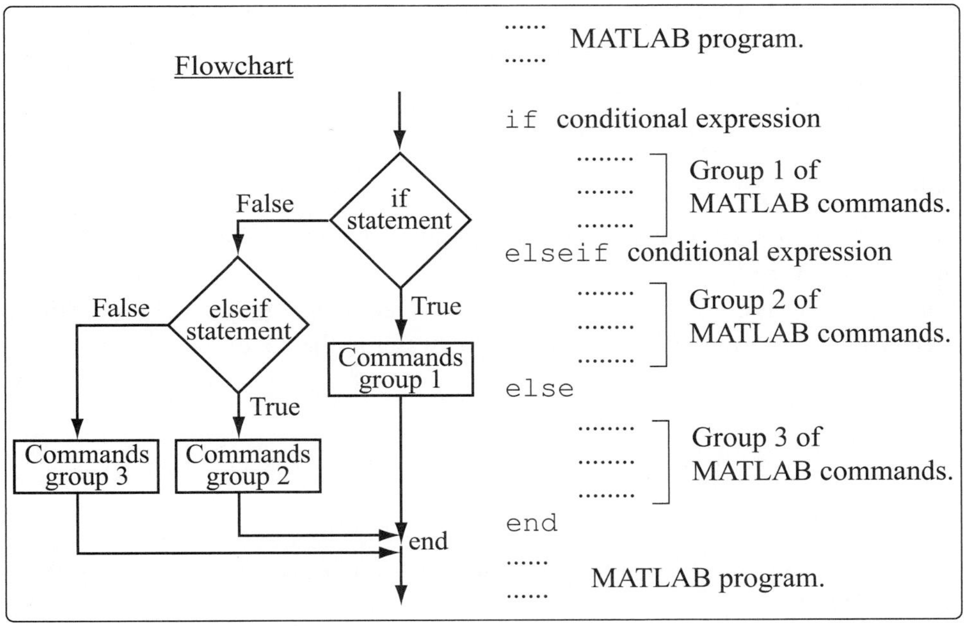

Figure 6-3: The structure of the `if-elseif-else-end` **conditional statement.**

It should be pointed out here that several `elseif` statements and associated groups of commands can be added. In this way more conditions can be included. Also, the `else` statement is optional. This means that in the case of several `elseif` statements and no `else` statement, if any of the conditional statements is true the associated commands are executed; otherwise nothing is executed.

The following example uses the `if-elseif-else-end` structure in a program.

Sample Problem 6-3: Water level in water tower

The tank in a water tower has the geometry shown in the figure (the lower part is a cylinder and the upper part is an inverted frustum of a cone). Inside the tank there is a float that indicates the level of the water. Write a MATLAB program that determines the volume of the water in the tank from the position (height *h*) of the float. The program asks the user to enter a value of *h* in m, and as output displays the volume of the water in m^3.

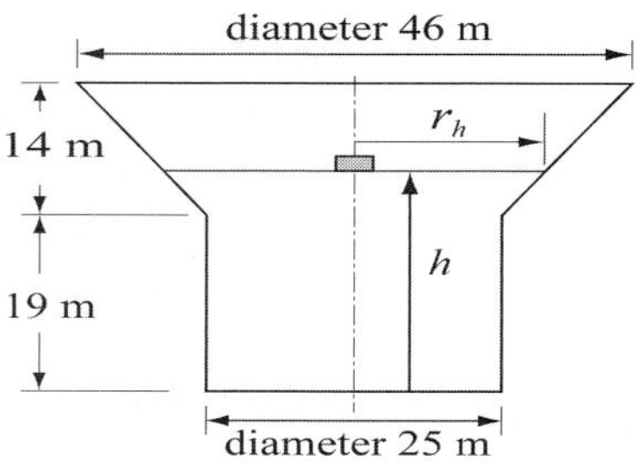

Solution

For $0 \le h \le 19$ m the volume of the water is given by the volume of a cylinder with height h: $V = \pi 12.5^2 h$.

For $19 < h \le 33$ m the volume of the water is given by adding the volume of a cylinder with $h = 19$ m, and the volume of the water in the cone:

$$V = \pi 12.5^2 \cdot 19 + \frac{1}{3}\pi(h - 19)(12.5^2 + 12.5 \cdot r_h + r_h^2)$$

where $r_h = 12.5 + \dfrac{10.5}{14}(h - 19)$.

The program is:

```
% The program calculates the volume of the water in the
water tower.
h=input('Please enter the height of the float in meter  ');
if h > 33
   disp('ERROR. The height cannot be larger than 33 m.')
elseif h < 0
   disp('ERROR. The height cannot be a negative number.')
elseif h <= 19
   v = pi*12.5^2*h;
   fprintf('The volume of the water is %7.3f cubic meter.\n',v)
```

```
else
   rh=12.5+10.5*(h-19)/14;
   v=pi*12.5^2*19+pi*(h-19)*(12.5^2+12.5*rh+rh^2)/3;
   fprintf('The volume of the water is %7.3f cubic meter.\n',v)
end
```

The following is the display in the Command Window when the program is used with three different values of water height.

```
Please enter the height of the float in meter  8
The volume of the water is 3926.991 cubic meter.
Please enter the height of the float in meter  25.7
The volume of the water is 14114.742 cubic meter.
Please enter the height of the float in meter  35
ERROR. The height cannot be larger than 33 m.
```

6.3 *THE* `switch-case` *STATEMENT*

The `switch-case` statement is another method that can be used to direct the flow of a program. It provides a means for choosing one group of commands for execution out of several possible groups. The structure of the statement is shown in Figure 6-4.

- The first line is the `switch` command, which has the form:

`switch` switch expression

The switch expression can be a scalar or a string. Usually it is a variable that has an assigned scalar or a string. It can also be, however, a mathematical expression that includes pre-assigned variables and can be evaluated.

- Following the `switch` command are one or several `case` commands. Each has a value (can be a scalar or a string) next to it (value1, value2, etc.) and an associated group of commands below it.
- After the last `case` command there is an optional `otherwise` command followed by a group of commands.
- The last line must be an `end` statement.

How does the switch-case statement work?

The value of the switch expression in the `switch` command is compared with the values that are next to each of the `case` statements. If a match is found, the group of commands that follow the `case` statement with the match are executed. (Only one group of commands—the one between the `case` that matches and either the

```
......
         MATLAB program.
......

switch  switch expression
      case  value1
      ........
                ]  Group 1 of commands.
      ........
      case  value2
      ........
                ]  Group 2 of commands.
      ........
      case  value3
      ........
                ]  Group 3 of commands.
      ........
      otherwise
      ........
                ]  Group 4 of commands.
      ........
end
......
         MATLAB program.
......
```

Figure 6-4: The structure of a `switch-case` **statement.**

`case`, `otherwise`, or `end` statement that is next—is executed).

- If there is more than one match, only the first matching case is executed.
- If no match is found and the `otherwise` statement (which is optional) is present, the group of commands between `otherwise` and `end` is executed.
- If no match is found and the `otherwise` statement is not present, none of the command groups is executed.
- A `case` statement can have more than one value. This is done by typing the values in the form: `{value1, value2, value3, ...}`. (This form, which is not covered in this book, is called a cell array.) The case is executed if at least one of the values matches the value of switch expression.

A Note: In MATLAB only the first matching case is executed. After the group of commands associated with the first matching case are executed, the program skips to the `end` *statement. This is different from the C language, where break statements are required.*

Sample Problem 6-4: Converting units of energy

Write a program in a script file that converts a quantity of energy (work) given in units of either joule, ft-lb, cal, or eV to the equivalent quantity in different units specified by the user. The program asks the user to enter the quantity of energy, its current units, and the desired new units. The output is the quantity of energy in the new units.

The conversion factors are: 1 J = 0.738 ft-lb = 0.239 cal = 6.24×10^{18} eV. Use the program to:

(*a*) Convert 150 J to ft-lb.

(*b*) Convert 2,800 cal to J.

(*c*) Convert 2.7 eV to cal.

Solution

The program includes two sets of `switch-case` statements and one `if-else-end` statement. The first `switch-case` statement is used to convert the input quantity from its initial units to units of joules. The second is used to convert the quantity from joules to the specified new units. The `if-else-end` statement is used to generate an error message if units are entered incorrectly.

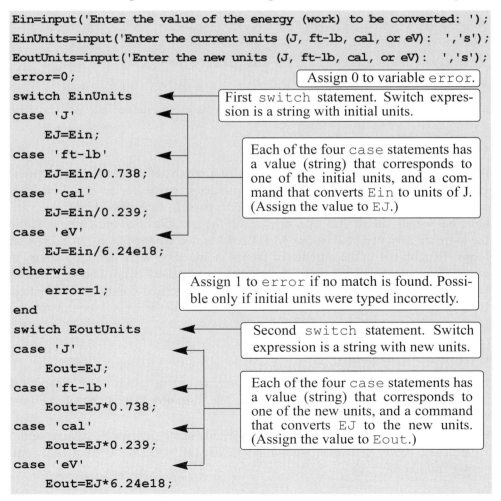

```
Ein=input('Enter the value of the energy (work) to be converted: ');
EinUnits=input('Enter the current units (J, ft-lb, cal, or eV):  ','s');
EoutUnits=input('Enter the new units (J, ft-lb, cal, or eV):  ','s');
error=0;
switch EinUnits
case 'J'
    EJ=Ein;
case 'ft-lb'
    EJ=Ein/0.738;
case 'cal'
    EJ=Ein/0.239;
case 'eV'
    EJ=Ein/6.24e18;
otherwise
    error=1;
end
switch EoutUnits
case 'J'
    Eout=EJ;
case 'ft-lb'
    Eout=EJ*0.738;
case 'cal'
    Eout=EJ*0.239;
case 'eV'
    Eout=EJ*6.24e18;
```

Assign 0 to variable `error`.

First `switch` statement. Switch expression is a string with initial units.

Each of the four `case` statements has a value (string) that corresponds to one of the initial units, and a command that converts `Ein` to units of J. (Assign the value to `EJ`.)

Assign 1 to `error` if no match is found. Possible only if initial units were typed incorrectly.

Second `switch` statement. Switch expression is a string with new units.

Each of the four `case` statements has a value (string) that corresponds to one of the new units, and a command that converts `EJ` to the new units. (Assign the value to `Eout`.)

```
otherwise
    error=1;
end
if error
    disp('ERROR current or new units are typed incorrectly.')
else
    fprintf('E = %g %s',Eout,EoutUnits)
end
```

Assign 1 to `error` if no match is found. Possible only if new units were typed incorrectly.

`If-else-end` statement.

If `error` is true (nonzero), display an error message.

If `error` is false (zero), display converted energy.

As an example, the script file (saved as EnergyConversion) is used next in the Command Window to make the conversion in part (*b*) of the problem statement.

```
>> EnergyConversion
Enter the value of the energy (work) to be converted: 2800
Enter the current units (J, ft-lb, cal, or eV):  cal
Enter the new units (J, ft-lb, cal, or eV):  J
E = 11715.5 J
```

6.4 LOOPS

A loop is another method to alter the flow of a computer program. In a loop, the execution of a command, or a group of commands, is repeated several times consecutively. Each round of execution is called a pass. In each pass at least one variable, but usually more than one, or even all the variables that are defined within the loop, are assigned new values. MATLAB has two kinds of loops. In `for-end` loops (Section 6.4.1) the number of passes is specified when the loop starts. In `while-end` loops (Section 6.4.2) the number of passes is not known ahead of time, and the looping process continues until a specified condition is satisfied. Both kinds of loops can be terminated at any time with the `break` command (see Section 6.6).

6.4.1 `for-end` Loops

In `for-end` loops the execution of a command, or a group of commands, is repeated a predetermined number of times. The form of a loop is shown in Figure 6-5.

- The loop index variable can have any variable name (usually `i`, `j`, `k`, `m`, and `n` are used, however, `i` and `j` should not be used if MATLAB is used with complex numbers).

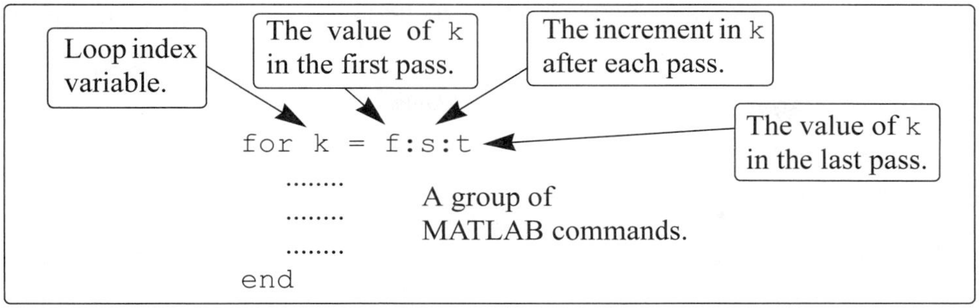

Figure 6-5: The structure of a `for-end` loop.

- In the first pass `k` = `f` and the computer executes the commands between the `for` and `end` commands. Then, the program goes back to the `for` command for the second pass. `k` obtains a new value equal to `k = f + s`, and the commands between the `for` and `end` commands are executed with the new value of `k`. The process repeats itself until the last pass, where `k = t`. Then the program does not go back to the `for`, but continues with the commands that follow the `end` command. For example, if `k` = 1:2:9, there are five loops, and the corresponding values of `k` are 1, 3, 5, 7, and 9.
- The increment `s` can be negative (i.e.; `k` = 25:–5:10 produces four passes with `k` = 25, 20, 15, 10).
- If the increment value `s` is omitted, the value is 1 (default) (i.e.; `k` = 3:7 produces five passes with `k` = 3, 4, 5, 6, 7).
- If `f` = `t`, the loop is executed once.
- If `f` > `t` and `s` > 0, or if `f` < `t` and `s` < 0, the loop is not executed.
- If the values of `k`, `s`, and `t` are such that `k` cannot be equal to `t`, then if `s` is positive, the last pass is the one where `k` has the largest value that is smaller than `t` (i.e., `k` = 8:10:50 produces five passes with `k` = 8, 18, 28, 38, 48). If `s` is negative, the last pass is the one where `k` has the smallest value that is larger than `t`.
- In the `for` command `k` can also be assigned a specific value (typed as a vector). Example: `for k = [7 9 -1 3 3 5]`.
- The value of `k` should not be redefined within the loop.
- Each `for` command in a program *must* have an `end` command.
- The value of the loop index variable (`k`) is not displayed automatically. It is possible to display the value in each pass (which is sometimes useful for debugging) by typing `k` as one of the commands in the loop.

- When the loop ends, the loop index variable (k) has the value that was last assigned to it.

A simple example of a for-end loop (in a script file) is:

```
for k=1:3:10
    x = k^2
end
```

When this program is executed, the loop is executed four times. The value of k in the four passes is k = 1, 4, 7, and 10, which means that the values that are assigned to x in the passes are x = 1, 16, 49, and 100, respectively. Since a semicolon is not typed at the end of the second line, the value of x is displayed in the Command Window at each pass. When the script file is executed, the display in the Command Window is:

```
>> x =
     1
x =
    16
x =
    49
x =
   100
```

Sample Problem 6-5: Sum of a series

(*a*) Use a for-end loop in a script file to calculate the sum of the first *n* terms of the series: $\sum_{k=1}^{n}\frac{(-1)^k k}{2^k}$. Execute the script file for $n = 4$ and $n = 20$.

(*b*) The function sin(*x*) can be written as a Taylor series by:

$$\sin x = \sum_{k=0}^{\infty}\frac{(-1)^k x^{2k+1}}{(2k+1)!}$$

Write a user-defined function file that calculates sin(*x*) by using the Taylor series. For the function name and arguments use y = Tsin(x,n). The input arguments are the angle x in degrees and n the number of terms in the series. Use the function to calculate sin(150°) using three and seven terms.

Solution

(*a*) A script file that calculates the sum of the first *n* terms of the series is shown below.

The summation is done with a loop. In each pass one term of the series is calcu-

```
n=input('Enter the number of terms ' );
S=0;                    Setting the sum to zero.
for k=1:n
    S=S+(-1)^k*k/2^k;   for-end loop.
end
fprintf('The sum of the series is: %f',S)
```

In each pass one element of the series is calculated and is added to the sum of the elements from the previous passes.

lated (in the first pass the first term, in the second pass the second term, and so on) and is added to the sum of the previous elements. The file is saved as Exp6_5a and then executed twice in the Command Window:

```
>> Exp6_5a
Enter the number of terms 4
The sum of the series is: -0.125000
>> Exp7_5a
Enter the number of terms 20
The sum of the series is: -0.222216
```

(*b*) A user-defined function file that calculates sin(*x*) by adding *n* terms of a Taylor series is shown below.

```
function y = Tsin(x,n)
% Tsin calculates the sin using Taylor formula.
% Input arguments:
% x The angle in degrees, n number of terms.

xr=x*pi/180;          Converting the angle from degrees to radians.
y=0;
for k=0:n-1
    y=y+(-1)^k*xr^(2*k+1)/factorial(2*k+1);   for-end loop.
end
```

The first element corresponds to $k = 0$, which means that in order to add n terms of the series, in the last loop $k = n - 1$. The function is used in the Command Window to calculate sin(150°) using three and seven terms:

```
>> Tsin(150,3)        Calculating sin(150°) with three terms of Taylor series.
ans =
      0.6523
```

```
>> Tsin(150,7)        Calculating sin(150°) with seven terms of Taylor series.
ans =
      0.5000                                     The exact value is 0.5.
```

A note about `for-end` loops and element-by-element operations:

In some situations the same end result can be obtained by either using `for-end` loops or using element-by-element operations. Sample Problem 6-5 illustrates how the `for-end` loop works, but the problem can also be solved by using element-by-element operations (see Problems 7 and 8 in Section 3.9). Element-by-element operations with arrays are one of the superior features of MATLAB that provide the means for computing in circumstances that otherwise require loops. In general, element-by-element operations are faster than loops and are recommended when either method can be used.

Sample Problem 6-6: Modify vector elements

A vector is given by V = [5, 17, –3, 8, 0, –7, 12, 15, 20, –6, 6, 4, –7, 16]. Write a program as a script file that doubles the elements that are positive and are divisible by 3 or 5, and, raises to the power of 3 the elements that are negative but greater than –5.

Solution

The problem is solved by using a `for-end` loop that has an `if-elseif-end` conditional statement inside. The number of passes is equal to the number of elements in the vector. In each pass one element is checked by the conditional statement. The element is changed if it satisfies the conditions in the problem statement. A program in a script file that carries out the required operations is:

```
V=[5, 17, -3, 8, 0, -7, 12, 15, 20 -6, 6, 4, -2, 16];
n=length(V);          Setting n to be equal to the number of elements in V.
for k=1:n
   if V(k)>0 & (rem(V(k),3)==0 | rem(V(k),5)==0)      for-end loop.
        V(k)=2*V(k);
   elseif V(k) < 0 & V(k) > -5
        V(k)=V(k)^3;                                   if-elseif-end statement.
   end
end
V
```

The file is saved as Exp7_6 and then executed in the Command Window:

```
>> Exp7_6
V =
    10    17   -27     8     0    -7    24    30    40    -6    12     4
-8    16
```

6.4.2 `while-end` Loops

`while-end` loops are used in situations when looping is needed but the number of passes is not known in advance. In `while-end` loops the number of passes is not specified when the looping process starts. Instead, the looping process continues until a stated condition is satisfied. The structure of a `while-end` loop is shown in Figure 6-6.

```
while  conditional expression
    ........
    ........        A group of
                    MATLAB commands.
    ........
end
```

Figure 6-6: The structure of a `while-end` loop.

The first line is a `while` statement that includes a conditional expression. When the program reaches this line the conditional expression is checked. If it is false (0), MATLAB skips to the `end` statement and continues with the program. If the conditional expression is true (1), MATLAB executes the group of commands that follow between the `while` and `end` commands. Then MATLAB jumps back to the `while` command and checks the conditional expression. This looping process continues until the conditional expression is false.

For a `while-end` loop to execute properly:

- The conditional expression in the `while` command must include at least one variable.
- The variables in the conditional expression must have assigned values when MATLAB executes the `while` command for the first time.
- At least one of the variables in the conditional expression must be assigned a new value in the commands that are between the `while` and the `end`. Otherwise, once the looping starts it will never stop since the conditional expression will remain true.

An example of a simple `while-end` loop is shown in the following program. In this program a variable x with an initial value of 1 is doubled in each pass as long as its value is equal to or smaller than 15.

```
x=1                    Initial value of x is 1.
while x<=15            The next command is executed only if x <= 15.
    x=2*x              In each pass x doubles.
end
```

When this program is executed the display in the Command Window is:

```
x =                    Initial value of x.
     1
x =
     2
x =                    In each pass x doubles.
     4
x =
     8
x =                    When x = 16, the conditional expression in the
    16                 while command is false and the looping stops.
```

Important note:

When writing a while-end loop, the programmer has to be sure that the variable (or variables) that are in the conditional expression and are assigned new values during the looping process will eventually be assigned values that make the conditional expression in the while command false. Otherwise the looping will continue indefinitely (indefinite loop). In the example above if the conditional expression is changed to x >= 0.5, the looping will continue indefinitely. Such a situation can be avoided by counting the passes and stopping the looping if the number of passes exceeds some large value. This can be done by adding the maximum number of passes to the conditional expression, or by using the break command (Section 6.6).

Since no one is free from making mistakes, a situation of indefinite looping can occur in spite of careful programming. If this happens, the user can stop the execution of an indefinite loop by pressing the **Ctrl + C** or **Ctrl + Break** keys.

Sample Problem 6-7: Taylor series representation of a function

The function $f(x) = e^x$ can be represented in a Taylor series by $e^x = \sum_{n=0}^{\infty} \frac{x^n}{n!}$.

Write a program in a script file that determines e^x by using the Taylor series representation. The program calculates e^x by adding terms of the series and stopping when the absolute value of the term that was added last is smaller than 0.0001. Use a `while-end` loop, but limit the number of passes to 30. If in the 30th pass the value of the term that is added is not smaller than 0.0001, the program stops and displays a message that more than 30 terms are needed.

Use the program to calculate e^2, e^{-4}, and e^{21}.

Solution

The first few terms of the Taylor series are:

$$e^x = 1 + x + \frac{x^2}{2!} + \frac{x^3}{3!} + \ldots$$

A program that uses the series to calculate the function is shown next. The program asks the user to enter the value of `x`. Then the first term, `an`, is assigned the number 1, and `an` is assigned to the sum `S`. Then, from the second term on, the program uses a `while` loop to calculate the *n*th term of the series and add it to the sum. The program also counts the number of terms `n`. The conditional expression in the `while` command is true as long as the absolute value of the *n*th `an` term is larger than 0.0001, and the number of passes `n` is smaller than 30. This means that if the 30th term is not smaller than 0.0001, the looping stops.

```
x=input('Enter x ' );
n=1; an=1; S=an;
while abs(an) >= 0.0001 & n <= 30          Start of the while loop.
    an=x^n/factorial(n);                   Calculating the nth term.
    S=S+an;                                Adding the nth term to the sum.
    n=n+1;                                 Counting the number of passes.
end                                        End of the while loop.
if n >= 30                                 if-else-end loop.
    disp('More than 30 terms are needed')
else
fprintf('exp(%f) = %f',x,S)
fprintf('\nThe number of terms used is: %i',n)
end
```

The program uses an `if-else-end` statement to display the results. If the looping stopped because the 30th term is not smaller than 0.0001, it displays a message indicating this. If the value of the function is calculated successfully, it displays the value of the function and the number of terms used. When the program executes, the number of passes depends on the value of `x`. The program (saved as `expox`) is used to calculate e^2, e^{-4}, and e^{21}:

```
>> expox
```

```
Enter x 2
exp(2.000000) = 7.389046
The number of terms used is: 12
>> expox
Enter x -4
exp(-4.000000) = 0.018307
The number of terms used is: 18
>> expox
Enter x 21
More than 30 terms are needed
```

Calculating exp(2).

12 terms used.

Calculating exp(–4).

18 terms used.

Trying to calculate exp(21).

6.5 *Nested Loops and Nested Conditional Statements*

Loops and conditional statements can be nested within other loops or conditional statements. This means that a loop and/or a conditional statement can start (and end) within another loop or conditional statement. There is no limit to the number of loops and conditional statements that can be nested. It must be remembered, however, that each `if`, `case`, `for`, and `while` statement must have a corresponding `end` statement. Figure 6-7 shows the structure of a nested `for-end`

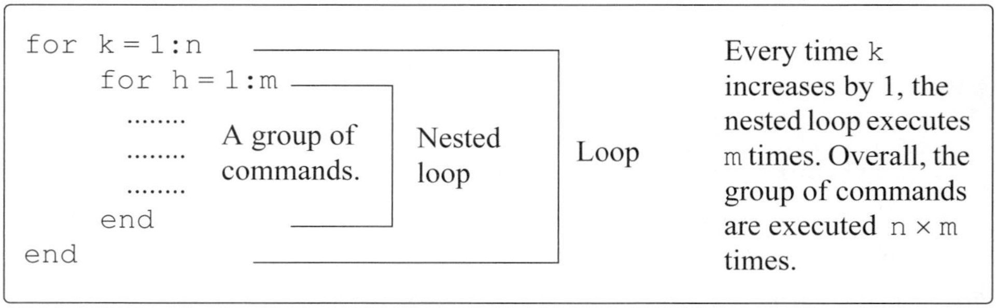

Figure 6-7: Structure of nested loops.

loop within another `for-end` loop. In the loops shown in this figure, if, for example, `n` = 3 and `m` = 4, then first `k` = 1 and the nested loop executes four times with `h` = 1, 2, 3, 4. Next `k` = 2 and the nested loop executes again four times with `h` = 1, 2, 3, 4. Finally `k` = 3 and the nested loop executes again four times. Every time a nested loop is typed, MATLAB automatically indents the new loop relative to the outside loop. Nested loops and conditional statements are demonstrated in the following sample problem.

Sample Problem 6-8: Creating a matrix with a loop

Write a program in a script file that creates an $n \times m$ matrix with elements that have the following values. The value of each element in the first row is the number of the column. The value of each element in the first column is the number of the row. The rest of the elements each has a value equal to the sum of the element above it and the element to the left. When executed, the program asks the user to enter values for n and m.

Solution

The program, shown below, has two loops (one nested) and a nested `if-elseif-else-end` structure. The elements in the matrix are assigned values row by row. The loop index variable of the first loop, `k`, is the address of the row, and the loop index variable of the second loop, `h`, is the address of the column.

```
n=input('Enter the number of rows ');
m=input('Enter the number of columns ');
A=[];                                   Define an empty matrix A
for k=1:n                               Start of the first for-end loop.
    for h=1:m                           Start of the second for-end loop.
        if k==1                         Start of the conditional statement.
            A(k,h)=h;                   Assign values to the elements of the first row.
        elseif h==1
            A(k,h)=k;                   Assign values to the elements of the first column.
        else
            A(k,h)=A(k,h-1)+A(k-1,h);   Assign values to other elements.
        end                             end of the if statement.
    end                                 end of the nested for-end loop.
end                                     end of the first for-end loop.
A
```

The program is executed in the Command Window to create a 4×5 matrix.

```
>> Chap6_exp8
Enter the number of rows 4
Enter the number of columns 5
```

```
A =
     1     2     3     4     5
     2     4     7    11    16
     3     7    14    25    41
     4    11    25    50    91
```

6.6 *THE* `break` *AND* `continue` *COMMANDS*

The `break` **command:**

- When inside a loop (`for` or `while`), the `break` command terminates the execution of the loop (the whole loop, not just the last pass). When the `break` command appears in a loop, MATLAB jumps to the `end` command of the loop and continues with the next command (it does not go back to the `for` command of that loop).
- If the `break` command is inside a nested loop, only the nested loop is terminated.
- When a `break` command appears outside a loop in a script or function file, it terminates the execution of the file.
- The `break` command is usually used within a conditional statement. In loops it provides a method to terminate the looping process if some condition is met —for example, if the number of loops exceeds a predetermined value, or an error in some numerical procedure is smaller than a predetermined value. When typed outside a loop, the `break` command provides a means to terminate the execution of a file, such as when data transferred into a function file is not consistent with what is expected.

The `continue` **command:**

- The `continue` command can be used inside a loop (`for` or `while`) to stop the present pass and start the next pass in the looping process.
- The `continue` command is usually a part of a conditional statement. When MATLAB reaches the `continue` command, it does not execute the remaining commands in the loop, but skips to the `end` command of the loop and then starts a new pass.

6.7 EXAMPLES OF MATLAB APPLICATIONS

Sample Problem 6-9: Withdrawing from a retirement account.

A person in retirement is depositing \$300,000 in a saving account that pays 5% interest per year. The person plans to withdraw money from the account once a year. He starts by withdrawing \$25,000 after the first year, and in future years he increases the amount he withdraws according to the inflation rate. For example, if the inflation rate is 3%, he withdraws \$25,750 after the second year. Calculate the number of years the money in the account will last assuming a constant yearly inflation rate of 2%. Make a plot that shows the yearly withdrawals and the balance of the account over the years.

Solution

The problem is solved by using a loop (a `while` loop since the number of passes is not known before the loop starts). In each pass the amount to be withdrawn and the account balance are calculated. The looping continues as long as the account balance is larger than or equal to the amount to be withdrawn. The following is a program in a script file that solves the problem. In the program, `year` is a vector in which each element is a year number, `W` is a vector with the amount withdrawn each year, and `AB` is a vector with the account balance each year.

```
rate=0.05; inf=0.02;
clear W AB year
year(1)=0;                         First element is year 0.
W(1)=0;                            Initial withdrawal amount.
AB(1)=300000;                      Initial account balance.
Wnext=25000;                       The amount to be withdrawn after a year.
ABnext=300000*(1 + rate);          The account balance after a year.
n=2;
    while ABnext >= Wnext          while checks if the next balance
        year(n)=n-1;               is larger than the next withdrawal.
        W(n)=Wnext;                Amount withdrawn in year n – 1.
        AB(n)=ABnext-W(n);         Account balance in year n – 1 after withdrawal.
        ABnext=AB(n)*(1+rate);     The account balance after additional year.
        Wnext=W(n)*(1+inf);
        n=n+1;                     The amount to be withdrawn
    end                            after an additional year.
fprintf('The money will last for %f years',year(n-1))
bar(year,[AB' W'],2.0)
```

The program is executed in the following Command Window:

```
>> Chap6_exp9
The money will last for 15 years.
```

The program also generates the following figure (axis labels and legend were added to the plot by using the Plot Editor).

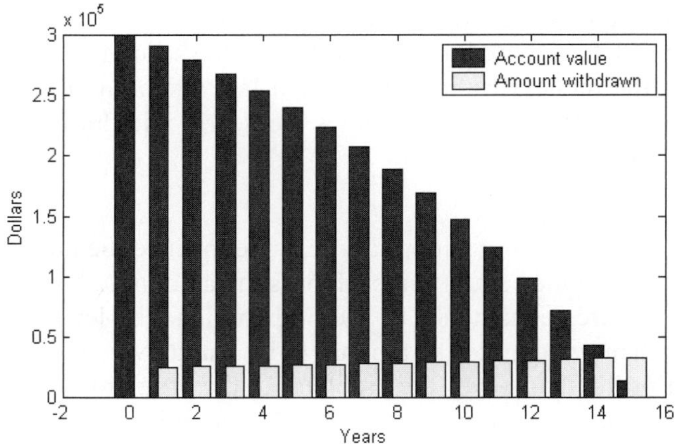

Sample Problem 6-10: Creating a random list

Six singers—John, Mary, Tracy, Mike, Katie, and David—are performing in a competition. Write a MATLAB program that generates a list of a random order in which the singers will perform.

Solution

An integer (1 through 6) is assigned to each name (1 to John, 2 to Mary, 3 to Tracy, 4 to Mike, 5 to Katie, and 6 to David). The program, shown below, first creates a list of the integers 1 through 6 in a random order. The integers are made the elements of six-element vector. This is done by using MATLAB's built-in function `randi` (see Section 3.7) for assigning integers to the elements of the vector. To make sure that all the integers of the elements are different from each other, the integers are assigned one by one. Each integer that is suggested by the `randi` function is compared with all the integers that have been assigned to previous elements. If a match is found, the integer is not assigned, and `randi` is used for suggesting a new integer. Since each singer name is associated with an integer, once the integer list is complete the `switch-case` statement is used to create the corresponding name list.

```
clear, clc
n=6;
```

```
L(1)=randi(n);                    Assign the first integer to L(1).
for p=2:n
    L(p)=randi(n);                Assign the next integer to L(p).
    r=0;                          Set r to zero.
    while r==0                    See explanation below.
      r=1;                        Set r to 1.
      for k=1:p-1                 for loop compares the integer assigned to L(p) to the
                                  integers that have been assigned to previous elements.
         if L(k)==L(p)            If a match if found, a
           L(p)=randi(n);         new integer is
           r=0;                   assigned to L(p) and
                                  r is set to zero.
           break                  The nested for loop is stopped. The pro-
         end                      gram goes back to the while loop. Since
      end                         r = 0 the nested loop inside the while
    end                           loop starts again and checks if the new
end                               integer that is assigned to L(p) is equal to
                                  an integer that is already in the vector L.
for i=1:n
    switch L(i)                   The switch-case state-
    case 1                        ment lists the names
        disp('John')              according to the values of
    case 2                        the integers in the elements
        disp('Mary')              of L.
    case 3
        disp('Tracy')
    case 4
        disp('Mike')
    case 5
        disp('Katie')
    case 6
        disp('David')
    end
end
```

The `while` loop checks that every new integer (element) that is to be added to the vector `L` is not equal any of the integers in elements already in the vector `L`. If a match is found, it keeps generating new integers until the new integer is different from all the integers that are already in `x`.

When the program is executed, the following is displayed in the Command Window. Obviously, a list in a different order will be displayed every time the program is executed.

```
The performing order is:
```

```
Katie
Tracy
David
Mary
John
Mike
```

Sample Problem 6-11: Flight of a model rocket

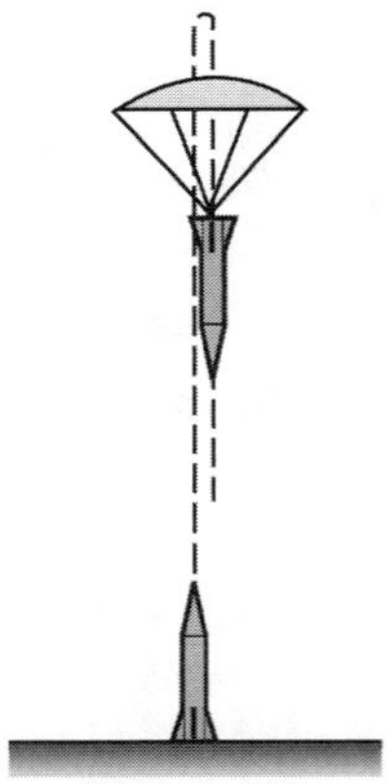

The flight of a model rocket can be modeled as follows. During the first 0.15 s the rocket is propelled upward by the rocket engine with a force of 16 N. The rocket then flies up while slowing down under the force of gravity. After it reaches the apex, the rocket starts to fall back down. When its downward velocity reaches 20 m/s a parachute opens (assumed to open instantly), and the rocket continues to drop at a constant speed of 20 m/s until it hits the ground. Write a program that calculates and plots the speed and altitude of the rocket as a function of time during the flight.

Solution

The rocket is assumed to be a particle that moves along a straight line in the vertical plane. For motion with constant acceleration along a straight line, the velocity and position as a function of time are given by:

$$v(t) = v_0 + at \quad \text{and} \quad s(t) = s_0 + v_0 t + \frac{1}{2}at^2$$

where v_0 and s_0 are the initial velocity and position, respectively. In the computer program the flight of the rocket is divided into three segments. Each segment is calculated in a `while` loop. In every pass the time increases by an increment.

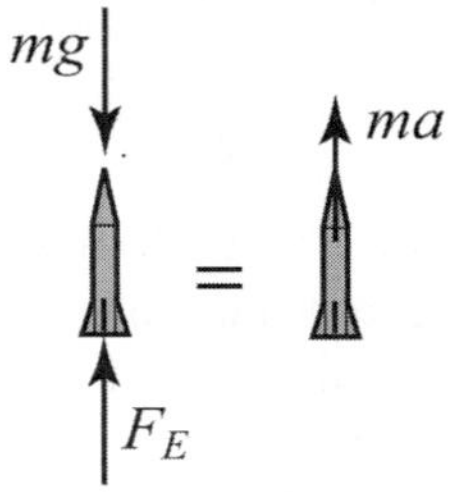

Segment 1: The first 0.15 s when the rocket engine is on. During this period, the rocket moves up with a constant acceleration. The acceleration is determined by drawing a free body and a mass acceleration diagram (shown on the right). From Newton's second law, the sum of the forces in the vertical direction is equal to the mass times the acceleration (equilibrium equation):

$$+\uparrow \Sigma F = F_E - mg = ma$$

Solving the equation for the acceleration gives:

$$a = \frac{F_E - mg}{m}$$

The velocity and height as a function of time are:

$$v(t) = 0 + at \quad \text{and} \quad h(t) = 0 + 0 + \frac{1}{2}at^2$$

where the initial velocity and initial position are both zero. In the computer program this segment starts at $t = 0$, and the looping continues as long as $t < 0.15$ s. The time, velocity, and height at the end of this segment are t_1, v_1, and h_1.

Segment 2: The motion from when the engine stops until the parachute opens. In this segment the rocket moves with a constant deceleration g. The speed and height of the rocket as functions of time are given by:

$$v(t) = v_1 - g(t - t_1) \quad \text{and} \quad h(t) = h_1 + v_1(t - t_1) - \frac{1}{2}g(t - t_1)^2$$

In this segment the looping continues until the velocity of the rocket is –20 m/s (negative since the rocket moves down). The time and height at the end of this segment are t_2 and h_2.

Segment 3: The motion from when the parachute opens until the rocket hits the ground. In this segment the rocket moves with constant velocity (zero acceleration). The height as a function of time is given by $h(t) = h_2 - v_{chute}(t - t_2)$, where v_{chute} is the constant velocity after the parachute opens. In this segment the looping continues as long as the height is greater than zero.

A program in a script file that carries out the calculations is shown below.

```
m=0.05; g=9.81; tEngine=0.15; Force=16; vChute=-20; Dt=0.01;
clear t v h
n=1;
t(n)=0; v(n)=0; h(n)=0;
% Segment 1
a1=(Force-m*g)/m;
while t(n) < tEngine & n < 50000          The first while loop.
    n=n+1;
    t(n)=t(n-1)+Dt;
    v(n)=a1*t(n);
    h(n)=0.5*a1*t(n)^2;
end
v1=v(n); h1=h(n); t1=t(n);
% Segment 2
while v(n) >= vChute & n < 50000          The second while loop.
    n=n+1;
    t(n)=t(n-1)+Dt;
    v(n)=v1-g*(t(n)-t1);
```

```
    h(n)=h1+v1*(t(n)-t1)-0.5*g*(t(n)-t1)^2;
end
v2=v(n); h2=h(n); t2=t(n);
% Segment 3
while h(n) > 0 & n < 50000                    The third while loop.
   n=n+1;
    t(n)=t(n-1)+Dt;
    v(n)=vChute;
    h(n)=h2+vChute*(t(n)-t2);
end
subplot(1,2,1)
plot(t,h,t2,h2,'o')
subplot(1,2,2)
plot(t,v,t2,v2,'o')
```

The accuracy of the results depends on the magnitude of the time increment `Dt`. An increment of 0.01 s appears to give good results. The conditional expression in the `while` commands also includes a condition for `n` (if `n` is larger than 50,000 the loop stops). This is done as a precaution to avoid an infinite loop in case there is an error in an of the statements inside the loop. The plots generated by the program are shown below (axis labels and text were added to the plots using the Plot Editor).

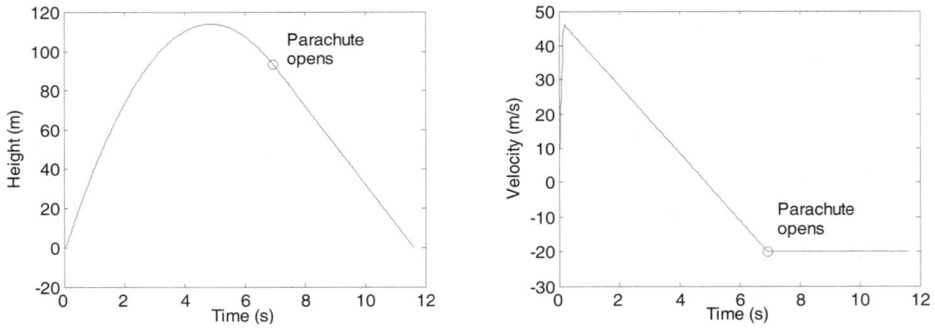

Note: The problem can be solved and programmed in different ways. The solution shown here is one option. For example, instead of using `while` loops, the times when the parachute opens and when the rocket hits the ground can be calculated first, and then `for-end` loops can be used instead of the `while` loop. If the times are determined first, it is possible also to use element-by-element calculations instead of loops.

Sample Problem 6-12: AC to DC converter

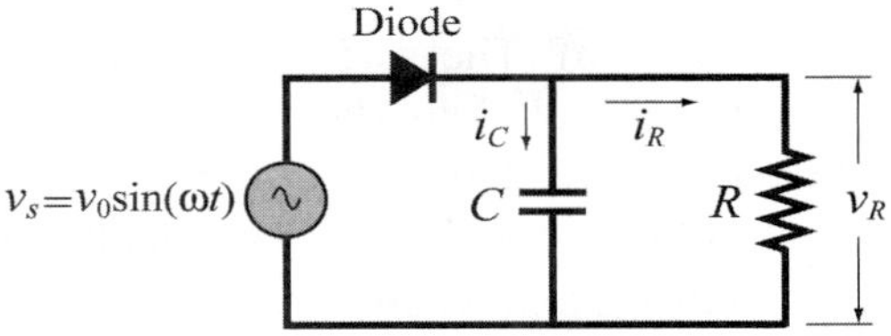

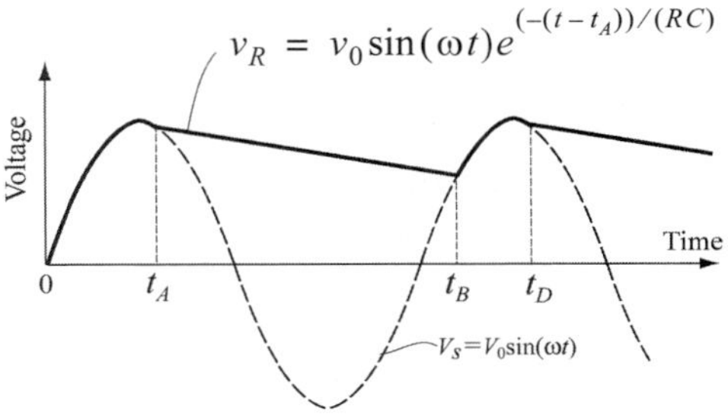

A half-wave diode rectifier is an electrical circuit that converts AC voltage to DC voltage. A rectifier circuit that consists of an AC voltage source, a diode, a capacitor, and a load (resistor) is shown in the figure. The voltage of the source is $v_s = v_0\sin(\omega t)$, where $\omega = 2\pi f$, in which f is the frequency. The operation of the circuit is illustrated in the lower diagram where the dashed line shows the source voltage and the solid line shows the voltage across the resistor. In the first cycle, the diode is on (conducting current) from $t = 0$ until $t = t_A$. At this time the diode turns off and the power to the resistor is supplied by the discharging capacitor. At $t = t_B$ the diode turns on again and continues to conduct current until $t = t_D$. The cycle continues as long as the voltage source is on. In this simplified analysis of this circuit, the diode is assumed to be ideal and the capacitor is assumed to have no charge initially (at $t = 0$). When the diode is on, the resistor's voltage and current are given by:

$$v_R = v_0\sin(\omega t) \text{ and } i_R = v_0\sin(\omega t)/R$$

The current in the capacitor is:

$$i_C = \omega C v_0\cos(\omega t)$$

When the diode is off, the voltage across the resistor is given by:

$$v_R = v_0\sin(\omega t_A)e^{(-(t-t_A))/(RC)}$$

The times when the diode switches off (t_A, t_D, and so on) are calculated from the condition $i_R = -i_C$. The diode switches on again when the voltage of the source reaches the voltage across the resistor (time t_B in the figure).

Write a MATLAB program that plots the voltage across the resistor v_R and the voltage of the source v_s as a function of time for $0 \le t \le 70$ ms. The resistance of the load is 1,800 Ω, the voltage source $v_0 = 12$ V, and $f = 60$ Hz. To examine the effect of capacitor size on the voltage across the load, execute the program twice, once with $C = 45$ μF and once with $C = 10$ μF.

Solution

A program that solves the problem is presented below. The program has two parts—one that calculates the voltage v_R when the diode is on, and the other when the diode is off. The `switch` command is used for switching between the two parts. The calculations start with the diode on (the variable `state='on'`), and when $i_R - i_C \le 0$ the value of `state` is changed to `'off'`, and the program switches to the commands that calculate v_R for this state. These calculations continue until $v_s \ge v_R$, when the program switches back to the equations that are valid when the diode is on.

```
V0=12; C=45e-6; R=1800; f=60;
Tf=70e-3; w=2*pi*f;
clear t VR Vs
t=0:0.05e-3:Tf;
n=length(t);
state='on'
for i=1:n
   Vs(i)=V0*sin(w*t(i));
   switch state
      case 'on'
      VR(i)=Vs(i);
      iR=Vs(i)/R;
      iC=w*C*V0*cos(w*t(i));
      sumI=iR+iC;
      if sumI <= 0
         state='off';
         tA=t(i);
      end
      case 'off'
      VR(i)=V0*sin(w*tA)*exp(-(t(i)-tA)/(R*C));
      if Vs(i) >= VR(i)
         state='on';
      end
   end
end
plot(t,Vs,':',t,VR,'k','linewidth',1)
xlabel('Time (s)'); ylabel('Voltage (V)')
```

Assign `'on'` to the variable `state`.

Calculate the voltage of the source at time t.

Diode is on.

Check if $i_R - i_C \le 0$.

If true, assign `'off'` to `state`.

Assign a value to t_A.

Diode is off.

Check if $v_s \ge v_R$.

If true, assign `'on'` to the variable `state`.

The two plots generated by the program are shown below. One plot shows the result with $C = 45$ µF and the other with $C = 10$ µF. It can be observed that with a larger capacitor the DC voltage is smoother (smaller ripple in the wave).

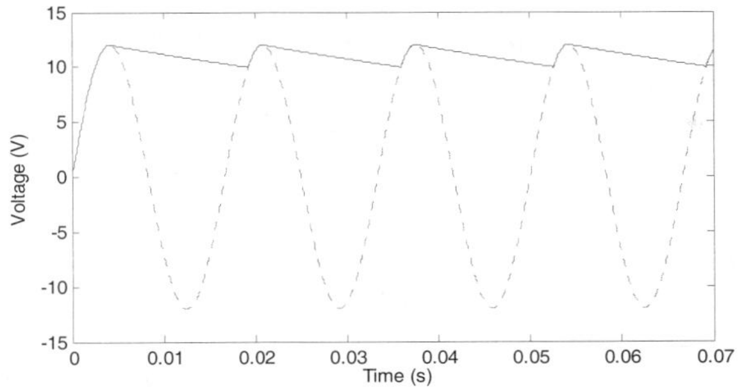

$C = 45$ µF

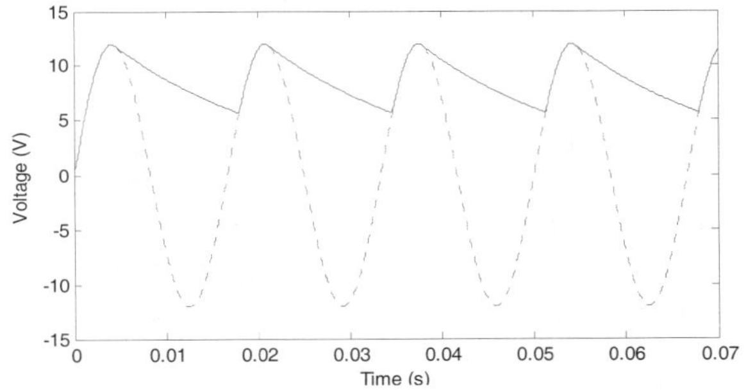

$C = 10$ µF

6.8 *Problems*

1. Evaluate the following expressions without using MATLAB. Check the answer with MATLAB.

   (*a*) $5 + 3 > 32/4$ (*b*) $y = 2 \times 3 > 10/5 + 1 > 2^2$

   (*c*) $y = 2 \times (3 > 10/5) + (1 > 2)^2$ (*d*) $5 \times 3 - 4 \times 4 <= \sim 2 \times 4 - 2 + \sim 0$

2. Given: $a = 6$, $b = 2$, $c = -5$. Evaluate the following expressions without using MATLAB. Check the answer with MATLAB.

   (*a*) $y = a + b > a - b < c$ (*b*) $y = -6 < c < -2$

   (*c*) $y = b + c >= c > a/b$ (*d*) $y = a + c == \sim(c + a \sim= a/b - b)$

3. Given: $v = [4\ -2\ -1\ 5\ 0\ 1\ -3\ 8\ 2]$ and $w = [0\ 2\ 1\ -1\ 0\ -2\ 4\ 3\ 2]$. Evaluate the following expressions without using MATLAB. Check the answer with MATLAB.

   (*a*) $\sim(\sim v)$ (*b*) $u == v$

   (*c*) $u - v < u$ (*d*) $u - (v < u)$

4. Use the vectors *v* and *w* from Problem 3. Use relational operators to create a vector *y* that is made up of the elements of *w* that are larger than or equal to the elements of *v*.

5. Evaluate the following expressions without using MATLAB. Check the answer with MATLAB.
   (*a*) 0&21 (*b*) ~-2>-1&11>=~0
   (*c*) 4-7/2&6<5|-3 (*d*) 3|-1&~2*-3|0

6. The maximum daily temperature (in °F) for Chicago and San Francisco during the month of August 2009 are given in the vectors below (data from the U.S. National Oceanic and Atmospheric Administration).
   TCH = [75 79 86 86 79 81 73 89 91 86 81 82 86 88 89 90 82 84 81 79 73 69 73 79 82 72 66 71 69 66 66]
   TSF = [69 68 70 73 72 71 69 76 85 87 74 84 76 68 79 75 68 68 73 72 79 68 68 69 71 70 89 95 90 66 69]
   Write a program in a script file to answer the following:
   (*a*) Calculate the average temperature for the month in each city.
   (*b*) How many days was the temperature above the average in each city?
   (*c*) How many days, and on which dates in the month, was the temperature in San Francisco lower than the temperature in Chicago?
   (*d*) How many days, and on which dates in the month, was the temperature the same in both cities?

7. Fibonacci numbers are the numbers in a sequence in which the first two elements are 0 and 1, and the value of each subsequent element is the sum of the previous two elements:
   $$0, 1, 1, 2, 3, 5, 8, 13, ...$$
   Write a MATLAB program in a script file that determines and displays the first 20 Fibonacci numbers.

8. Use loops to create a 4×3 matrix in which the value of each element is the sum of its row number and its column number divided by the square of its column number. For example, the value of element (2,3) is $(2+3)/3^2 = 0.5555$.

9. The elements of the symmetric Pascal matrix are obtained from:
   $$P_{ij} = \frac{(i+j-2)!}{(i-1)!(j-1)!}$$
   Write a MATLAB program that creates an $n \times n$ symmetric Pascal matrix. Use the program to create 4×4 and 7×7 Pascal matrices.

10. A Fibonacci sequence is a sequence of numbers beginning with 0 and 1, where the value of each subsequent element is the sum of the previous two elements:

$$a_{i+1} = a_i + a_{i-1}, \text{ i.e. } 0, 1, 1, 2, 3, 5, 8, 13, ...$$

Related sequences can be constructed with other beginning numbers. Write a MATLAB program in a script file that construct an $n \times n$ matrix such that the first row contains the first n elements of a sequence, the second row contains the $n+1$ through $2n$th elements and so on. The first line of the script should show the order n of the matrix followed by the values of the first two elements. These two elements can be any two integers, except they cannot both be zero. A property of matrices thus constructed is that their determinants are always zero. Run the program for $n = 4$ and $n = 6$ and for different values of the first two elements. Verify that the determinant is zero in each case (use MATLAB's built-in function `det`).

11. Write a program in a script file that determines the real roots of a quadratic equation $ax^2 + bx + c = 0$. Name the file `quadroots`. When the file runs, it asks the user to enter the values of the constants a, b, and c. To calculate the roots of the equation the program calculates the discriminant D, given by:

$$D = b^2 - 4ac$$

If $D > 0$, the program displays message "The equation has two roots," and the roots are displayed in the next line.
If $D = 0$, the program displays message "The equation has one root," and the root is displayed in the next line.
If $D < 0$, the program displays message "The equation has no real roots."
Run the script file in the Command Window three times to obtain solutions to the following three equations:
(a) $2x^2 + 8x + 8 = 0$
(b) $-5x^2 + 3x - 4 = 0$
(c) $-2x^2 + 7x + 4 = 0$

12. Write a program in a script file that finds the smallest odd integer that is divisible by 11 and whose square root is greater than 132. Use a loop in the program. The loop should start from 1 and stop when the number is found. The program prints the message "The required number is:" and then prints the number.

13. Write a program (using a loop) that determines the expression:

$$\sqrt{12}\sum_{n=0}^{m}\frac{(-1/3)^n}{2n+1}$$

Run the program with $m = 5$, $m = 10$, and $m = 20$. Compare the result with π. (Use format long.)

14. Write a program (using a loop) that determines the expression:

$$2\prod_{n=1}^{m}\frac{(2n)^2}{(2n)^2-1} = 2\left(\frac{4}{3}\cdot\frac{16}{15}\cdot\frac{36}{35}\cdot\ldots\right)$$

Run the program with $m = 100$, $m = 100{,}000$, and $m = 1{,}0000{,}000$. Compare the result with π. (Use format long.)

15. A vector is given by x = [–3.5 –5 6.2 11 0 8.1 –9 0 3 –1 3 2.5]. Using conditional statements and loops, write a program that creates two vectors from x—one (call it P) that contains the positive elements of x, and a second (call it N) that contains the negative elements of x. In both P and N, the elements are in the same order as in x.

16. A vector is given by x = [–3.5 5 –6.2 11.1 0 7 –9.5 2 15 –1 3 2.5]. Using conditional statements and loops, write a program that rearranges the elements of x in order from the smallest to the largest. Do not use MATLAB's built-in function `sort`.

17. The following is a list of 20 exam scores. Write a computer program that calculates the average of the top 8 scores.

Exam scores: 73, 91, 37, 81, 63, 66, 50, 90, 75, 43, 88, 80, 79, 69, 26, 82, 89, 99, 71, 59

18. The Taylor series expansion for $\sin(x)$ is

$$\sin(x) = x - \frac{x^3}{3!} + \frac{x^5}{5!} - \frac{x^7}{7!} + \ldots = \sum_{n=0}^{\infty}\frac{(-1)^n}{(2n+1)!}x^{2n+1}$$

where x is in radians. Write a MATLAB program that determines $\sin(x)$ using the Taylor series expansion. The program asks the user to type a value for an angle in degrees. Then the program uses a loop for adding the terms of the Taylor series. If a_n is the nth term in the series, then the sum S_n of the n terms is $S_n = S_{n-1} + a_n$. In each pass calculate the estimated error E given by $E = \left|\frac{S_n - S_{n-1}}{S_{n-1}}\right|$. Stop adding terms when $E \le 0.000001$. The program displays the value of $\sin(x)$. Use the program for calculating:

(a) $\sin(45°)$ (b) $\sin(195°)$.

Compare the values with those obtained by using a calculator.

19. Write a MATLAB program in a script file that finds a positive integer n such that the sum of all the integers $1 + 2 + 3 + \ldots + n$ is a number between 100 and 1000 whose three digits are identical. As output the program displays the integer n and the corresponding sum.

20. The following are formulas for calculating the training heart rate (*THR*) for men and women

For men (Karvonen formula): $THR = [(220 - AGE) - RHR] \times INTEN + RHR$

For women: $THR = [(206 - 0.88 \times AGE) - RHR] \times INTEN + RHR$

where *AGE* is the person's age, *RHR* the resting heart rate, and *INTEN* the fitness level (0.55 for low, 0.65 for medium, and 0.8 for high fitness). Write a program in a script file that determines the *THR*. The program asks users to enter their gender (male or female), age (number), resting heart rate (number), and fitness level (low, medium, or high). The program then displays the training heart rate. Use the program for determining the training heart rate for the following two individuals:

(*a*) A 21-years-old male, resting heart rate of 62, and low fitness level.

(*b*) A 19-years-old female, resting heart rate of 67, and high fitness level.

21. Write a program that determines the center and the radius of a circle that passes through three given points. The program asks the user to enter the coordinates of the points one at a time. The program displays the coordinate of the center and the radius, and makes a plot of the circle and the three points displayed on the plot with asterisk markers. Execute the program to find the circle that passes through the points (13, 15), (4, 18), and (19, 3).

22. Body Mass Index (*BMI*) is a measure of obesity. In standard units it is calculated by the formula

$$BMI = 703\frac{W}{H^2}$$

where *W* is weight in pounds, and *H* is height in inches. The obesity classification is:

BMI	Classification
Below 18.5	Underweight
18.5 to 24.9	Normal
25 to 29.9	Overweight
30 and above	Obese

Write a program in a script file that calculates the *BMI* of a person. The program asks the person to enter his or her weight (lb) and height (in.). The program displays the result in a sentence that reads: "Your BMI value is XXX, which classifies you as SSSS," where XXX is the BMI value rounded to the nearest tenth, and SSSS is the corresponding classification. Use the program for determining the obesity of the following two individuals:

(*a*) A person 6 ft 2 in. tall with a weight of 180 lb.

(*b*) A person 5 ft 1 in. tall with a weight of 150 lb.

23. Write a program in a script file that calculates the cost of a telephone call according to the following price schedule:

Time the call made	**Duration of call**		
	1–10 min	**10–30 min**	**More than 30 min**
Day: 8 A.M. to 6 P.M.	\$0.10/min	\$1.00 + \$0.08/min for additional min above 10.	\$2.60 + \$0.06/min for additional min above 30.
Evening: 6 P.M. to 12 A.M.	\$0.07/min	\$0.70 + \$0.05/min for additional min above 10.	\$1.70 + \$0.04/min for additional min above 30.
Night: 12 A.M. to 8 A.M.	\$0.04/min	\$0.40 + \$0.03/min for additional min above 10.	\$1.00 + \$0.02/min for additional min above 13.

The program asks the user to enter the time the call is made (day, evening, or night) and the duration of the call (a number that can have one digit to the right of the decimal point). If the duration of the call is not an integer, the program rounds up the duration to the next integer. The program then displays the cost of the call.

Run the program three times for the following calls:
(*a*) 8.3 min at 1:32 P.M. (*b*) 34.5 min at 8:00 P.M. (*c*) 29.6 min at 1:00 A.M.

24. Write a program that determines the change given back to a customer in a self-service checkout machine of a supermarket for purchases of up to \$20. The program generates a random number between 0.01 and 20.00 and displays the number as the amount to be paid. The program then asks the user to enter payment, which can be one \$1 bill, one \$5 bill, one \$10 bill, or one \$20 bill. If the payment is less than the amount to be paid, an error message is displayed. If the payment is sufficient, the program calculates the change and lists the bills and/or the coins that make up the change, which has to be composed of the least number each of bills and coins. For example, if the amount to be paid is \$2.33 and a \$10 bill is entered as payment, then the change is one \$5 bill, two \$1 bills, two quarters, one dime, one nickel, and two pennies.

25. The concentration of a drug in the body C_P can be modeled by the equation

$$Cp = \frac{D_G}{V_d}\frac{k_a}{(k_a - k_e)}(e^{-k_e t} - e^{-k_a t})$$

where D_G is the dosage administered (mg), V_d is the volume of distribution (L), k_a is the absorption rate constant (h^{-1}), k_e is the elimination rate constant (h^{-1}), and t is the time (h) since the drug was administered. For a certain drug, the following quantities are given: $D_G = 150$ mg, $V_d = 50$ L, $k_a = 1.6\ h^{-1}$, and $k_e = 0.4\ h^{-1}$.

(*a*) A single dose is administered at $t = 0$. Calculate and plot C_P versus t for 10 hours.

(*b*) A first dose is administered at $t = 0$, and subsequently four more doses are administered at intervals of 4 hours (i.e., at $t = 4, 8, 12, 16$). Calculate and plot C_P versus t for 24 hours.

26. One numerical method for calculating the square root of a number is the Babylonian method. In this method $\sqrt{P}$ is calculated in iterations. The solution process starts by choosing a value x_1 as a first estimate of the solution. Using this value, a second, more accurate solution x_2 can be calculated with $x_2 = (x_1 + P/x_1)/2$, which is then used for calculating a third, still more accurate solution x_3, and so on. The general equation for calculating the value of the solution x_{i+1} from the solution x_i is $x_{i+1} = (x_i + P/x_i)/2$. Write a MATLAB program that calculates the square root of a number. In the program use $x = P$ for the first estimate of the solution. Then, by using the general equation in a loop, calculate new, more accurate solutions. Stop the looping when the estimated relative error E defined by $E = \left|\frac{x_{i+1} - x_i}{x_i}\right|$ is smaller than 0.00001. Use the program to calculate:

(*a*) $\sqrt{110}$ (*b*) $\sqrt{93,443}$ (*c*) $\sqrt{23.25}$

27. A twin primes is a pair of prime numbers such that the difference between them is 2 (for example, 17 and 19). Write a computer program that finds all the twin primes between 10 and 500. The program displays the results in a two-column matrix in which each row is a twin prime.

28. Write a program in a script file that converts a measure of volume given in units of either m^3, L, ft^3, or gat (U.S. gallons) to the equivalent quantity in different units specified by the user. The program asks the user to enter the amount of volume, its current units, and the desired new units. The output is the specification of volume in the new units. Use the program to:
(*a*) Convert 3.5 m^3 to gal.
(*b*) Convert 200 L to ft^3.
(*c*) Convert 480 ft^3 to m^3.

29. In a one-dimensional random walk the position x of a walker is computed by

$$x_j = x_j + s$$

where s is a random number. Write a program that calculates the number of steps required for the walker to reach a boundary $x = \pm B$. Use MATLAB's built-in function `randn(1,1)` to calculate s. Run the program 100 times (by using a loop) and calculate the average number of steps when $B = 10$.

30. The Sierpinski triangle can be implemented in MATLAB by plotting points iteratively according to one of the following three rules which are selected randomly with equal probability.

Rule 1: $x_{n+1} = 0.5x_n, \quad y_{n+1} = 0.5y_n$

Rule 2: $x_{n+1} = 0.5x_n + 0.25, \quad y_{n+1} = 0.5y_n + \frac{\sqrt{3}}{4}$

Rule 3: $x_{n+1} = 0.5x_n + 0.5, \quad y_{n+1} = 0.5y_n$

Write a program in a script file that calculates the x and y vectors and then plots y versus x as individual points (use `plot(x,y,'^')`). Start with $x_1 = 0$ and $y_1 = 0$. Run the program four times with 10, 100, 1,000, and 10,000 iterations.

31. There are 12 teams in a league, numbered 1 through 12. Six games are planned for the weekend. Write a MATLAB program that randomly assign the teams for each game. Display the results in a two-column table where each row contains the two teams that play each other.

32. The temperature dependence of the heat capacity C_p of many gases can be described in terms of a cubic equation:

$$C_p = a + bT + cT^2 + dT^3$$

The following table gives the coefficients of the cubic equation for four gases. C_p is in J/(g mol)(°C) and T is in °C.

Gas	a	b	c	d
SO_2	38.91	3.904×10^{-2}	-3.105×10^{-5}	8.606×10^{-9}
SO_3	48.50	9.188×10^{-2}	-8.540×10^{-5}	32.40×10^{-9}
O_2	29.10	1.158×10^{-2}	-0.6076×10^{-5}	1.311×10^{-9}
N_2	29.00	0.2199×10^{-2}	-0.5723×10^{-5}	-2.871×10^{-9}

Write a program that does the following:

- Prints the four gases on the screen and asks the user to select which gas to find the heat capacity for.
- Asks the user for a temperature.
- Asks the user if another temperature is needed (enter yes or no). If the answer is yes, the user is asked to enter another temperature. This process continues until the user enters no.
- Display a table containing the temperatures entered and the corresponding heat capacities.

(*a*) Use the program for determining the heat capacity of SO_3 at $100°$ and $180°$.

(*b*) Use the program for finding the heat capacity of N_2 at $220°$ and $300°$.

33. The overall grade in a course is determined from the grades of 5 quizzes, 3 midterms, and a final, using the following scheme:
Quizzes: Quizzes are graded on a scale from 0 to 10. The grade of the lowest quiz is dropped and the average of the 4 quizzes with the higher grades constitutes 25% of the course grade.
Midterms: Midterms are graded on a scale from 0 to 100. If the average of the midterm scores is higher than the score on the final, the average of the midterms is 35% of the course grade. If the final grade is higher than the average of the midterms, then the lowest midterm is dropped and the average of the two midterms with the higher grades is 35% of the course grade.
Final: Finals are graded on a scale from 0 to 10. The final is 40% of the course grade.
Write a computer program in a script file that determines the course grade for a student. The program first asks the user to enter the five quiz grades (in a vector), the three midterm grades (in a vector), and the grade of the final. Then the program calculates a numerical course grade (a number between 0 and 100). Finally, the program assigns a letter grade according to the following key: *A* for $\text{Grade} \geq 90$, *B* for $80 \leq \text{Grade} \leq 90$, *C* for $70 \leq \text{Grade} \leq 80$, *D* for $60 \leq \text{Grade} \leq 70$, and *E* for a grade lower than 60. Execute the program for the following cases:
(*a*) Quiz grades: 7, 9, 4, 8 , 7. Midterm grades: 93, 83, 87. Final grade: 89.
(*b*) Quiz grades: 8, 6, 9, 6 , 9. Midterm grades: 81, 75, 79. Final grade: 72.

34. The handicap differential (*HCD*) for a round of golf is calculated from the formula:

$$HCD = \frac{(Score - Course\ \ Rating)}{Course\ \ Slope} \times 113$$

The course rating and the slope are measures of how difficult a particular course is. A golfers handicap is calculated from a certain number *N* of their best (lowest) handicap scores according to the following table.

# Rounds played	*N*	# Rounds played	*N*
5-6	1	15-16	6
7-8	2	17	7
9-10	3	18	8
11-12	4	19	9
13-14	5	20	10

For example, if 13 rounds have been played, only the best five handicaps are used. A handicap cannot be computed for fewer than five rounds. If more than 20 rounds have been played, only the 20 most recent results are used.

Once the lowest *N* handicap differentials have been identified, they are averaged and then rounded down to the nearest tenth. The result is the player's handicap. Write a program in a script file that calculates a persons handicap. The program asks the user to enter the golfers record in a three columns matrix where the first column is the course rating, the second is the course slope, and the third is the players score. Each row corresponds to one round. The program displays the person's handicap. Execute the program for players with the following records.

(*a*)

Rating	Slope	Score
71.6	122	85
72.8	118	87
69.7	103	83
70.3	115	81
70.9	116	79
72.3	117	91
71.6	122	89
70.3	115	83
72.8	118	92
70.9	109	80
73.1	132	94
68.2	115	78
74.2	135	103
71.9	121	84

(*b*)

Rating	Slope	Score
72.2	119	71
71.6	122	73
74.0	139	78
68.2	125	69
70.2	130	74
69.6	109	69
66.6	111	74

Chapter 7

User-Defined Functions and Function Files

A simple function in mathematics, $f(x)$, associates a unique number to each value of x. The function can be expressed in the form $y = f(x)$, where $f(x)$ is usually a mathematical expression in terms of x. A value of y (output) is obtained when a value of x (input) is substituted in the expression. Many functions are programmed inside MATLAB as built-in functions, and can be used in mathematical expressions simply by typing their name with an argument (see Section 1.5); examples are `sin(x)`, `cos(x)`, `sqrt(x)`, and `exp(x)`. Frequently, in computer programs, there is a need to calculate the value of functions that are not built-in. When a function expression is simple and needs to be calculated only once, it can be typed as part of the program. However, when a function needs to be evaluated many times for different values of arguments, it is convenient to create a "user-defined" function. Once a user-defined function is created (saved) it can be used just like the built-in functions.

A user-defined function is a MATLAB program that is created by the user, saved as a function file, and then can be used like a built-in function. The function can be a simple, single mathematical expression or a complicated and involved series of calculations. In many cases it is actually a subprogram within a computer program. The main feature of a function file is that it has an input and an output. This means that the calculations in the function file are carried out using the input data, and the results of the calculations are transferred out of the function file by the output. The input and the output can be one or several variables, and each can be a scalar, vector, or an array of any size. Schematically, a function file can be illustrated by:

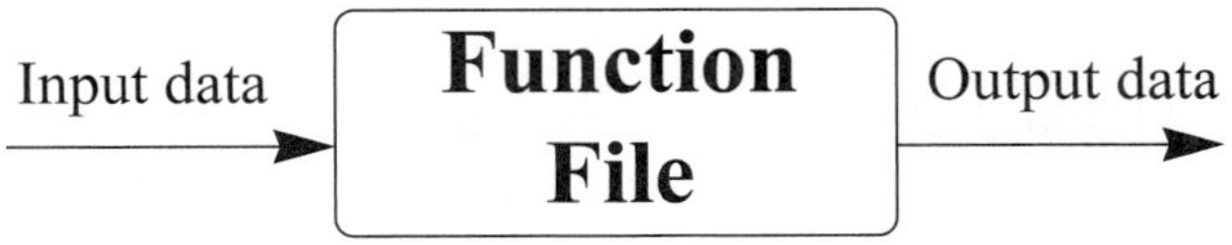

A very simple example of a user-defined function is a function that calculates the maximum height that a ball reaches when thrown upward with a certain velocity. For a velocity v_0, the maximum height h_{max} is given by $h_{max} = \frac{v_0^2}{2g}$, where g is the gravitational acceleration. In function form this can be written as $h_{max}(v_0) = \frac{v_0^2}{2g}$. In this case the input to the function is the velocity (a number), and the output is the maximum height (a number). For example, in SI units (g = 9.81 m/s^2) if the input is 15 m/s, the output is 11.47 m.

In addition to being used as math functions, user-defined functions can be used as subprograms in large programs. In this way large computer programs can be made up of smaller "building blocks" that can be tested independently. Function files are similar to subroutines in Basic and Fortran, procedures in Pascal, and functions in C.

The fundamentals of user-defined functions are explained in Sections 7.1 through 7.7. In addition to user-defined functions that are saved in separate function files and called for use in a computer program, MATLAB provides an option to define and use a user-defined math function within a computer program (not in a separate file). This can be done by using anonymous and/or inline functions, which are presented in Section 7.8. There are built-in and user-defined functions that have to be supplied with other functions when they are called. These functions, which in MATLAB are called function functions, are introduced in Section 7.9. The last two sections cover subfunctions and nested functions. Both are methods for incorporating two or more user-defined functions in a single function file.

7.1 CREATING A FUNCTION FILE

Function files are created and edited, like script files, in the Editor/Debugger Window. This window is opened from the Command Window. In the **File** menu, select **New**, and then select **Function**. Once the Editor/Debugger Window opens, it looks like that shown in Figure 7-1. The editor contains several pre-typed lines that outline the structure of a function file. The first line is the function definition line, which is followed by comments the describe the function. Next comes the program (the empty lines 4 and 5 in Figure 7-1), and the last line is an `end` statement, which is optional. The structure of a function file is described in detail in the next section.

Note: The Editor/Debugger Window can also be opened (as was described in Chapter 1) by selecting **Script** after **New**. The window that opens is empty, without any pre-typed lines. The window can be used for writing a script file or a

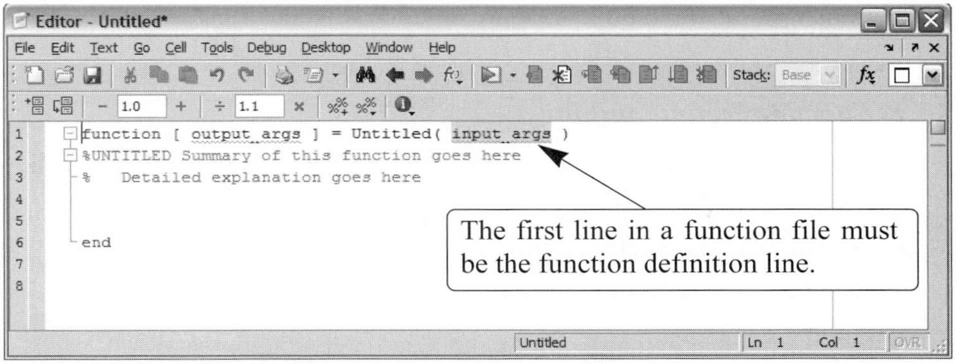

Figure 7-1: The Editor/Debugger Window.

function file. If the Editor/Debugger Window is opened by selecting **Function** after **New**, it can also be used for writing a script file or a function file.

7.2 STRUCTURE OF A FUNCTION FILE

The structure of a typical complete function file is shown in Figure 7-2. This particular function calculates the monthly payment and the total payment of a loan. The inputs to the function are the amount of the loan, the annual interest rate, and the duration of the loan (number of years). The output from the function is the monthly payment and the total payment.

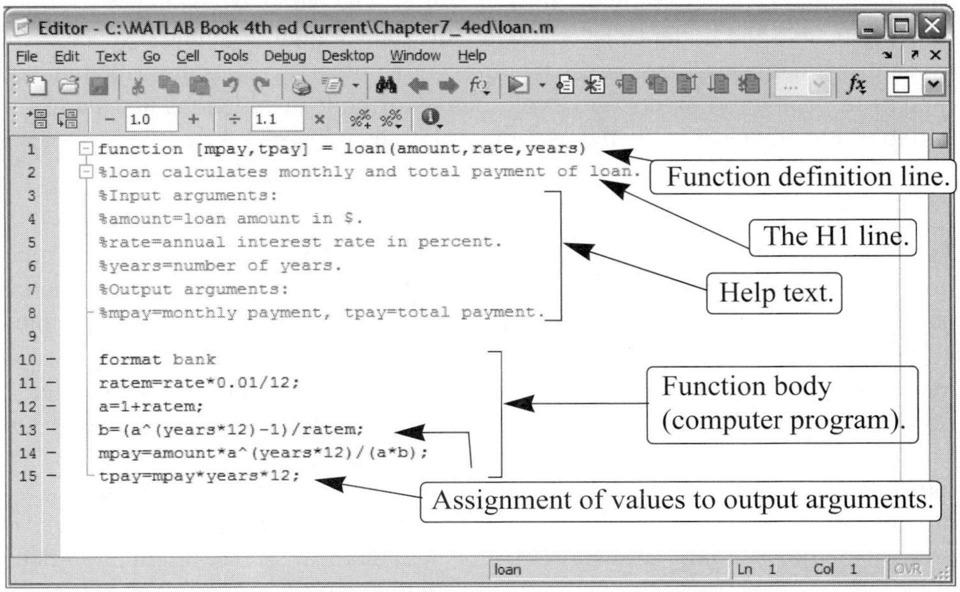

Figure 7-2: Structure of a typical function file.

The various parts of the function file are described in detail in the following sections.

7.2.1 Function Definition Line

The first executable line in a function file *must* be the function definition line. Otherwise the file is considered a script file. The function definition line:

- Defines the file as a function file.
- Defines the name of the function.
- Defines the number and order of the input and output arguments.

The form of the function definition line is:

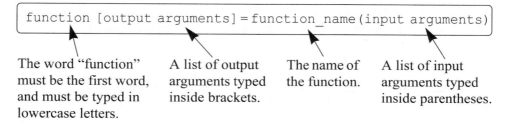

```
function [output arguments]=function_name(input arguments)
```

The word "function" must be the first word, and must be typed in lowercase letters.

A list of output arguments typed inside brackets.

The name of the function.

A list of input arguments typed inside parentheses.

The word "function," typed in lowercase letters, must be the first word in the function definition line. On the screen the word function appears in blue. The function name is typed following the equal sign. The name can be made up of letters, digits, and the underscore character (the name cannot include a space). The rules for the name are the same as the rules for naming variables described in Section 1.6.2. It is good practice to avoid names of built-in functions and names of variables already defined by the user or predefined by MATLAB.

7.2.2 Input and Output Arguments

The input and output arguments are used to transfer data into and out of the function. The input arguments are listed inside parentheses following the function name. Usually, there is at least one input argument, although it is possible to have a function that has no input arguments. If there are more than one, the input arguments are separated with commas. The computer code that performs the calculations within the function file is written in terms of the input arguments and assumes that the arguments have assigned numerical values. This means that the mathematical expressions in the function file must be written according to the dimensions of the arguments, since the arguments can be scalars, vectors, or arrays. In the example shown in Figure 7-2 there are three input arguments (`amount,rate,years`), and in the mathematical expressions they are assumed to be scalars. The actual values of the input arguments are assigned when the function is used (called). Similarly, if the input arguments are vectors or arrays, the mathematical expressions in the function body must be written to follow linear algebra or element-by-element calculations.

The output arguments, which are listed inside brackets on the left side of the assignment operator in the function definition line, transfer the output from the function file. Function files can have zero, one, or several output arguments. If there are more than one, the output arguments are separated with commas. If there is only one output argument, it can be typed without brackets. *In order for the function file to work, the output arguments must be assigned values in the computer program that is in the function body.* In the example in Figure 7-2 there are two output arguments, `mpay` and `tpay`. When a function does not have an output argument, the assignment operator in the function definition line can be omitted. A function without an output argument can, for example, generate a plot or write data to a file.

It is also possible to transfer strings into a function file. This is done by typing the string as part of the input variables (text enclosed in single quotes). Strings can be used to transfer names of other functions into the function file.

Usually, all the input to, and the output from, a function file transferred through the input and output arguments. In addition, however, all the input and output features of script files are valid and can be used in function files. This means that any variable that is assigned a value in the code of the function file will be displayed on the screen unless a semicolon is typed at the end of the command. In addition, the `input` command can be used to input data interactively, and the `disp`, `fprintf`, and `plot` commands can be used to display information on the screen, save to a file, or plot figures just as in a script file. The following are examples of function definition lines with different combinations of input and output arguments.

Function definition line	**Comments**
function [mpay,tpay] = loan(amount,rate,years)	Three input arguments, two output arguments.
function [A] = RectArea(a,b)	Two input arguments, one output argument.
function A = RectArea(a,b)	Same as above; one output argument can be typed without the brackets.
function [V, S] = SphereVolArea(r)	One input variable, two output variables.
function trajectory(v,h,g)	Three input arguments, no output arguments.

7.2.3 The H1 Line and Help Text Lines

The H1 line and help text lines are comment lines (lines that begin with the percent, %, sign) following the function definition line. They are optional but are frequently used to provide information about the function. The H1 line is the first comment line and usually contains the name and a short definition of the function. When a user types (in the Command Window) `lookfor a_word`, MATLAB searches for `a_word` in the H1 lines of all the functions, and if a match is found, the H1 line that contains the match is displayed.

The help text lines are comment lines that follow the H1 line. These lines contain an explanation of the function and any instructions related to the input and output arguments. The comment lines that are typed between the function definition line and the first non-comment line (the H1 line and the help text) are displayed when the user types `help function_name` in the Command Window. This is true for MATLAB built-in functions as well as the user-defined functions. For example, for the function `loan` in Figure 7-2, if `help loan` is typed in the Command Window (make sure the current directory or the search path includes the directory where the file is saved), the display on the screen is:

```
>> help loan
loan calculates monthly and total payment of loan.
Input arguments:
amount=loan amount in $.
rate=annual interest rate in percent.
years=number of years.
Output arguments:
mpay=monthly payment, tpay=total payment.
```

A function file can include additional comment lines in the function body. These lines are ignored by the `help` command.

7.2.4 Function Body

The function body contains the computer program (code) that actually performs the computations. The code can use all MATLAB programming features. This includes calculations, assignments, any built-in or user-defined functions, flow control (conditional statements and loops) as explained in Chapter 6, comments, blank lines, and interactive input and output.

7.3 Local and Global Variables

All the variables in a function file are local (the input and output arguments and any variables that are assigned values within the function file). This means that the variables are defined and recognized only inside the function file. When a function file is executed, MATLAB uses an area of memory that is separate from the workspace (the memory space of the Command Window and the script files). In a function file the input variables are assigned values each time the function is called. These variables are then used in the calculations within the function file. When the function file finishes its execution the values of the output arguments are transferred to the variables that were used when the function was called. All of this means that a function file can have variables with the same names as variables in the Command Window or in script files. The function file does not recognize variables with the same names as have been assigned values outside the function. The assignment of values to these variables in the function file will not change their assignment elsewhere.

Each function file has its own local variables, which are not shared with other functions or with the workspace of the Command Window and the script files. It is possible, however, to make a variable common (recognized) in several different function files, and perhaps in the workspace too. This is done by declaring the variable global with the `global` command, which has the form:

```
global variable_name
```

Several variables can be declared global by listing them, separated with spaces, in the global command. For example:

```
global GRAVITY_CONST FrictionCoefficient
```

- The variable has to be declared global in every function file that the user wants it to be recognized in. The variable is then common only to these files.
- The `global` command must appear before the variable is used. It is recommended to enter the `global` command at the top of the file.
- The `global` command has to be entered in the Command Window, or in a script file, for the variable to be recognized in the workspace.
- The variable can be assigned, or reassigned, a value in any of the locations in which it is declared common.
- The use of long descriptive names (or all capital letters) is recommended for global variables in order to distinguish them from regular variables.

7.4 Saving a Function File

A function file must be saved before it can be used. This is done, as with a script file, by choosing **Save as . . .** from the **File** menu, selecting a location (many students save to a flash drive), and entering the file name. It is highly recommended that the file be saved with a name that is identical to the function name in the function definition line. In this way the function is called (used) by using the function name. (If a function file is saved with a different name, the name it is saved under must be used when the function is called.) Function files are saved with the extension .m. Examples:

Function definition line	File name
function [mpay,tpay] = loan(amount,rate,years)	loan.m
function [A] = RectArea(a,b)	RectArea.m
function [V, S] = SphereVolArea(r)	SphereVolArea.m
function trajectory(v,h,g)	trajectory.m

7.5 USING A USER-DEFINED FUNCTION

A user-defined function is used in the same way as a built-in function. The function can be called from the Command Window, from a script file, or from another function. To use the function file, the folder where it is saved must either be in the current folder or be in the search path (see Sections 1.8.3 and 1.8.4).

A function can be used by assigning its output to a variable (or variables), as a part of a mathematical expression, as an argument in another function, or just by typing its name in the Command Window or in a script file. In all cases the user must know exactly what the input and output arguments are. An input argument can be a number, a computable expression, or a variable that has an assigned value. The arguments are assigned according to their position in the input and output argument lists in the function definition line.

Two of the ways that a function can be used are illustrated below with the user-defined `loan` function in Figure 7-2, which calculates the monthly and total payments (two output arguments) of a loan. The input arguments are the loan amount, annual interest rate, and the length (number of years) of the loan. In the first illustration the `loan` function is used with numbers as input arguments:

```
>> [month total]=loan(25000,7.5,4)

month =
        600.72
total =
      28834.47
```

First argument is loan amount, second is interest rate, and third is number of years.

In the second illustration the `loan` function is used with two pre-assigned variables and a number as the input arguments:

```
>> a=70000;  b=6.5;
>> [x y]=loan(a,b,30)
```

Define variables `a` and `b`.

Use `a`, `b`, and the number 30 for input arguments and `x` (monthly pay) and `y` (total pay) for output arguments.

```
x =
        440.06
y =
     158423.02
```

7.6 EXAMPLES OF SIMPLE USER-DEFINED FUNCTIONS

Sample Problem 7-1: User-defined function for a math function

Write a function file (name it chp7one) for the function $f(x) = \frac{x^4\sqrt{3x+5}}{(x^2+1)^2}$. The input to the function is x and the output is $f(x)$. Write the function such that x can be a vector. Use the function to calculate:

(*a*) $f(x)$ for $x = 6$.

(*b*) $f(x)$ for $x = 1, 3, 5, 7, 9,$ and 11.

Solution

The function file for the function $f(x)$ is:

```
function y=chp7one(x)                    Function definition line.
y=(x.^4.*sqrt(3*x+5))./(x.^2+1).^2;      Assignment to output argument.
```

Note that the mathematical expression in the function file is written for element-by-element calculations. In this way if x is a vector, y will also be a vector. The function is saved and then the search path is modified to include the directory where the file was saved. As shown below, the function is used in the Command Window.

(*a*) Calculating the function for $x = 6$ can be done by typing chp7one(6) in the Command Window, or by assigning the value of the function to a new variable:

```
>> chp7one(6)

ans =
    4.5401

>> F=chp7one(6)

F =
    4.5401
```

(*b*) To calculate the function for several values of x, a vector with the values of x is created and then used for the argument of the function.

```
>> x=1:2:11

x =
     1     3     5     7     9    11
```

```
>> chp7one(x)
ans =
    0.7071    3.0307    4.1347    4.8971    5.5197    6.0638
```

Another way is to type the vector *x* directly in the argument of the function.

```
>> H=chp7one([1:2:11])
H =
    0.7071    3.0307    4.1347    4.8971    5.5197    6.0638
```

Sample Problem 7-2: Converting temperature units

Write a user-defined function (name it `FtoC`) that converts temperature in degrees F to temperature in degrees C. Use the function to solve the following problem. The change in the length of an object, ΔL, due to a change in the temperature, ΔT, is given by: $\Delta L = \alpha L \Delta T$, where α is the coefficient of thermal expansion. Determine the change in the area of a rectangular (4.5 m by 2.25 m) aluminum ($\alpha = 23 \cdot 10^{-6}$ 1/°C) plate if the temperature changes from 40°F to 92°F.

Solution

A user-defined function that converts degrees F to degrees C is:

```
function C=FtoC(F)                        Function definition line.
%FtoC converts degrees F to degrees C
C=5*(F-32)./9;                            Assignment to output argument.
```

A script file (named Chapter7Example2) that calculates the change of the area of the plate due to the temperature is:

```
a1=4.5; b1=2.25; T1=40; T2=92; alpha=23e-6;
deltaT=FtoC(T2)-FtoC(T1);    Using the FtoC function to calculate the
                             temperature difference in degrees C.
a2=a1+alpha*a1*deltaT;                  Calculating the new length.
b2=b1+alpha*b1*deltaT;                  Calculating the new width.
AreaChange=a2*b2-a1*b1;                 Calculating the change in the area.
fprintf('The change in the area is %6.5f meters
square.',AreaChange)
```

Executing the script file in the Command Window gives the solution:

```
>> Chapter7Example2
The change in the area is 0.01346 meters square.
```

7.7 COMPARISON BETWEEN SCRIPT FILES AND FUNCTION FILES

Students who are studying MATLAB for the first time sometimes have difficulty understanding exactly the differences between script and function files since, for many of the problems that they are asked to solve using MATLAB, either type of file can be used. The similarities and differences between script and function files are summarized below.

- Both script and function files are saved with the extension .m (that is why they are sometimes called M-files).
- The first executable line in a function file is (must be) the function definition line.
- The variables in a function file are local. The variables in a script file are recognized in the Command Window.
- Script files can use variables that have been defined in the workspace.
- Script files contain a sequence of MATLAB commands (statements).
- Function files can accept data through input arguments and can return data through output arguments.
- When a function file is saved, the name of the file should be the same as the name of the function.

7.8 ANONYMOUS AND INLINE FUNCTIONS

User-defined functions written in function files can be used for simple mathematical functions, for large and complicated math functions that require extensive programming, and as subprograms in large computer programs. In cases when the value of a relatively simple mathematical expression has to be determined many times within a program, MATLAB provides the option of using anonymous functions. An anonymous function is a user-defined function that is defined and written within the computer code (not in a separate function file) and is then used in the code. Anonymous functions can be defined in any part of MATLAB (in the Command Window, in script files, and inside regular user-defined functions).

Anonymous functions were introduced in MATLAB 7. They replace inline functions that were used for the same purpose in previous versions of MATLAB. Both anonymous and inline functions can be used in MATLAB R2010b). Anonymous functions, however, have several advantages over inline functions, and it is expected that inline functions will gradually be phased out. Anonymous functions are covered in detail in Section 7.8.1, and inline functions are described in the section that follows.

7.8.1 Anonymous Functions

An anonymous function is a simple (one-line) user-defined function that is defined without creating a separate function file (M-file). Anonymous functions can be constructed in the Command Window, within a script file, or inside a regular user-defined function.

An anonymous function is created by typing the following command:

```
name = @ (arglist) expr
```

name	@	(arglist)	expr
The name of the anonymous function.	The @ symbol.	A list of input arguments (independent variables).	Mathematical expression.

A simple example is: `cube = @ (x) x^3`, which calculates the cube of the input argument.

- The command creates the anonymous function and assigns a handle for the function to the variable name on the left-hand side of the = sign. (Function handles provide means for using the function and passing it to other functions; see Section 7.9.1.)
- The `expr` consists of a single valid mathematical MATLAB expression.
- The mathematical expression can have one or several independent variables. The independent variable(s) is (are) entered in the `(arglist)`. Multiple independent variables are separated with commas. An example of an anonymous function that has two independent variables is: `circle = @ (x,y) 16*x^2+9*y^2`
- The mathematical expression can include any built-in or user-defined functions.
- The expression must be written according to the dimensions of the arguments (element-by-element or linear algebra calculations).
- The expression can include variables that are already defined when the anonymous function is defined. For example, if three variables `a`, `b`, and `c` are defined (have assigned numerical values), then they can be used in the expression of the anonymous function `parabola = @ (x) a*x^2+b*x+c`.

Important note: MATLAB captures the values of the predefined variables when the anonymous function is defined. This means that if new values are subsequently assigned to the predefined variables, the anonymous function is not changed. The anonymous function has to be redefined in order for the new values of the predefined variables to be used in the expression.

Using an anonymous function:

- Once an anonymous function is defined, it can be used by typing its name and a value for the argument (or arguments) in parentheses (see examples that follow).
- Anonymous functions can also be used as arguments in other functions (see Section 7.9.1).

Example of an anonymous function with one independent variable:

The function $f(x) = \dfrac{e^{x^2}}{\sqrt{x^2+5}}$ can be defined (in the Command Window) as an anonymous function for x as a scalar by:

```
>> FA = @ (x) exp(x^2)/sqrt(x^2+5)
FA =
    @(x)exp(x^2)/sqrt(x^2+5)
```

If a semicolon is not typed at the end, MATLAB responds by displaying the function. The function can then be used for different values of x, as shown below.

```
>> FA(2)
ans =
   18.1994
>> z = FA(3)
z =
  2.1656e+003
```

If x is expected to be an array, with the function calculated for each element, then the function must be modified for element-by-element calculations.

```
>> FA = @ (x) exp(x.^2)./sqrt(x.^2+5)
FA =
    @(x)exp(x.^2)./sqrt(x.^2+5)
>> FA([1 0.5 2])
ans =
    1.1097    0.5604   18.1994
```

Using a vector as input argument.

Example of an anonymous function with several independent variables:

The function $f(x, y) = 2x^2 - 4xy + y^2$ can be defined as an anonymous function by:

```
>> HA = @ (x,y) 2*x^2 - 4*x*y + y^2
HA =
    @(x,y)2*x^2-4*x*y+y^2
```

Then the anonymous function can be used for different values of x and y. For example, typing `HA(2,3)` gives:

```
>> HA(2,3)
ans =
    -7
```

Another example of using an anonymous function with several arguments is shown in Sample Problem 6-3.

Sample Problem 7-3: Distance between points in polar coordinates

Write an anonymous function that calculates the distance between two points in a plane when the position of the points is given in polar coordinates. Use the anonymous function to calculate the distance between point A (2, π/6) and point B (5, 3π/4).

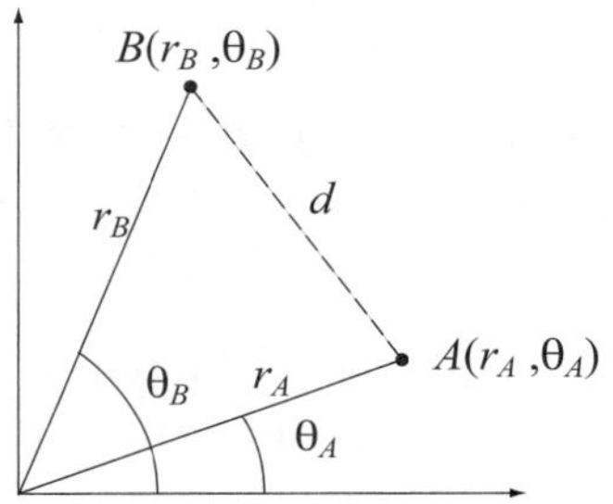

Solution

The distance between two points in polar coordinates can be calculated by using the Law of Cosines:

$$d = \sqrt{r_A^2 + r_B^2 - 2r_A r_B \cos(\theta_A - \theta_B)}$$

The formula for the distance is entered as an anonymous function with four input arguments $(r_A, \theta_A, r_B, \theta_B)$. Then the function is used for calculating the distance between points A and B.

```
>> d= @ (rA,thetA,rB,thetB) sqrt(rA^2+rB^2-2*rA*rB*cos(thetB-thetA))
                                        List of input arguments.
d =
    @(rA,thetA,rB,thetB)sqrt(rA^2+rB^2-2*rA*rB*cos(thetB-
thetA))
>> DistAtoB = d(2,pi/6,5,3*pi/4)
DistAtoB =
    5.8461     The arguments are typed in the order defined in the function.
```

7.8.2 Inline Functions

Similar to an anonymous function, an inline function is a simple user-defined function that is defined without creating a separate function file (M-file). As already mentioned, anonymous functions replace the inline functions used in earlier versions of MATLAB. Inline functions are created with the `inline` command according to the following format:

```
name = inline('math expression typed as a string')
```

A simple example is `cube = inline('x^3')`, which calculates the cube of the input argument.

- The mathematical expression can have one or several independent variables.
- Any letter except i and j can be used for the independent variables in the expression.
- The mathematical expression can include any built-in or user-defined functions.
- The expression must be written according to the dimension of the argument (element-by-element or linear algebra calculations).
- The expression *cannot* include pre assigned variables.
- Once the function is defined it can be used by typing its name and a value for the argument (or arguments) in parentheses (see example below).
- The `inline` function can be used as an argument in other functions.

For example, the function: $f(x) = \frac{e^{x^2}}{\sqrt{x^2+5}}$ can be defined as an inline function for x by:

```
>> FA=inline('exp(x.^2)./sqrt(x.^2+5)')
FA =
     Inline function:
     FA(x) = exp(x.^2)./sqrt(x.^2+5)
>> FA(2)
ans =
   18.1994
>> FA([1 0.5 2])
ans =
    1.1097    0.5604   18.1994
```

Expression written with element-by-element operations.

Using a scalar as the argument.

Using a vector as the argument.

An inline function that has two or more independent variables can be written by using the following format:

```
name = inline('mathematical expression','arg1',
                                       'arg2','arg3')
```

In the format shown here the order of the arguments to be used when calling the function is defined. If the independent variables are not listed in the command, MATLAB arranges the arguments in alphabetical order. For example, the function $f(x, y) = 2x^2 - 4xy + y^2$ can be defined as an inline function by:

```
>> HA=inline('2*x^2-4*x*y+y^2')
HA =
     Inline function:
     HA(x,y) = 2*x^2-4*x*y+y^2
```

Once defined, the function can be used with any values of x and y. For example, HA(2,3) gives:

```
>> HA(2,3)
ans =
    -7
```

7.9 Function Functions

There are many situations where a function (Function *A*) works on (uses) another function (Function *B*). This means that when Function *A* is executed it has to be provided with Function *B*. A function that accepts another function is called in MATLAB a function function. For example, MATLAB has a built-in function called `fzero` (Function *A*) that finds the zero of a math function $f(x)$ (Function *B*), i.e., the value of x where $f(x) = 0$. The program in the function `fzero` is written such that it can find the zero of any $f(x)$. When `fzero` is called, the specific function to be solved is passed into `fzero`, which finds the zero of the $f(x)$. (The function `fzero` is described in detail in Chapter 9.)

A function function, which accepts another function (imported function), includes in its input arguments a name that represents the imported function. The imported function name is used for the operations in the program (code) of the function function. When the function function is used (called), the specific function that is imported is listed in its input argument. In this way different functions can be imported (passed) into the function function. There are two methods for listing the name of an imported function in the argument list of a function function. One is by using a function handle (Section 7.9.1), and the other is by typing the name of the function that is being passed in as a string expression (Section 7.9.2). The method that is used affects the way that the operations in the function

function are written (this is explained in more detail in the next two sections). Using function handles is easier and more efficient, and should be the preferred method.

7.9.1 Using Function Handles for Passing a Function into a Function Function

Function handles are used for passing (importing) user-defined functions, built-in functions, and anonymous functions into function functions that can accept them. This section first explains what a function handle is, then shows how to write a user-defined function function that accepts function handles, and finally shows how to use function handles for passing functions into function functions.

Function handle:

A function handle is a MATLAB value that is associated with a function. It is a MATLAB data type and can be passed as an argument into another function. Once passed, the function handle provides means for calling (using) the function it is associated with. Function handles can be used with any kind of MATLAB function. This includes built-in functions, user-defined functions (written in function files), and anonymous functions.

- For built-in and user-defined functions, a function handle is created by typing the symbol @ in front of the function name. For example, `@cos` is the function handle of the built-in function `cos`, and `@FtoC` is the function handle of the user-defined function `FtoC` that was created in Sample Problem 7-2.
- The function handle can also be assigned to a variable name. For example, `cosHandle=@cos` assigns the handle `@cos` to `cosHandle`. Then the name `cosHandle` can be used for passing the handle.
- As anonymous functions (see Section 7.8.1), their name is already a function handle.

Writing a function function that accepts a function handle as an input argument:

As already mentioned, the input arguments of a function function (which accepts another function) includes a name (dummy function name) that represents the imported function. This dummy function (including a list of input arguments enclosed in parentheses) is used for the operations of the program inside the function function.

- The function that is actually being imported must be in a form consistent with the way that the dummy function is being used in the program. This means that both must have the same number and type of input and output arguments.

The following is an example of a user-defined function function, named `funplot`, that makes a plot of a function (any function $f(x)$ that is imported into it) between the points $x = a$ and $x = b$. The input arguments are `(Fun,a,b)`, where Fun is a dummy name that represents the imported function, and a and b are the end points of the domain. The function funplot also has a numerical output xyout, which is a 3×2 matrix with the values of x and $f(x)$ at the three points $x = a$, $x = (a+b)/2$, and $x = b$. Note that in the program, the dummy function Fun has one input argument (x) and one output argument y, which are both vectors.

```
function xyout=funplot(Fun,a,b)        A name for the function that is passed in.
% funplot makes a plot of the function Fun which is passed in
% when funplot is called in the domain [a, b].
% Input arguments are:
% Fun:  Function handle of the function to be plotted.
% a:  The first point of the domain.
% b:  The last point of the domain.
% Output argument is:
% xyout: The values of x and y at x=a, x=(a+b)/2, and x=b
% listed in a 3 by 2 matrix.
x=linspace(a,b,100);
y=Fun(x);          Using the imported function to calculate f(x) at 100 points.
xyout(1,1)=a; xyout(2,1)=(a+b)/2; xyout(3,1)=b;
xyout(1,2)=y(1);
xyout(2,2)=Fun((a+b)/2);      Using the imported function to
xyout(3,2)=y(100);            calculate f(x) at the midpoint.
plot(x,y)
xlabel('x'), ylabel('y')
```

As an example, the function $f(x) = e^{-0.17x}x^3 - 2x^2 + 0.8x - 3$ over the domain $[0.5, 4]$ is passed into the user-defined function funplot. This is done in two ways: first, by writing a user-defined function for $f(x)$, and then by writing $f(x)$ as an anonymous function.

Passing a user-defined function into a function function:

First, a user-defined function is written for $f(x)$. The function, named Fdemo, calculates $f(x)$ for a given value of x and is written using element-by-element operations.

```
function y=Fdemo(x)
y=exp(-0.17*x).*x.^3-2*x.^2+0.8*x-3;
```

Next, the function Fdemo is passed into the user-defined function function funplot, which is called in the Command Window. Note that a handle of the user-defined function Fdemo is entered (the handle is @Fdemo) for the input argument Fun in the user-defined function funplot.

```
>> ydemo=funplot(@Fdemo,0.5,4)
ydemo =
    0.5000   -2.9852
    2.2500   -3.5548
    4.0000    0.6235
```

Enter a handle of the user-defined function Fdemo.

In addition to the display of the numerical output, when the command is executed, the plot shown in Figure 7-3 is displayed in the Figure Window.

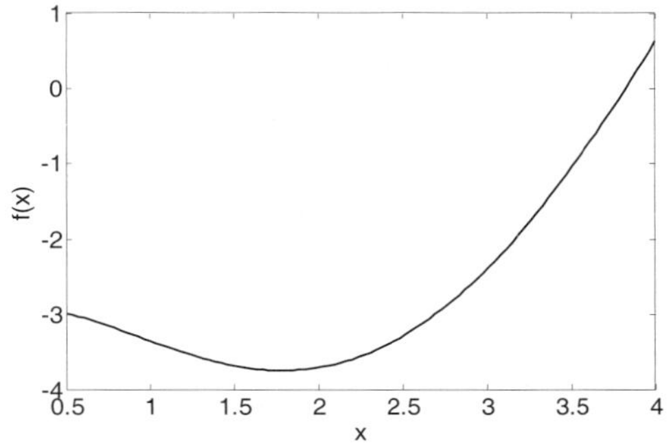

Figure 7-3: A plot of the function $f(x) = e^{-0.17x}x^3 - 2x^2 + 0.8x - 3$.

Passing an anonymous function into a function function:

To use an anonymous function, the function $f(x) = e^{-0.17x}x^3 - 2x^2 + 0.8x - 3$ first has to be written as an anonymous function, and then passed into the user-defined function funplot. The following shows how both of these steps are done in the Command Window. Note that the name of the anonymous function FdemoAnony is entered without the @ sign for the input argument Fun in the user-defined function funplot (since the name is already the handle of the anonymous function).

```
>> FdemoAnony=@(x) exp(-0.17*x).*x.^3-2*x.^2+0.8*x-3
FdemoAnony =
    @(x) exp(-0.17*x).*x.^3-2*x.^2+0.8*x-3

>> ydemo=funplot(FdemoAnony,0.5,4)
ydemo =
    0.5000   -2.9852
    2.2500   -3.5548
    4.0000    0.6235
```

Create an anonymous function for $f(x)$.

Enter the name of the anonymous function (FdemoAnony).

In addition to the display of the numerical output in the Command Window, the plot shown in Figure 7-3 is displayed in the Figure Window.

7.9.2 Using a Function Name for Passing a Function into a Function Function

A second method for passing a function into a function function is by typing the name of the function that is being imported as a string in the input argument of the function function. The method which was used before the introduction of function handles, can be used for importing user-defined functions. As mentioned, function handles are easier to use and more efficient and should be the preferred method. Importing user-defined functions by using their name is covered in the present edition of the book for the benefit of readers who need to understand programs written before MATLAB 7. New programs should use function handles.

When a user-defined function is imported by using its name, the value of the imported function inside the function function has to be calculated with the `feval` command. This is different from the case where a function handle is used, which means that there is a difference in the way that the code in the function function is written that depends on how the imported function is passed in.

The `feval` command:

The `feval` (short for "function evaluate") command evaluates the value of a function for a given value (or values) of the function's argument (or arguments). The format of the command is:

```
variable = feval('function name', argument value)
```

The value that is determined by `feval` can be assigned to a variable, or if the command is typed without an assignment, MATLAB displays `ans =` and the value of the function.

- The function name is typed as string.
- The function can be a built-in or a user-defined function.
- If there is more than one input argument, the arguments are separated with commas.
- If there is more than one output argument, the variables on the left-hand side of the assignment operator are typed inside brackets and separated with commas.

Two examples using the `feval` command with built-in functions follow.

```
>> feval('sqrt',64)
ans =
     8
>> x=feval('sin',pi/6)
```

```
x =
    0.5000
```

The following shows the use of the `feval` command with the user-defined function `loan` which was created earlier in the chapter (Figure 7-2). This function has three input arguments and two output arguments.

```
>> [M,T]=feval('loan',50000,3.9,10)      A $50,000 loan, 3.9% interest, 10 years.
M =
        502.22                            Monthly payment.
T =
      60266.47                            Total payment.
```

Writing a function function that accepts a function by typing its name as an input argument:

As already mentioned, when a user-defined function is imported by using its name, the value of the function inside the function function has to be calculated with the `feval` command. This is demonstrated in the following user-defined function function that is called `funplotS`. The function is the same as the function `funplot` from Section 7.9.1, except that the command `feval` is used for the calculations with the imported function.

```
function xyout=funplotS(Fun,a,b)          A name for the function that is passed in.
% funplotS makes a plot of the function Fun which is passed in
% when funplotS is called in the domain [a, b].
% Input arguments are:
% Fun: The function to be plotted. Its name is entered as
string expression.
% a:  The first point of the domain.
% b:  The last point of the domain.
% Output argument is:
% xyout: The values of x and y at x=a, x=(a+b)/2, and x=b
% listed in a 3 by 2 matrix.
x=linspace(a,b,100);
y=feval(Fun,x);     Using the imported function to calculate f(x) at 100 points.
xyout(1,1)=a; xyout(2,1)=(a+b)/2; xyout(3,1)=b;
xyout(1,2)=y(1);
xyout(2,2)=feval(Fun,(a+b)/2);    Using the imported function to
xyout(3,2)=y(100);                calculate f(x) at the midpoint.
```

```
plot(x,y)
xlabel('x'), ylabel('y')
```

Passing a user-defined function into another function by using a string expression:

The following demonstrates how to pass a user-defined function into a function function by typing the name of the imported function as a string in the input argument. The function $f(x) = e^{-0.17x}x^3 - 2x^2 + 0.8x - 3$ from Section 7.9.1, created as a user-defined function named `Fdemo`, is passed into the user-defined function `funplotS`. Note that the name `Fdemo` is typed in a string for the input argument `Fun` in the user-defined function `funplotS`.

```
>> ydemoS=funplotS('Fdemo',0.5,4)
ydemoS =
    0.5000   -2.9852
    2.2500   -3.5548
    4.0000    0.6235
```

The name of the imported function is typed as a string.

In addition to the display of the numerical output in the Command Window, the plot shown in Figure 7-3 is displayed in the Figure Window.

7.10 *SUBFUNCTIONS*

A function file can contain more than one user-defined function. The functions are typed one after the other. Each function begins with a function definition line. The first function is called the primary function and the rest of the functions are called subfunctions. The subfunctions can be typed in any order. The name of the function file that is saved should correspond to the name of the primary function. Each of the functions in the file can call any of the other functions in the file. Outside functions, or programs (script files), can call only the primary function. Each of the functions in the file has its own workspace, which means that in each the variables are local. In other words, the primary function and the subfunctions cannot access each other's variables (unless variables are declared to be global).

Subfunctions can help in writing user-defined functions in an organized manner. The program in the primary function can be divided into smaller tasks, each of which is carried out in a subfunction. This is demonstrated in Sample Problem 7-4.

Sample Problem 7-4: Average and standard deviation

Write a user-defined function that calculates the average and the standard deviation of a list of numbers. Use the function to calculate the average and the standard deviation of the following list of grades:
80 75 91 60 79 89 65 80 95 50 81

Solution

The average x_{ave} (mean) of a given set of n numbers $x_1, x_2, \ldots, x_n$ is given by:

$$x_{ave} = (x_1 + x_2 + \ldots + x_n)/n$$

The standard deviation is given by:

$$\sigma = \sqrt{\frac{\sum_{i=1}^{i=n}(x_i - x_{ave})^2}{n-1}}$$

A user-defined function, named `stat`, is written for solving the problem. To demonstrate the use of subfunctions, the function file includes `stat` as a primary function, and two subfunctions called `AVG` and `StandDiv`. The function `AVG` calculates x_{ave}, and the function `StandDiv` calculates σ. The subfunctions are called by the primary function.The following listing is saved as one function file called `stat`.

```
function [me SD] = stat(v)                    The primary function.
n=length(v);
me=AVG(v,n);
SD=StandDiv(v,me,n);

function av=AVG(x,num)                        Subfunction.
av=sum(x)/num;

function Sdiv=StandDiv(x,xAve,num)            Subfunction.
xdif=x-xAve;
xdif2=xdif.^2;
Sdiv= sqrt(sum(xdif2)/(num-1));
```

The user-defined function `stat` is then used in the Command Window for calculating the average and the standard deviation of the grades:

```
>> Grades=[80 75 91 60 79 89 65 80 95 50 81];
>> [AveGrade StanDeviation] = stat(Grades)
AveGrade =
   76.8182
StanDeviation =
   13.6661
```

7.11 *NESTED FUNCTIONS*

A nested function is a user-defined function that is written inside another user-defined function. The portion of the code that corresponds to the nested function starts with a function definition line and ends with an `end` statement. An `end` statement must also be entered at the end of the function that contains the nested function. (Normally, a user-defined function does not require a terminating `end` statement. However, an `end` statement is required if the function contains one or more nested functions.) Nested functions can also contain nested functions. Obviously, having many levels of nested functions can be confusing. This section considers only two levels of nested functions.

One nested function:

The format of a user-defined function `A` (called the primary function) that contains one nested function `B` is:

```
function y=A(a1,a2)
.......
    function z=B(b1,b2)
    .......
    end
.......
end
```

- Note the `end` statements at the ends of functions `B` and `A`.
- The nested function `B` can access the workspace of the primary function `A`, and the primary function `A` can access the workspace of the function `B`. This means that a variable defined in the primary function `A` can be read and redefined in nested function `B` and vice versa.
- Function `A` can call function `B`, and function `B` can call function `A`.

Two (or more) nested functions at the same level:

The format of a user-defined function `A` (called the primary function) that contains two nested functions `B` and `C` at the same level is:

```
function y=A(a1,a2)
.......
    function z=B(b1,b2)
    .......
    end
.......
    function w=C(c1,c2)
    .......
    end
.......
end
```

- The three functions can access the workspace of each other.
- The three functions can call each other.

As an example, the following user-defined function (named `statNest`), with two nested functions at the same level, solves Sample Problem 7-4. Note that the nested functions are using variables (`n` and `me`) that are defined in the primary function.

```
function [me SD]=statNest(v)                    The primary function.
n=length(v);
me=AVG(v);

    function av=AVG(x)                          Nested function.
    av=sum(x)/n;
    end

    function Sdiv=StandDiv(x)                   Nested function.
    xdif=x-me;
    xdif2=xdif.^2;
    Sdiv= sqrt(sum(xdif2)/(n-1));
    end

SD=StandDiv(v);
end
```

Using the user-defined function `statNest` in the Command Window for calculating the average of the grade data gives:

```
>> Grades=[80 75 91 60 79 89 65 80 95 50 81];
>> [AveGrade StanDeviation] = statNest(Grades)
AveGrade =
   76.8182
StanDeviation =
   13.6661
```

Two levels of nested functions:

Two levels of nested functions are created when nested functions are written inside nested functions. The following shows an example for the format of a user-defined function with four nested functions in two levels.

```
function y=A(a1,a2)                  (Primary function A.)
.......
    function z=B(b1,b2)              (B is nested function in A.)
    .......
        function w=C(c1,c2)          (C is nested function in B.)
        .......
        end
    end
    function u=D(d1,d2)              (D is nested function in A.)
    .......
        function h=E(e1,e2)          (E is nested function in D.)
        .......
        end
    end
.......
end
```

The following rules apply to nested functions:

- A nested function can be called from a level above it. (In the preceding example, function A can call B or D, but not C or E.)
- A nested function can be called from a nested function at the same level within the primary function. (In the preceding example, function B can call D, and D can call B.)
- A nested function can be called from a nested function at any lower level.
- A variable defined in the primary function is recognized and can be redefined by a function that is nested at any level within the primary function.
- A variable defined in a nested function is recognized and can be redefined by any of the functions that contain the nested function.

7.12 EXAMPLES OF MATLAB APPLICATIONS

Sample Problem 7-5: Exponential growth and decay

A model for exponential growth or decay of a quantity is given by

$$A(t) = A_0 e^{kt}$$

where $A(t)$ and A_0 are the quantity at time t and time 0, respectively, and k is a constant unique to the specific application.

Write a user-defined function that uses this model to predict the quantity $A(t)$ at time t from knowledge of A_0 and $A(t_1)$ at some other time t_1. For function name and arguments use `At = expGD(A0,At1,t1,t)`, where the output argument `At` corresponds to $A(t)$, and for input arguments use `A0,At1,t1,t,` corresponding to A_0, $A(t_1)$, t_1, and t, respectively.

Use the function file in the Command Window for the following two cases:

(*a*) The population of Mexico was 67 million in the year 1980 and 79 million in 1986. Estimate the population in 2000.

(*b*) The half-life of a radioactive material is 5.8 years. How much of a 7-gram sample will be left after 30 years?

Solution

To use the exponential growth model, the value of the constant k has to be determined first by solving for k in terms of A_0, $A(t_1)$, and t_1:

$$k = \frac{1}{t_1}\ln\frac{A(t_1)}{A_0}$$

Once k is known, the model can be used to estimate the population at any time.

The user-defined function that solves the problem is:

```
function At=expGD(A0,At1,t1,t)                Function definition line.
% expGD calculates exponential growth and decay
% Input arguments are:
% A0: Quantity at time zero.
% At1: Quantity at time t1.
% t1: The time t1.
% t: time t.
% Output argument is:
% At: Quantity at time t.
k=log(At1/A0)/t1;                             Determination of k.
At=A0*exp(k*t);                               Determination of A(t).
                                              (Assignment of value to output variable.)
```

Once the function is saved, it is used in the Command Window to solve the two cases. For case *a*) $A_0 = 67$, $A(t_1) = 79$, $t_1 = 6$, and $t = 20$:

```
>> expGD(67,79,6,20)
ans =
       116.03
```

Estimation of the population in the year 2000.

For case *b*) $A_0 = 7$, $A(t_1) = 3.5$ (since t_1 corresponds to the half-life, which is the time required for the material to decay to half of its initial quantity), $t_1 = 5.8$, and $t = 30$.

```
>> expGD(7,3.5,5.8,30)
ans =
         0.19
```

The amount of material after 30 years.

Sample Problem 7-6: Motion of a projectile

Create a function file that calculates the trajectory of a projectile. The inputs to the function are the initial velocity and the angle at which the projectile is fired. The outputs from the function are the maximum height and distance. In addition, the function generates a plot of the trajectory. Use the function to calculate the trajectory of a projectile that is fired at a velocity of 230 m/s at an angle of $39°$.

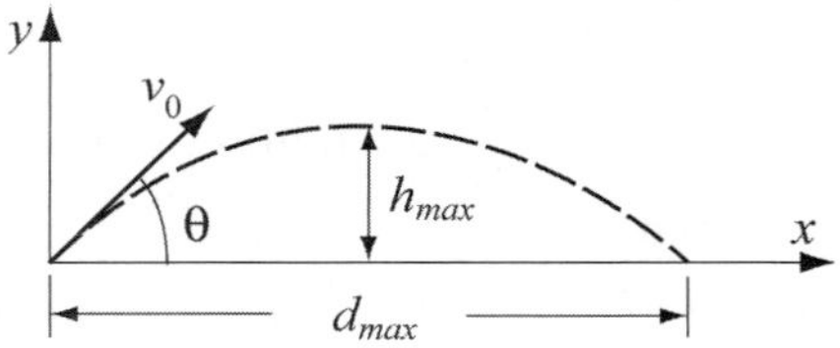

Solution

The motion of a projectile can be analyzed by considering the horizontal and vertical components. The initial velocity v_0 can be resolved into horizontal and vertical components

$$v_{0x} = v_0\cos(\theta) \quad \text{and} \quad v_{0y} = v_0\sin(\theta)$$

In the vertical direction the velocity and position of the projectile are given by:

$$v_y = v_{0y} - gt \quad \text{and} \quad y = v_{0y}t - \frac{1}{2}gt^2$$

The time it takes the projectile to reach the highest point $(v_y = 0)$ and the corresponding height are given by:

$$t_{hmax} = \frac{v_{0y}}{g} \quad \text{and} \quad h_{max} = \frac{v_{0y}^2}{2g}$$

The total flying time is twice the time it takes the projectile to reach the highest point, $t_{tot} = 2t_{hmax}$. In the horizontal direction the velocity is constant, and the position of the projectile is given by:

$$x = v_{0x}t$$

In MATLAB notation the function name and arguments are entered as `[hmax,dmax] = trajectory(v0,theta)`. The function file is:

```
function [hmax,dmax]=trajectory(v0,theta)      Function definition line.
% trajectory calculates the max height and distance of a
projectile, and makes a plot of the trajectory.
% Input arguments are:
% v0: initial velocity in (m/s).
% theta: angle in degrees.
% Output arguments are:
% hmax: maximum height in (m).
% dmax: maximum distance in (m).
% The function creates also a plot of the trajectory.
g=9.81;
v0x=v0*cos(theta*pi/180);
v0y=v0*sin(theta*pi/180);
thmax=v0y/g;
hmax=v0y^2/(2*g);
ttot=2*thmax;
dmax=v0x*ttot;
% Creating a trajectory plot
tplot=linspace(0,ttot,200);   Creating a time vector with 200 elements.
x=v0x*tplot;                  Calculating the x and y coordi-
y=v0y*tplot-0.5*g*tplot.^2;   nates of the projectile at each time.
plot(x,y)                     Note the element-by-element multiplication.
xlabel('DISTANCE (m)')
ylabel('HEIGHT (m)')
title('PROJECTILE''S TRAJECTORY')
```

After the function is saved, it is used in the Command Window for a projectile that is fired at a velocity of 230 m/s and an angle of 39°.

```
>> [h d]=trajectory(230,39)
h =
  1.0678e+003
d =
  5.2746e+003
```

In addition, the following figure is created in the Figure Window:

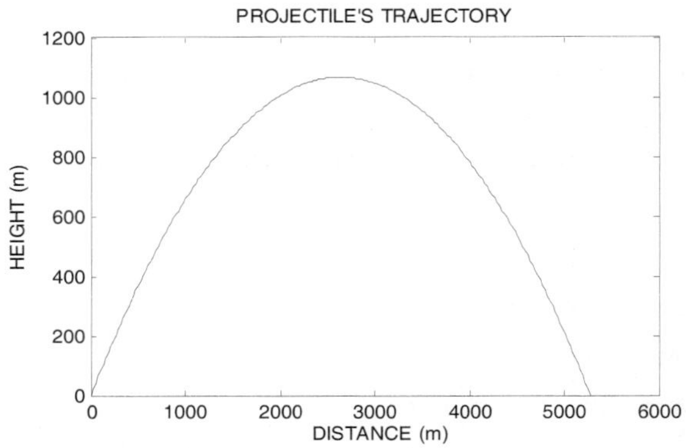

7.13 PROBLEMS

1. The fuel efficiency of an automobile is measured in mi/gal (miles per U.S. gallon) or in km/L (kilometers per liter). Write a MATLAB user-defined function that converts fuel efficiency values from km/L to mi/gal. For the function name and arguments use `mpg=kmlTOmpg(kml)`. The input argument `kml` is the efficiency in km/L, and the output argument `mpg` is the efficiency in mi/gal. Use the function in the Command Window to:
   (*a*) Determine the fuel efficiency in mi/gal of a car that consumes 9 km/L.
   (b) Determine the fuel efficiency in mi/gal of a car that consumes 14 km/L.

2. Write a user-defined MATLAB function for the following math function:
   $$y(x) = -0.2x^4 + e^{-0.5x}x^3 + 7x^2$$
   The input to the function is x and the output is y. Write the function such that x can be a vector (use element-by-element operations).
   (*a*) Use the function to calculate $y(-2.5)$, and $y(3)$.
   (*b*) Use the function to make a plot of the function $y(x)$ for $-3 \le x \le 4$.

3. Write a user-defined MATLAB function, with two input and two output arguments, that determines the height in centimeters and mass in kilograms of a person from his height in inches and weight in pounds. For the function name and arguments use `[cm, kg] = STtoSI(in, lb)`. The input arguments are the height in inches and weight in pounds, and the output arguments are the height in centimeters and mass in kilograms. Use the function in the Command Window to:
   (*a*) Determine in SI units the height and mass of a 5 ft 8 in. person who weighs 175 lb.
   (b) Determine your own height and weight in SI units.

4. Write a user-defined MATLAB function that converts speed given in units of miles per hour to speed in units of meters per second. For the function name and arguments use `mps = mphTOmets(mph)`. The input argument is the speed in mi/h, and the output argument is the speed in m/s. Use the function to convert 55 mi/h to units of m/s.

5. Write a user-defined MATLAB function for the following math function:

$$r(\theta) = 2\cos\theta\sin\theta\sin(\theta/4)$$

The input to the function is θ (in radians) and the output is r. Write the function such that θ can be a vector.
(*a*) Use the function to calculate $r(3\pi/4)$ and $r(7\pi/4)$.
(*b*) Use the function to plot (polar plot) $r(\theta)$ for $0 \le \theta \le 2\pi$.

6. Write a user-defined MATLAB function that determines the area of a triangle when the lengths of the sides are given. For the function name and arguments use `[Area] = triangle(a,b,c)`. Use the function to determine the areas of triangles with the following sides:
(*a*) $a = 3$, $b = 8$, $c = 10$. (*b*) $a = 7$, $b = 7$, $c = 5$.

7. A cylindrical vertical fuel tank has hemispheric end caps as shown. The radius of the cylinder and the caps is r = 15 in., and the height of the cylindrical middle section is 40 in.

Write a user-defined function (for the function name and arguments use `V = Volfuel(h)`) that gives the volume of fuel in the tank (in gallons) as a function of the height h (measured from the bottom). Use the function to make a plot of the volume as a function of h for $0 \le h \le 70$ in.

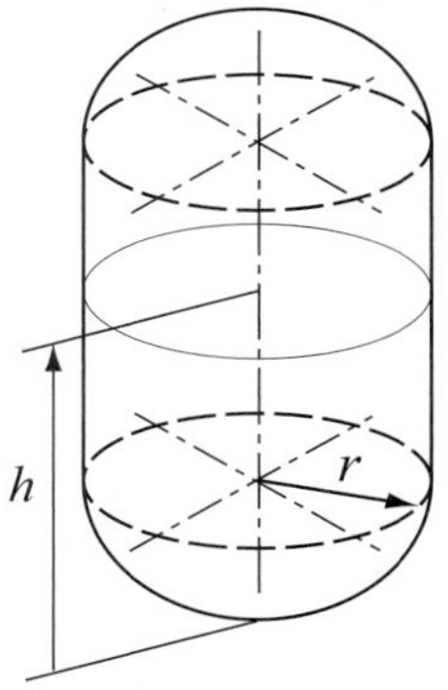

8. The surface area S of a ring in shape of a torus with an inner radius r and a diameter d is given by:

$$S = \pi^2(2r + d)d$$

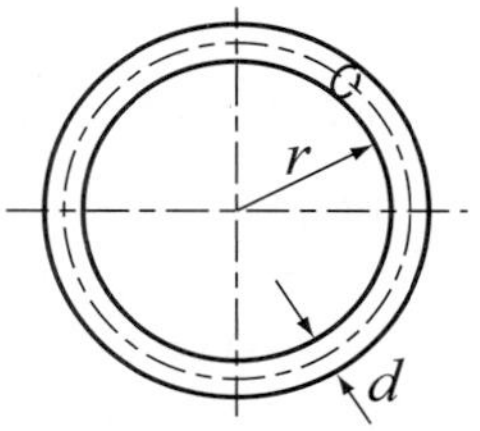

The ring is to be plated with a thin layer of coating. The weight of the coating W can be calculated approximately as $W = \gamma S t$, where γ is the specific weight of the coating material and t is its thickness. Write an anonymous function that calculates the weight of the coating. The function should have four input arguments, r, d, t, and γ. Use the anonymous function to calculate the weight of a gold coating (γ = 0.696 lb/in.^{3}) of a ring with r = 0.35 in., d = 0.12 in., and t = 0.002 in.

9. The monthly deposit into a savings account S needed to reach an investment goal B can be calculated by the formula

$$M = S\frac{\frac{r}{1200}}{\left(1+\frac{r}{1200}\right)^{12N}-1}$$

where M is the monthly deposit, S is the saving goal, N is the number of years, and r is the annual interest rate (%). Write a MATLAB user-defined function that calculates the monthly deposit into a savings account. For the function name and arguments use `M = invest(S,r,N)`. The input arguments are `S` (the investment goal), `r` (the annual interest rate, %), and `N` (duration of the savings in years). The output `M` is the amount of the monthly deposit. Use the function to calculate the monthly deposit for a 10-year investment if the investment goal is $25,000 and the annual interest rate is 4.25%.

10. The heat index, *HI* (in degrees F), is an apparent temperature. For temperatures higher than 80°F and humidity higher than 40% it is calculated by:

$$HI = C_1 + C_2T + C_3R + C_4TR + C_5T^2 + C_6R^2 + C_7T^2R + C_8TR^2 + C_9R^2T^2$$

where T is temperature in degrees F, R is the relative humidity in percent, $C_1 = -42.379$, $C_2 = 2.04901523$, $C_3 = 10.14333127$, $C_4 = -0.22475541$, $C_5 = -6.83783 \times 10^{-3}$, $C_6 = -5.481717 \times 10^{-2}$, $C_7 = 1.22874 \times 10^{-3}$, $C_8 = 8.5282 \times 10^{-4}$, and $C_9 = -1.99 \times 10^{-6}$. Write a user-defined function for calculating *HI* for given T and R. For the function name and arguments use `HI=HeatIn(T,R)`. The input arguments are `T` in °F and, `R` in %, and the output argument is `HI` in °F (rounded to the nearest integer). Use the function to determine the heat index for the following conditions:

(*a*) $T = 95$ °F, $R = 80$ %.

(*b*) $T = 100$ °F, $R = 100$ % (condition in a sauna).

11. The body fat percentage (*BFP*) of a person can be estimated by the formula

$$BFP = 1.2 \times BMI + 0.23 \times Age - 10.8 \times Gender - 0.54$$

where *BMI* is the body mass index, given by $BMI = 703\frac{W}{H^2}$, in which W is the weight in pounds and H is the height in inches, *Age* is the person's age, and $Gender = 1$ for a male and $Gender = 0$ for a female.

Write a MATLAB user-defined function that calculates the body fat percentage. For the function name and arguments use `BFP = Body-Fat(w,h,age,gen)`. The input arguments are the weight, height, age, and gender (1 for male, 0 for female), respectively. The output argument is the *BEF* value. Use the function to calculate the body fat percentage of:

a) A 35-years-old, 6 ft 2 in. tall, 220 lb male.

b) A 22-years-old, 5 ft 7 in. tall, 135 lb female.

12. Write a user-defined function that calculates grade point average (GPA) on a scale of 0 to 4, where $A = 4$, $B = 3$, $C = 3$, $D = 1$, and $E = 0$. For the function name and arguments use `av = GPA(g,h)`. The input argument `g` is a vector whose elements are letter grades *A*, *B*, *C*, *D*, or *E* entered as strings. The input argument `h` is a vector with the corresponding credit hours. The output argument `av` is the calculated GPA. Use the function to calculate the GPA for a student with the following record:

Grade	B	A	C	E	A	B	D	B
Credit Hours	3	4	3	4	3	4	3	2

For this case the input arguments are:
`g=['BACEABDB']` and `h=[3 4 3 4 3 4 3 2]`.

13. The factorial $n!$ of a positive number (integer) is defined by $n! = n \cdot (n-1) \cdot (n-2) \cdot \ldots \cdot 3 \cdot 2 \cdot 1$, where $0! = 1$. Write a user-defined function that calculates the factorial $n!$ of a number. For function name and arguments use `y=fact(x)`, where the input argument `x` is the number whose factorial is to be calculated, and the output argument `y` is the value $x!$. The function displays an error message if a negative or non-integer number is entered when the function is called. Use `fact` with the following numbers:
(*a*) 12! (*b*) 0! (*c*) –7! (*d*) 6.7!

14. Write a user-defined MATLAB function that determines the vector connecting two points (*A* and *B*). For the function name and arguments use `V=vector(A,B)`. The input arguments to the function are vectors `A` and `B`, each with the Cartesian coordinates of points *A* and *B*. The output `V` is the vector from point *A* to point *B*. If points *A* and *B* have two coordinates each (they are in the *x y* plane), then `V` is a two-element vector. If points *A* and *B* have three coordinates each (general points in space), then `V` is a three-element vector. Use the function `vector` for determining the following vectors.
(*a*) The vector from point (0.5, 1.8) to point (–3, 16).
(*b*) The vector from point (–8.4, 3.5, –2.2) to point (5, –4.6, 15).

15. Write a user-defined MATLAB function that determines the dot product of two vectors. For the function name and arguments use `D=dotpro(u,v)`. The input arguments to the function are two vectors, which can be two- or three-dimensional. The output `D` is the result (a scalar). Use the function `dotpro` for determining the dot product of:
(*a*) Vectors $a = 3i + 11j$ and $b = 14i - 7.3j$.
(*b*) Vectors $c = -6i + 14.2j + 3k$ and $d = 6.3i - 8j - 5.6k$.

16. Write a user-defined MATLAB function that determines the unit vector in the direction of the line that connects two points (A and B) in space. For the function name and arguments use `n = unitvec(A,B)`. The input to the function are two vectors `A` and `B`, each with the Cartesian coordinates of the corresponding point. The output is a vector with the components of the unit vector in the direction from A to B. If points A and B have two coordinates each (they are in the x y plane), then `n` is a two-element vector. If points A and B have three coordinate each (general points in space), then `n` is a three-element vector. Use the function to determine the following unit vectors:
    (*a*) In the direction from point (1.2, 3.5) to point (12, 15).
    (*b*) In the direction from point (–10, –4, 2.5) to point (–13, 6, –5).

17. Write a user-defined MATLAB function that determines the cross product of two vectors. For the function name and arguments use `w=crosspro(u,v)`. The input arguments to the function are the two vectors, which can be two- or three-dimensional. The output `w` is the result (a vector). Use the function `crisper` for determining the cross product of:
    (*a*) Vectors $a = 3i + 11j$ and $b = 14i - 7.3j$.
    (*b*) Vectors $c = -6i + 14.2j + 3k$ and $d = 6.3i - 8j - 5.6k$.

18. The area of a triangle ABC can be calculated by:

$$A = \frac{1}{2}|\boldsymbol{AB} \times \boldsymbol{AC}|$$

where $\boldsymbol{AB}$ is the vector from point A to point B and $\boldsymbol{AC}$ is the vector from point A to point C. Write a user-defined MATLAB function that determines the area of a triangle given its vertices' coordinates. For the function name and arguments use `[Area] = TriArea(A,B,C)`. The input arguments `A`, `B`, and `C`, are vectors, each with the coordinates of the corresponding vertex. Write the code of `TriArea` such that it has two subfunctions—one that determines the vectors $\boldsymbol{AB}$ and $\boldsymbol{AC}$ and an other that executes the cross product. (If available, use the user-defined functions from Problems 15 and 17. The function should work for a triangle in the x y plane (each vertex is defined by two coordinates) or for a triangle in space (each vertex is defined by three coordinates). Use the function to determine the areas of triangles with the following vertices:
    (*a*) $A = (1, 2)$, $B = (10, 3)$, $C = (6, 11)$.
    (*b*) $A = (-1.5, -4.2, -3)$, $B = (-5.1, 6.3, 2)$, $C = (12.1, 0, -0.5)$.

19. Write a user-defined function that plots a circle given the coordinates of the center and the radius. For the function name and arguments use `circleplot(x,y,R)`. The input arguments are the x and y coordinates of the center and the radius. This function has no output arguments. Use the function to plot the following circles:
    (*a*) $x = 3.5$, $y = 2.0$, $R = 8.5$. (*b*) $x = -4.0$, $y = -1.5$, $R = 10$.

20. Write a user-defined function that plots a circle that passes through three given points. For the function name and arguments use `cirpnts(P)`. The input arguments is a 3×2 matrix in which the two elements of a row are the x and y coordinates of one point. This function has no output arguments. The figure that is created by the function displays the circle and the three points marked with asterisks. Use the function to plot a circle that passes through the points (6, 1.5), (2, 4), (–3, –1.8).

21. In polar coordinates a two-dimensional vector is given by its radius and angle (r, θ). Write a user-defined MATLAB function that adds two vectors that are given in polar coordinates. For the function name and arguments use `[r th]= AddVecPol(r1,th1,r2,th2)`, where the input arguments are (r_1, θ_1) and (r_2, θ_2), and the output arguments are the radius and angle of the result. Use the function to carry out the following additions:

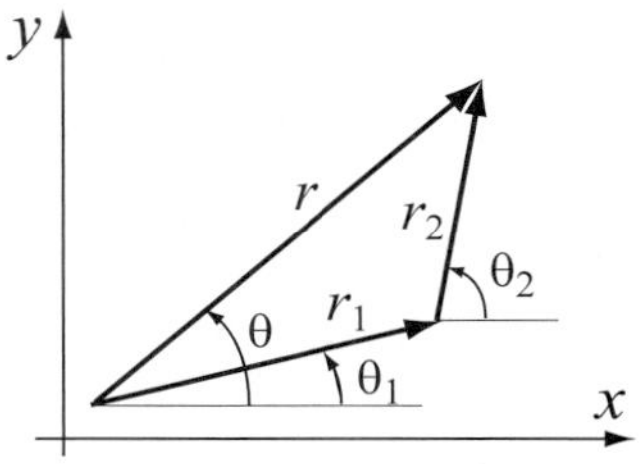

(*a*) $r_1 = (5, 23°)$, $r_2 = (12, 40°)$.   (*b*) $r_1 = (6, 80°)$, $r_2 = (15, 125°)$.

22. Write a user-defined function that plots an ellipse with axes that are parallel to the x and y axes, given the coordinates of its center and the length of the axes. For the function name and arguments use `ellipseplot(xc,yc,a,b)`. The input arguments `xc` and `yc` are the coordinates of the center, and `a` and `b` are half the lengths of the horizontal and vertical axes (see figure), respectively. This function has no output arguments. Use the function to plot the following ellipses:

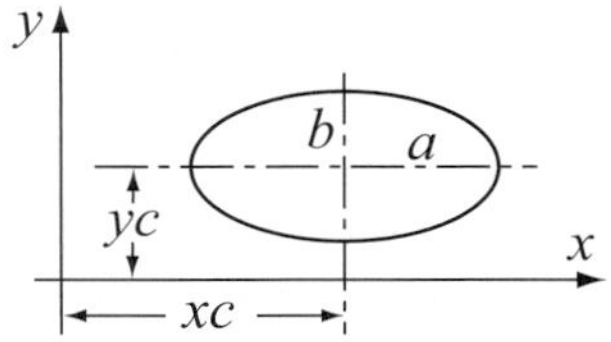

(*a*) $xc = 3.5$, $yc = 2.0$, $a = 8.5$, $b = 3$.

(*b*) $xc = -5$, $yc = 1.5$, $a = 4$, $b = 8$.

23. Write a user-defined function that finds all the prime numbers between two numbers m and n. Name the function `pr=prime(m,n)`, where the input arguments `m` and `n` are positive integers, and the output argument `pr` is a vector with the prime numbers. If $m > n$ is entered when the function is called, the error message "The value of n must be larger than the value of m." is displayed. If a negative number or a number that is not an integer is entered when the function is called, the error message "The input argument must be a positive integer." is displayed. Use the function with:

(*a*) `prime(12,80)`   (*b*) `prime(21,63.5)`

(*c*) `prime(100,200)`   (*d*) `prime(90,50)`

24. The geometric mean *GM* of a set of *n* positive numbers $x_1, x_2, \ldots, x_n$ is defined by:

$$GM = (x_1 \cdot x_2 \cdot \ldots \cdot x_n)^{1/n}$$

Write a user-defined function that calculates the geometric mean of a set of numbers. For function name and arguments use `GM=Geomean(x)`, where the input argument `x` is a vector of numbers (any length) and the output argument `GM` is their geometric mean. The geometric mean is useful for calculating the average return of a stock. The following table gives the returns for IBM stock over the last ten years (a return of 16% means 1.16). Use the user-defined function `Geomean` to calculate the average return of the stock.

Year	1997	1998	1999	2000	2001	2002	2003	2004	2005	2006
Return	1.38	1.76	1.17	0.79	1.42	0.64	1.2	1.06	0.83	1.18

25. Write a user-defined function that determines the polar coordinates of a point from the Cartesian coordinates in a two-dimensional plane. For the function name and arguments use `[th rad]=CartToPolar(x,y)`. The input arguments are the *x* and *y* coordinates of the point, and the output arguments are the angle θ and the radial distance to the point. The angle θ is in degrees and is measured relative to the positive *x* axis, such that it is a positive number in quadrants I and II, and a negative number in quadrant III and IV. Use the function to determine the polar coordinates of points (14, 9), (–11, –20), (–15, 4), and (13.5, –23.5).

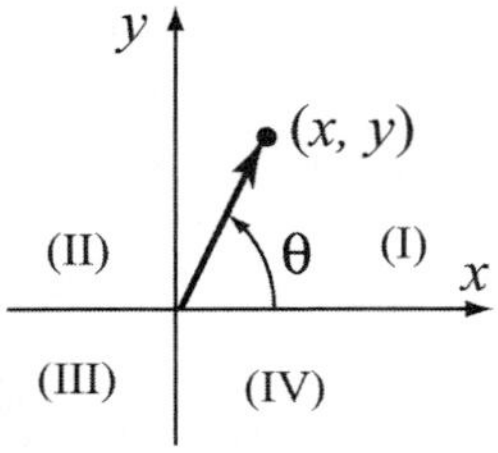

26. Write a user-defined function that sorts the elements of a vector from the largest to the smallest. For the function name and arguments use `y=downsort(x)`. The input to the function is a vector `x` of any length, and the output `y` is a vector in which the elements of `x` are arranged in a descending order. Do not use the MATLAB built-in function `sort`, `max`, or `min`. Test your function on a vector with 14 numbers (integers) randomly distributed between –30 and 30. Use the MATLAB `randi` function to generate the initial vector.

27. Write a user-defined function that sorts the elements of a matrix. For the function name and arguments use `B = matrixsort(A)`, where `A` is any size matrix and `B` is a matrix of the same size with the elements of `A` rearranged in descending order row after row with the (1,1) element the largest and the (*m*,*n*) element the smallest. If available, use the user-defined function `downsort` from Problem 26 as a subfunction within `matrixsort`.

Test your function on a 4×7 matrix with elements (integers) randomly distributed between –30 and 30. Use MATLAB's `randi` function to generate the initial matrix.

28. Write a user-defined MATLAB function that calculates the determinant of a 3×3 matrix by using the formula:

$$det = A_{11}\begin{vmatrix} A_{22} & A_{23} \\ A_{32} & A_{33} \end{vmatrix} - A_{12}\begin{vmatrix} A_{21} & A_{23} \\ A_{31} & A_{33} \end{vmatrix} + A_{13}\begin{vmatrix} A_{21} & A_{22} \\ A_{31} & A_{32} \end{vmatrix}$$

For the function name and arguments use `d3 = det3by3(A)`, where the input argument `A` is the matrix and the output argument `d3` is the value of the determinant. Write the code of `det3by3` such that it has a subfunction that calculates the 2×2 determinant. Use `det3by3` for calculating the determinants of:

(a) $\begin{bmatrix} 1 & 3 & 2 \\ 6 & 5 & 4 \\ 7 & 8 & 9 \end{bmatrix}$ (b) $\begin{bmatrix} -2.5 & 7 & 1 \\ 5 & -3 & -2.6 \\ 4 & 2 & -1 \end{bmatrix}$

29. A two-dimensional state of stress at a point in a loaded material is defined by three components of stress σ_{xx}, σ_{yy}, and τ_{xy}. The maximum and minimum normal stresses (principal stresses) at the point, σ_{max} and σ_{min}, are calculated from the stress components by:

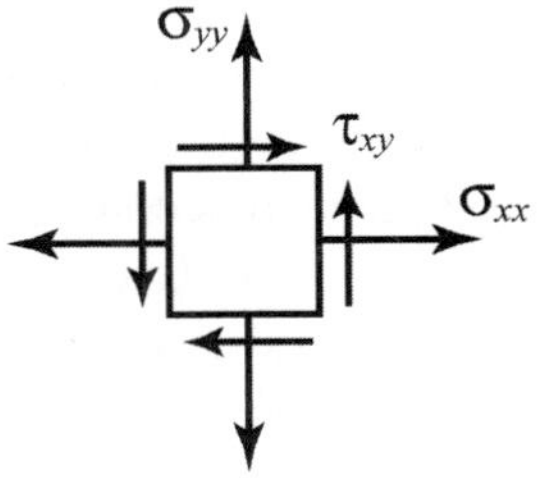

$$\sigma_{\substack{max \\ min}} = \frac{\sigma_{xx} + \sigma_{yy}}{2} \pm \sqrt{\left(\frac{\sigma_{xx} - \sigma_{yy}}{2}\right)^2 + \tau_{xy}^2}$$

Write a user-defined MATLAB function that determines the principal stresses from the stress components. For the function name and arguments use `[Smax,Smin] = princstress(Sxx,Syy,Sxy)`. The input arguments are the three stress components, and the output arguments are the maximum and minimum stresses.

Use the function to determine the principal stresses for the following states of stress:

(a) $\sigma_{xx} = -190$ MPa, $\sigma_{yy} = 145$ MPa, and $\tau_{xy} = 110$ MPa.

(b) $\sigma_{xx} = 14$ ksi, $\sigma_{yy} = -15$ ksi, and $\tau_{xy} = 8$ ksi.

30. The dew point temperature T_d and the relative humidity RH can be calculated (approximately) from the dry-bulb T and wet-bulb T_w temperatures by (http://www.wikipedia.org)

$$e_s = 6.112\exp\left(\frac{17.67T}{T + 243.5}\right) \qquad e_w = 6.112\exp\left(\frac{17.67T_w}{T_w + 243.5}\right)$$

$$e = e_w - p_{sta}(T - T_w)0.00066(1 + 0.00115T_w)$$

$$RH = 100\frac{e}{e_s} \quad T_d = \frac{243.5\ln(e/6.112)}{17.67 - \ln(e/6.112)}$$

where the temperatures are in degrees Celsius, RH is in %, and p_{sta} is the barometric pressure in units of millibars.

Write a user-defined MATLAB function that calculates the dew point temperature and relative humidity for given dry-bulb and wet-bulb temperatures and barometric pressure. For the function name and arguments use `[Td,RH] = DewptRhum(T,Tw,BP)`, where the input arguments are T, T_w and p_{sta}, and the output arguments are T_d and RH. The values of the output arguments should be rounded to the nearest tenth. Use the user-defined function `dewpoint` for calculating the dew point temperature and relative humidity for the following cases:

(*a*) T = 25 °C, T_w = 19 °C, p_{sta} = 985 mbar.

(*b*) T = 36 °C, T_w = 31 °C, p_{sta} = 1020 mbar.

31. Write a user-defined MATLAB function that calculates a student's final grade in a course using the scores from three midterm exams, a final exam, and six homework assignments. The midterms are graded on a scale from 0 to 100 and each accounts for 15% of the course grade. The final exam is graded on a scale from 0 to 100 and accounts for 40% of the course grade. The six homework assignments are each graded on a scale from 0 to 10. The homework assignment with the lowest grade is dropped, and the average of the remaining assignments accounts for 15% of the course grade. In addition, the following adjustment is made when the grade is calculated. If the average grade for the three midterms is higher than the grade for the final exam, then the grade of the final exam is not used and the average grade of the three midterms accounts for 85% of the course grade. The program calculates a course grade that is a number between 0 and 100.

For the function name and arguments use `g = fgrade(R)`. The input argument `R` is a matrix in which the elements in each row are the grades of one student. The first six columns are the homework grades (numbers between 0 and 10), the next three columns are the midterm grades (numbers between 0 and 100), and the last column is the final exam grade (a number between 0 and 100). The output from the function, `g`, is a column vector with the student grades for the course. Each row has the course grade of the student with the grades in the corresponding row of the matrix `R`.

The function can be used to calculate the grades of any number of students. For one student the matrix `R` has one row. Use the function for the following cases:

(*a*) Use the Command Window to calculate the course grade of one student with the following grades: 8, 9, 6, 10, 9, 7, 76, 86, 91, 80.

(b) Write a program in a script file. The program asks the user to enter the students' grades in an array (one student per row). The program then calculates the course grades by using the function `fgrade`. Run the script file in the Command Window to calculate the grades of the following four students:

Student A: 7, 10, 6, 9, 10, 9, 91, 71, 81, 88.
Student B: 5, 5, 6, 1, 8, 6, 59, 72, 66, 59.
Student C: 6, 8, 10, 4, 5, 9, 72, 78, 84 78.
Student D: 7, 7, 8, 8, 9, 8, 83, 82, 81 84.

32. In a lottery the player has to select several numbers out of a list. Write a MATLAB program that generates a list of n integers that are uniformly distributed between the numbers a and b. All the selected numbers on the list must be different.

(*a*) Use the function to generate a list of seven numbers from the numbers 1 through 59.

(*b*) Use the function to generate a list of eight numbers from the numbers 50 through 65.

(*c*) Use the function to generate a list of nine numbers from the numbers –25 through –2.

33. The solution of the nonlinear equation $x^5 - P = 0$ gives the fifth root of the number P. A numerical solution of the equation can be calculated with Newton's method. The solution process starts by choosing a value x_1 as a first estimate of the solution. Using this value, a second, more accurate solution x_2 can be calculated with $x_2 = x_1 - \frac{x_1^5 - P}{5x_1^4}$, which is then used for calculating a third, still more accurate solution x_3, and so on. The general equation for calculating the value of the solution x_{i+1} from the solution x_i is $x_{i+1} = x_i - \frac{x_i^5 - P}{5x_i^4}$. Write a user-defined function that calculates the fifth root of a number. For function name and arguments use `y=fifthroot(P)`, where the input argument `P` is the number whose fifth root is to be determined, and the output argument `y` is the value $\sqrt[5]{P}$. In the program use $x = P$ for the first estimate of the solution. Then, by using the general equation in a loop, calculate new, more accurate solutions. Stop the looping when the estimated relative error E defined by $E = \left|\frac{x_{i+1} - x_i}{x_i}\right|$ is smaller than 0.00001. Use the function `cubic` to calculate:

(*a*) $\sqrt[5]{120}$ (*b*) $\sqrt[5]{16807}$ (*c*) $\sqrt[5]{-15}$

34. Write a user-defined function that determines the coordinate y_c of the centroid of the T-shaped cross-sectional area shown in the figure. For the function name and arguments use `yc = centroidT(w,h,t,d)`, where the input arguments `w`, `h`, `t` and `d`, are the dimensions shown in the figure, and the output argument `yc` is the coordinate y_c.

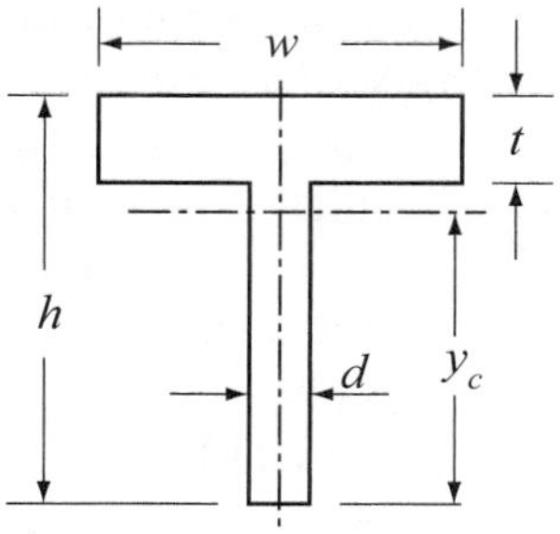

Use the function to determine y_c for an area with $w = 240$ mm, $h = 380$ mm, $d = 42$ mm, and $t = 60$ mm.

35. The area moment of inertia I_{x_o} of a rectangle about the axis x_o passing through its centroid is $I_{x_o} = \frac{1}{12}bh^3$. The moment of inertia about an axis x that is parallel to x_o is given by $I_x = I_{x_o} + Ad_x^2$, where A is the area of the rectangle, and d_x is the distance between the two axes.

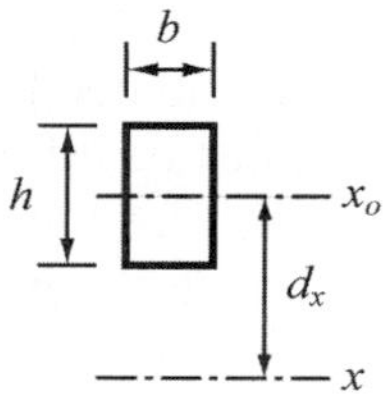

Write a MATLAB user-defined function that determines the area moment of inertia I_{x_c} of a "T" beam about the axis that passes through its centroid (see drawing). For the function name and arguments use `Ixc = IxcTBeam(w,h,t,d)`, where the input arguments `w`, `h`, `t`, and `d` are the dimensions shown in the figure, and the output argument `Ixc` is I_{x_c}. For finding the coordinate y_c of the of the centroid use the user-defined function `centroidT` from Problem 34 as a subfunction inside `IxcTBeam`.

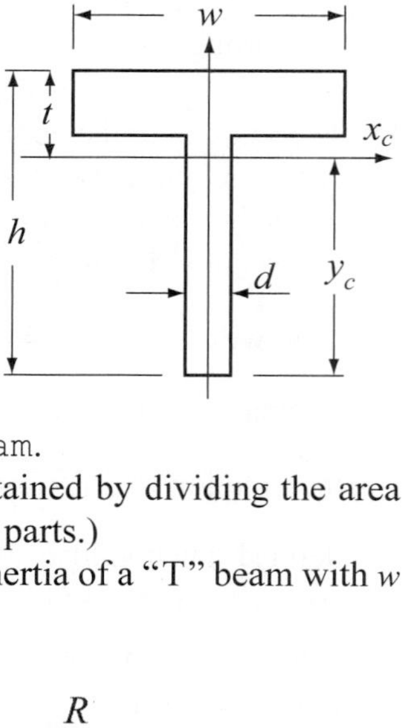

(The moment of inertia of a composite area is obtained by dividing the area into parts and adding the moments of inertia of the parts.)

Use the function to determine the moment of inertia of a "T" beam with $w = 240$ mm, $h = 380$ mm, $d = 42$ mm, and $t = 60$ mm.

36. In a low-pass RC filter (a filter that passes signals with low frequencies), the ratio of the magnitudes of the voltages is given by:

$$RV = \left|\frac{V_o}{V_i}\right| = \frac{1}{\sqrt{1 + (\omega RC)^2}}$$

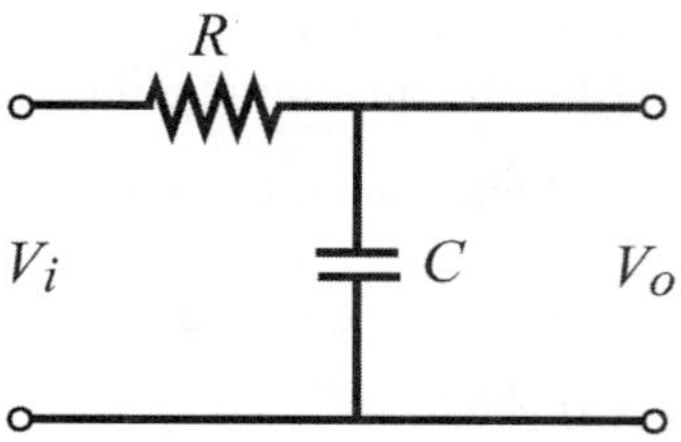

where ω is the frequency of the input signal.

Write a user-defined MATLAB function that calculates the magnitude ratio. For the function name and arguments use `RV = lowpass(R,C,w)`. The input arguments are `R`, the size of the resistor in Ω (ohms); `C`, the size of the capacitor in F (farads); and `w`, the frequency of the input signal in rad/s. Write the function such that `w` can be a vector.

Write a program in a script file that uses the `lowpass` function to generate a plot of *RV* as a function of ω for $10^{-2} \le \omega \le 10^{6}$ rad/s. The plot has a logarithmic scale on the horizontal axis (ω). When the script file is executed, it asks the user to enter the values of *R* and *C*. Label the axes of the plot.

Run the script file with $R = 1200\,\Omega$, and $C = 8\,\mu\text{F}$.

37. A bandpass filter passes signals with frequencies that are within a certain range. In this filter the ratio of the magnitudes of the voltages is given by

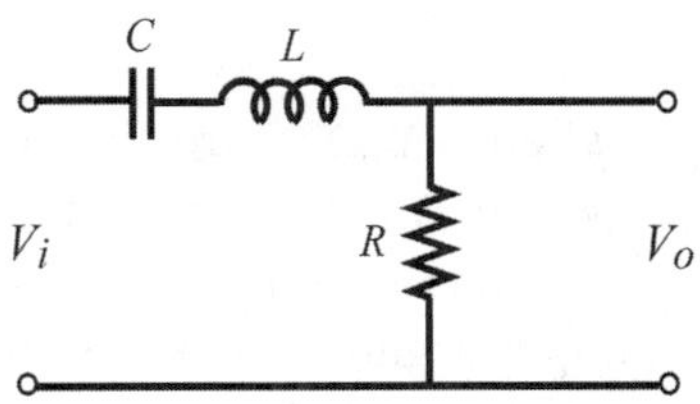

$$RV = \left|\frac{V_o}{V_i}\right| = \frac{\omega RC}{\sqrt{(1-\omega^2 LC)^2 + (\omega RC)^2}}$$

where ω is the frequency of the input signal.

Write a user-defined MATLAB function that calculates the magnitude ratio. For the function name and arguments use `RV = bandpass(R,C,L,w)`. The input arguments are `R` the size of the resistor in Ω (ohms); `C`, the size of the capacitor in F (farads); `L`, the inductance of the coil in H (henrys); and `w`, the frequency of the input signal in rad/s. Write the function such that `w` can be a vector.

Write a program in a script file that uses the `bandpass` function to generate a plot of *RV* as a function of ω for $10^{-2} \le \omega \le 10^{7}$ rad/s. The plot has a logarithmic scale on the horizontal axis (ω). When the script file is executed, it asks the user to enter the values of *R, L,* and *C*. Label the axes of the plot.

Run the script file for the following two cases:

(*a*) $R = 1100\,\Omega$, $C = 9\ \mu\text{F}$, $L = 7$ mH.

(*b*) $R = 500\,\Omega$, $C = 300\ \mu\text{F}$, $L = 400$ mH.

38. The first derivative $\frac{df(x)}{dx}$ of a function $f(x)$ at $x = x_0$ can be approximated with the four-point central difference formula

$$\frac{df(x)}{dx} = \frac{f(x_0 - 2h) - f(x_0 - h) + f(x_0 + h) - f(x_0 + 2h)}{12h}$$

where h is a small number relative to x_0. Write a user-defined function function (see Section 7.9) that calculates the derivative of a math function $f(x)$ by using the four-point central difference formula. For the user-defined function name use `dfdx=Funder(Fun,x0)`, where `Fun` is a name for the function that is passed into `Funder`, and `x0` is the point where the derivative is calculated. Use $h = x_0/10$ in the four-point central difference formula. Use the user-defined function `Funder` to calculate the following:

(a) The derivative of $f(x) = x^2e^x$ at $x_0 = 0.25$.

(b) The derivative of $f(x) = \frac{2^x}{x}$ at $x_0 = 2$.

In both cases compare the answer obtained from `Funder` with the analytical solution (use format long).

39. The new coordinates (X_r, Y_r) of a point in the x y plane that is rotated about the z axis at an angle θ (positive is clockwise) are given by

$$X_r = X_0\cos\theta - Y_0\sin\theta$$
$$Y_r = X_0\sin\theta + Y_0\cos\theta$$

where (X_0, Y_0) are the coordinates of the point before the rotation. Write a user-defined function that calculates (X_r, Y_r) given (X_0, Y_0) and θ. For function name and arguments use `[xr,yr]=rotation(x,y,q)`, where the input arguments are the initial coordinates and the rotation angle in degrees, and the output arguments are the new coordinates.

(a) Use `rotation` to determine the new coordinates of a point originally at $(6.5, 2.1)$ that is rotated about the z-axis by $25°$.

(b) Consider the function $y = (x-7)^2 + 1.5$ for $5 \le x \le 9$. Write a program in a script file that makes a plot of the function. Then use `rotation` to rotate all the points that make up the first plot and make a plot of the rotated function. Make both plots in the same figure and set the range of both axes at 0 to 10.

Chapter 8
Polynomials, Curve Fitting, and Interpolation

Polynomials are mathematical expressions that are frequently used for problem solving and modeling in science and engineering. In many cases an equation that is written in the process of solving a problem is a polynomial, and the solution of the problem is the zero of the polynomial. MATLAB has a wide selection of functions that are specifically designed for handling polynomials. How to use polynomials in MATLAB is described in Section 8.1.

Curve fitting is a process of finding a function that can be used to model data. The function does not necessarily pass through any of the points, but models the data with the smallest possible error. There are no limitations to the type of the equations that can be used for curve fitting. Often, however, polynomial, exponential, and power functions are used. In MATLAB curve fitting can be done by writing a program, or by interactively analyzing data that is displayed in the Figure Window. Section 8.2 describes how to use MATLAB programming for curve fitting with polynomials and other functions. Section 8.4 describes the basic fitting interface that is used for interactive curve fitting and interpolation.

Interpolation is the process of estimating values between data points. The simplest kind of interpolation is done by drawing a straight line between the points. In a more sophisticated interpolation, data from additional points is used. How to interpolate with MATLAB is discussed in Sections 8.3 and 8.4.

8.1 POLYNOMIALS

Polynomials are functions that have the form:

$$f(x) = a_n x^n + a_{n-1} x^{n-1} + \ldots + a_1 x + a_0$$

The coefficients $a_n, a_{n-1}, \ldots, a_1, a_0$ are real numbers, and n, which is a nonnegative integer, is the degree, or order, of the polynomial.

Examples of polynomials are:

$f(x) = 5x^5 + 6x^2 + 7x + 3$ polynomial of degree 5.

$f(x) = 2x^2 - 4x + 10$ polynomial of degree 2.

$f(x) = 11x - 5$ polynomial of degree 1.

A constant (e.g., $f(x) = 6$) is a polynomial of degree 0.

In MATLAB, polynomials are represented by a row vector in which the elements are the coefficients $a_n, a_{n-1}, \ldots, a_1, a_0$. The first element is the coefficient of the x with the highest power. The vector has to include all the coefficients, including the ones that are equal to 0. For example:

Polynomial	MATLAB representation
$8x + 5$	p = [8 5]
$2x^2 - 4x + 10$	d = [2 –4 10]
$6x^2 - 150$, MATLAB form: $6x^2 + 0x - 150$	h = [6 0 –150]
$5x^5 + 6x^2 - 7x$, MATLAB form: $5x^5 + 0x^4 + 0x^3 + 6x^2 - 7x + 0$	c = [5 0 0 6 –7 0]

8.1.1 *Value of a Polynomial*

The value of a polynomial at a point x can be calculated with the function `polyval` which has the form:

`polyval(p,x)`

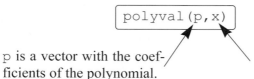

`p` is a vector with the coefficients of the polynomial.

`x` is a number, or a variable that has an assigned value, or a computable expression.

`x` can also be a vector or a matrix. In such a case the polynomial is calculated for each element (element-by-element), and the answer is a vector, or a matrix, with the corresponding values of the polynomial.

Sample Problem 8-1: Calculating polynomials with MATLAB

For the polynomial $f(x) = x^5 - 12.1x^4 + 40.59x^3 - 17.015x^2 - 71.95x + 35.88$:

(*a*) Calculate $f(9)$.

(*b*) Plot the polynomial for $-1.5 \le x \le 6.7$.

Solution

The problem is solved in the Command Window.

(*a*) The coefficients of the polynomials are assigned to vector `p`. The function polyval is then used to calculate the value at $x = 9$.

```
>> p = [1 -12.1 40.59 -17.015 -71.95 35.88];
>> polyval(p,9)
ans =
   7.2611e+003
```

(*b*) To plot the polynomial, a vector x is first defined with elements ranging from –1.5 to 6.7. Then a vector y is created with the values of the polynomial for every element of x. Finally, a plot of y vs. x is made.

```
>> x=-1.5:0.1:6.7;
>> y=polyval(p,x);
>> plot(x,y)
```

Calculating the value of the polynomial for each element of the vector x.

The plot created by MATLAB is presented below (axis labels were added with the Plot Editor).

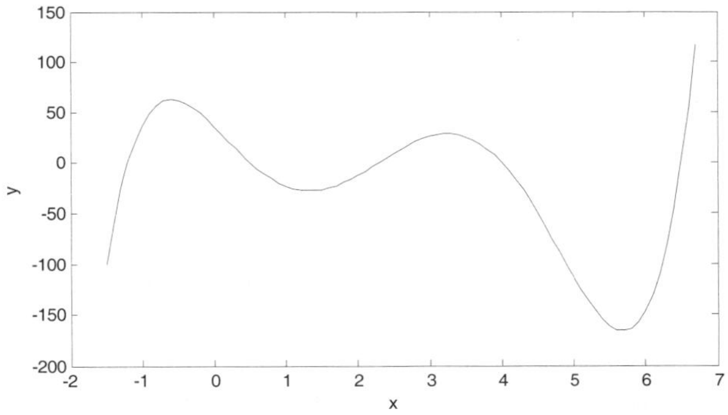

8.1.2 Roots of a Polynomial

The roots of a polynomial are the values of the argument for which the value of the polynomial is equal to zero. For example, the roots of the polynomial $f(x) = x^2 - 2x - 3$ are the values of x for which $x^2 - 2x - 3 = 0$, which are $x = -1$ and $x = 3$.

MATLAB has a function, called roots, that determines the root, or roots, of a polynomial. The form of the function is:

```
r = roots(p)
```

r is a column vector with the roots of the polynomial.

p is a row vector with the coefficients of the polynomial.

For example, the roots of the polynomial in Sample Problem 8-1 can be determined by:

```
>> p= 1 -12.1 40.59 -17.015 -71.95 35.88];
>> r=roots(p)
r =
    6.5000
    4.0000
    2.3000
   -1.2000
    0.5000
```

When the roots are known, the polynomial can actually be written as:
$f(x) = (x+1.2)(x-0.5)(x-2.3)(x-4)(x-6.5)$

The `roots` command is very useful for finding the roots of a quadratic equation. For example, to find the roots of $f(x) = 4x^2 + 10x - 8$, type:

```
>> roots([4 10 -8])
ans =
   -3.1375
    0.6375
```

When the roots of a polynomial are known, the `poly` command can be used for determining the coefficients of the polynomial. The form of the `poly` command is:

```
p = poly(r)
```

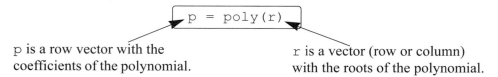

`p` is a row vector with the coefficients of the polynomial.

`r` is a vector (row or column) with the roots of the polynomial.

For example, the coefficients of the polynomial in Sample Problem 8-1 can be obtained from the roots of the polynomial (see above) by:

```
>> r=6.5 4 2.3 -1.2 0.5];
>> p=poly(r)
p =
   1.0000  -12.1000   40.5900  -17.0150  -71.9500   35.8800
```

8.1.3 Addition, Multiplication, and Division of Polynomials

Addition:

Two polynomials can be added (or subtracted) by adding (subtracting) the vectors of the coefficients. If the polynomials are not of the same order (which means that the vectors of the coefficients are not of the same length), the shorter vector has to be modified to be of the same length as the longer vector by adding zeros (called padding) in front. For example, the polynomials

$f_1(x) = 3x^6 + 15x^5 - 10x^3 - 3x^2 + 15x - 40$ and $f_2(x) = 3x^3 - 2x - 6$ can be added by:

```
>> p1=[3 15 0 -10 -3 15 -40];
>> p2=[3 0 -2 -6];
>> p=p1+[0 0 0 p2]
p =
     3    15     0    -7    -3    13   -46
```

Three 0s are added in front of p2, since the order of p1 is 6 and the order of p2 is 3.

Multiplication:

Two polynomials can be multiplied using the MATLAB built-in function conv, which has the form:

```
c = conv(a,b)
```

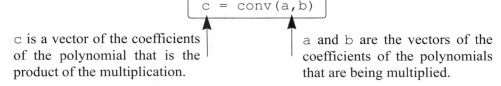

c is a vector of the coefficients of the polynomial that is the product of the multiplication.

a and b are the vectors of the coefficients of the polynomials that are being multiplied.

- The two polynomials do not have to be of the same order.
- Multiplication of three or more polynomials is done by using the conv function repeatedly.

For example, multiplication of the polynomials $f_1(x)$ and $f_2(x)$ above gives:

```
>> pm=conv(p1,p2)
pm =
     9    45    -6   -78   -99    65   -54   -12   -10   240
```

which means that the answer is:

$9x^9 + 45x^8 - 6x^7 - 78x^6 - 99x^5 + 65x^4 - 54x^3 - 12x^2 - 10x + 240$

Division:

A polynomial can be divided by another polynomial with the MATLAB built-in function deconv, which has the form:

```
[q,r] = deconv(u,v)
```

q is a vector with the coefficients of the quotient polynomial.
r is a vector with the coefficients of the remainder polynomial.

u is a vector with the coefficients of the numerator polynomial.
v is a vector with the coefficients of the denominator polynomial.

For example, dividing $2x^3 + 9x^2 + 7x - 6$ by $x + 3$ is done by:

```
>> u=[2 9 7 -6];
>> v=[1 3];
```

```
>> [a b]=deconv(u,v)
a =
     2     3    -2
b =
     0     0     0     0
```

The answer is: $2x^2 + 3x - 2$.

Remainder is zero.

An example of division that gives a remainder is $2x^6 - 13x^5 + 75x^3 + 2x^2 - 60$ divided by $x^2 - 5$:

```
>> w=[2 -13 0 75 2 0 -60];
>> z=[1 0 -5];
>> [g h]=deconv(w,z)
g =
    2   -13    10    10    52
h =
    0     0     0     0     0    50   200
```

The quotient is: $2x^4 - 13x^3 + 10x^2 + 10x + 52$.

The remainder is: $50x + 200$.

The answer is: $2x^4 - 13x^3 + 10x^2 + 10x + 52 + \frac{50x + 200}{x^2 - 5}$.

8.1.4 Derivatives of Polynomials

The built-in function `polyder` can be used to calculate the derivative of a single polynomial, a product of two polynomials, or a quotient of two polynomials, as shown in the following three commands.

`k = polyder(p)` Derivative of a single polynomial. `p` is a vector with the coefficients of the polynomial. `k` is a vector with the coefficients of the polynomial that is the derivative.

`k = polyder(a,b)` Derivative of a product of two polynomials. `a` and `b` are vectors with the coefficients of the polynomials that are multiplied. `k` is a vector with the coefficients of the polynomial that is the derivative of the product.

`[n d] = polyder(u,v)` Derivative of a quotient of two polynomials. `u` and `v` are vectors with the coefficients of the numerator and denominator polynomials. `n` and `d` are vectors with the coefficients of the numerator and denominator polynomials in the quotient that is the derivative.

The only difference between the last two commands is the number of output arguments. With two output arguments MATLAB calculates the derivative of the quotient of two polynomials. With one output argument the derivative is of the product.

For example, if $f_1(x) = 3x^2 - 2x + 4$, and $f_2(x) = x^2 + 5$, the derivatives of $3x^2 - 2x + 4$, $(3x^2 - 2x + 4)(x^2 + 5)$, and $\frac{3x^2 - 2x + 4}{x^2 + 5}$ can be determined by:

```
>> f1= 3 -2 4];
>> f2=[1 0 5];
>> k=polyder(f1)
k =
     6    -2
>> d=polyder(f1,f2)
d =
    12    -6    38   -10
>> [n d]=polyder(f1,f2)
n =
     2    22   -10
d =
     1     0    10     0    25
```

Creating the vectors of coefficients of f_1 and f_2.

The derivative of f_1 is: $6x - 2$.

The derivative of $f_1 * f_2$ is: $12x^3 - 6x^2 + 38x - 10$.

The derivative of $\frac{3x^2 - 2x + 4}{x^2 + 5}$ is: $\frac{2x^2 + 22x - 10}{x^4 + 10x^2 + 25}$.

8.2 CURVE FITTING

Curve fitting, also called regression analysis, is a process of fitting a function to a set of data points. The function can then be used as a mathematical model of the data. Since there are many types of functions (linear, polynomial, power, exponential, etc.), curve fitting can be a complicated process. Many times one has some idea of the type of function that might fit the given data and will need only to determine the coefficients of the function. In other situations, where nothing is known about the data, it is possible to make different types of plots that provide information about possible forms of functions that might fit the data well. This section describes some of the basic techniques for curve fitting and the tools that MATLAB has for this purpose.

8.2.1 Curve Fitting with Polynomials; The `polyfit` Function

Polynomials can be used to fit data points in two ways. In one the polynomial passes through all the data points, and in the other the polynomial does not necessarily pass through any of the points, but overall gives a good approximation of the data. The two options are described below.

Polynomials that pass through all the points:

When n points (x_i, y_i) are given, it is possible to write a polynomial of degree $n - 1$ that passes through all the points. For example, if two points are given it is possible to write a linear equation in the form of $y = mx + b$ that passes through the points. With three points the equation has the form of $y = ax^2 + bx + c$. With n

points the polynomial has the form $a_{n-1}x^{n-1} + a_{n-2}x^{n-2} + \ldots + a_1x + a_0$. The coefficients of the polynomial are determined by substituting each point in the polynomial and then solving the n equations for the coefficients. As will be shown later in this section, polynomials of high degree might give a large error if they are used to estimate values between data points.

Polynomials that do not necessarily pass through any of the points:

When n points are given, it is possible to write a polynomial of degree less than $n-1$ that does not necessarily pass through any of the points, but overall approximates the data. The most common method of finding the best fit to data points is the method of least squares. In this method the coefficients of the polynomial are determined by minimizing the sum of the squares of the residuals at all the data points. The residual at each point is defined as the difference between the value of the polynomial and the value of the data. For example, consider the case of finding the equation of a straight line that best fits four data points as shown in Figure 8-1. The points are (x_1,y_1), (x_2,y_2), (x_3,y_3), and (x_4,y_4), and the polynomial of the

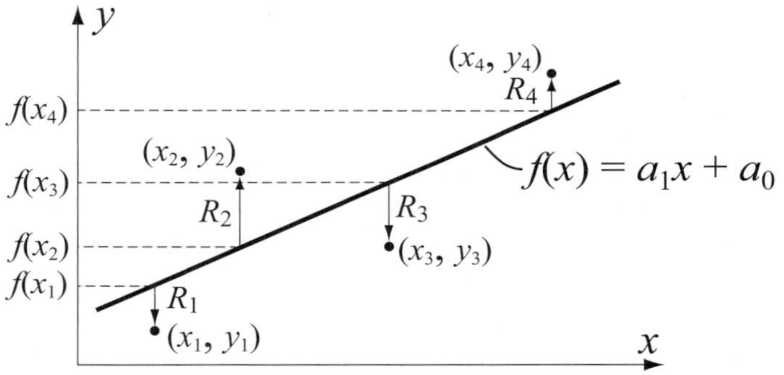

Figure 8-1: Least squares fitting of first-degree polynomial to four points.

first degree can be written as $f(x) = a_1x + a_0$. The residual, R_i, at each point is the difference between the value of the function at x_i and y_i, $R_i = f(x_i) - y_i$. An equation for the sum of the squares of the residuals R_i of all the points is given by

$$R = [f(x_1) - y_1]^2 + [f(x_2) - y_2]^2 + [f(x_3) - y_3]^2 + [f(x_4) - y_4]^2$$

or, after substituting the equation of the polynomial at each point, by:

$$R = [a_1x_1 + a_0 - y_1]^2 + [a_1x_2 + a_0 - y_2]^2 + [a_1x_3 + a_0 - y_3]^2 + [a_1x_4 + a_0 - y_4]^2$$

At this stage R is a function of a_1 and a_0. The minimum of R can be determined by taking the partial derivative of R with respect to a_1 and a_0 (two equations) and equating them to zero.

$$\frac{\partial R}{\partial a_1} = 0 \quad \text{and} \quad \frac{\partial R}{\partial a_0} = 0$$

This results in a system of two equations with two unknowns, a_1 and a_0. The solution of these equations gives the values of the coefficients of the polynomial that best fits the data. The same procedure can be followed with more points and higher-order polynomials. More details on the least squares method can be found in books on numerical analysis.

Curve fitting with polynomials is done in MATLAB with the `polyfit` function, which uses the least squares method. The basic form of the `polyfit` function is:

```
p = polyfit(x,y,n)
```

`p` is the vector of the coefficients of the polynomial that fits the data.

`x` is a vector with the horizontal coordinates of the data points (independent variable).
`y` is a vector with the vertical coordinates of the data points (dependent variable).
`n` is the degree of the polynomial.

For the same set of m points, the `polyfit` function can be used to fit polynomials of any order up to $m-1$. If $n = 1$ the polynomial is a straight line, if $n = 2$ the polynomial is a parabola, and so on. The polynomial passes through all the points if $n = m-1$ (the order of the polynomial is one less than the number of points). It should be pointed out here that a polynomial that passes through all the points, or polynomials with higher order, do not necessarily give a better fit overall. High-order polynomials can deviate significantly between the data points.

Figure 8-2 shows how polynomials of different degrees fit the same set of data points. A set of seven points is given by (0.9, 0.9), (1.5, 1.5), (3, 2.5), (4, 5.1),

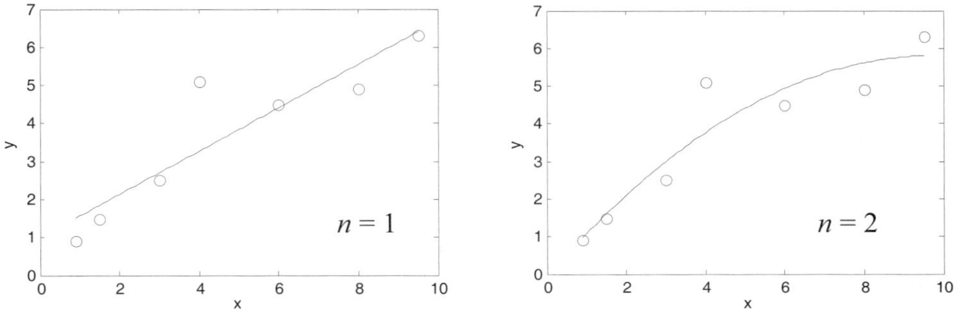

Figure 8-2: Fitting data with polynomials of different order.

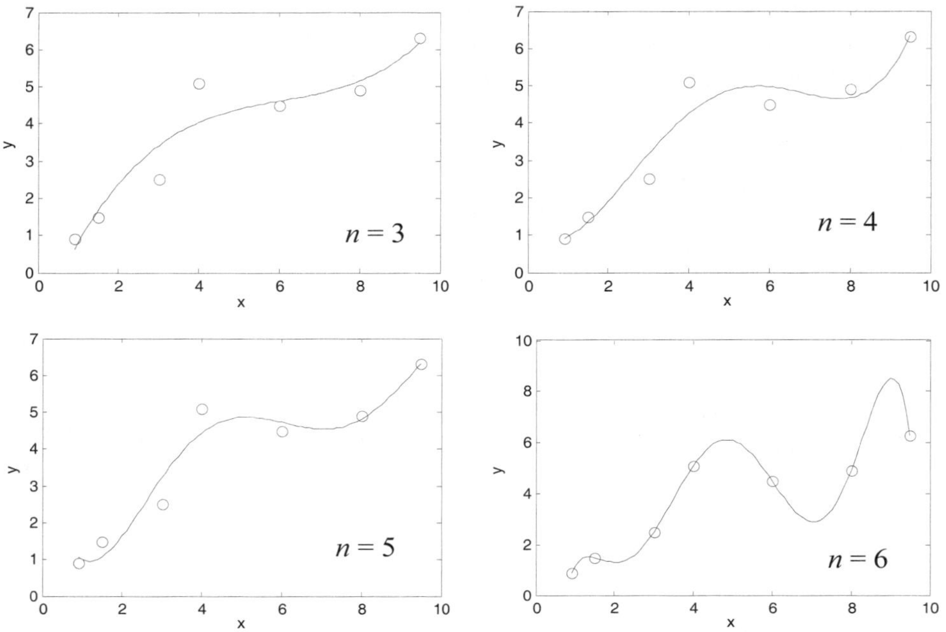

Figure 8-2: Fitting data with polynomials of different order. (Continued)

(6, 4.5), (8, 4.9), and (9.5, 6.3). The points are fitted using the `polyfit` function with polynomials of degrees 1 through 6. Each plot in Figure 8-2 shows the same data points, marked with circles, and a curve-fitted line that corresponds to a polynomial of the specified degree. It can be seen that the polynomial with $n = 1$ is a straight line, and with $n = 2$ is a slightly curved line. As the degree of the polynomial increases, the line develops more bends such that it passes closer to more points. When $n = 6$, which is one less than the number of points, the line passes through all the points. However, between some of the points, the line deviates significantly from the trend of the data.

The script file used to generate one of the plots in Figure 8-2 (the polynomial with $n = 3$) is shown below. Note that in order to plot the polynomial (the line) a new vector `xp` with small spacing is created. This vector is then used with

`x=[0.9 1.5 3 4 6 8 9.5];` `y=[0.9 1.5 2.5 5.1 4.5 4.9 6.3];`	Create vectors `x` and `y` with the coordinates of the data points.
`p=polyfit(x,y,3)`	Create a vector `p` using the `polyfit` function.
`xp=0.9:0.1:9.5;`	Create a vector `xp` to be used for plotting the polynomial.
`yp=polyval(p,xp)`	Create a vector `yp` with values of the polynomial at each `xp`.
`plot(x,y,'o',xp,yp)`	A plot of the seven points and the polynomial.
`xlabel('x'); ylabel('y')`	

the function `polyval` to create a vector `yp` with the value of the polynomial for each element of `xp`.

When the script file is executed, the following vector `p` is displayed in the Command Window.

```
p =
      0.0220    -0.4005     2.6138    -1.4158
```

This means that the polynomial of the third degree in Figure 8-2 has the form $0.022x^3 - 0.4005x^2 + 2.6138x - 1.4148$.

8.2.2 Curve Fitting with Functions Other than Polynomials

Many situations in science and engineering require fitting functions that are not polynomials to given data. Theoretically, any function can be used to model data within some range. For a particular data set, however, some functions provide a better fit than others. In addition, determining the best-fitting coefficients can be more difficult for some functions than for others. This section covers curve fitting with power, exponential, logarithmic, and reciprocal functions, which are commonly used. The forms of these functions are:

$$y = bx^m \qquad \text{(power function)}$$

$$y = be^{mx} \text{ or } y = b10^{mx} \qquad \text{(exponential function)}$$

$$y = m\ln(x) + b \text{ or } y = m\log(x) + b \qquad \text{(logarithmic function)}$$

$$y = \frac{1}{mx + b} \qquad \text{(reciprocal function)}$$

All of these functions can easily be fitted to given data with the `polyfit` function. This is done by rewriting the functions in a form that can be fitted with a linear polynomial ($n = 1$), which is

$$y = mx + b$$

The logarithmic function is already in this form, and the power, exponential and reciprocal equations can be rewritten as:

$$\ln(y) = m\ln(x) + \ln b \qquad \text{(power function)}$$

$$\ln(y) = mx + \ln(b) \text{ or } \log(y) = mx + \log(b) \qquad \text{(exponential function)}$$

$$\frac{1}{y} = mx + b \qquad \text{(reciprocal function)}$$

These equations describe a linear relationship between $\ln(y)$ and $\ln(x)$ for the power function, between $\ln(y)$ and x for the exponential function, between y and $\ln(x)$ or $\log(x)$ for the logarithmic function, and between $1/y$ and x for the reciprocal function. This means that the `polyfit(x,y,1)` function can be used to determine the best-fit constants m and b for best fit if, instead of `x` and `y`, the following arguments are used.

Function		`polyfit` **function form**
power	$y = bx^m$	`p=polyfit(log(x),log(y),1)`
exponential	$y = be^{mx}$ or $y = b10^{mx}$	`p=polyfit(x,log(y),1)` or `p=polyfit(x,log10(y),1)`
logarithmic	$y = m\ln(x) + b$ or $y = m\log(x) + b$	`p=polyfit(log(x),y,1)` or `p=polyfit(log10(x),y,1)`
reciprocal	$y = \dfrac{1}{mx + b}$	`p=polyfit(x,1./y,1)`

The result of the `polyfit` function is assigned to `p`, which is a two-element vector. The first element, `p(1)`, is the constant `m`, and the second element, `p(2)`, is `b` for the logarithmic and reciprocal functions, $\ln(b)$ or $\log(b)$ for the exponential function, and $\ln(b)$ for the power function ($b = e^{p(2)}$ or $b = 10^{p(2)}$ for the exponential function, and $b = e^{p(2)}$ for the power function).

For given data it is possible to estimate, to some extent, which of the functions has the potential for providing a good fit. This is done by plotting the data using different combinations of linear and logarithmic axes. If the data points in one of the plots appear to fit a straight line, the corresponding function can provide a good fit according to the list below.

***x* axis**	***y* axis**	**Function**
linear	linear	linear $y = mx + b$
logarithmic	logarithmic	power $y = bx^m$
linear	logarithmic	exponential $y = be^{mx}$ or $y = b10^{mx}$
logarithmic	linear	logarithmic $y = m\ln(x) + b$ or $y = m\log(x) + b$
linear	linear (plot $1/y$)	reciprocal $y = \dfrac{1}{mx + b}$

Other considerations in choosing a function:

- Exponential functions cannot pass through the origin.
- Exponential functions can fit only data with all positive y's or all negative y's.
- Logarithmic functions cannot model $x = 0$ or negative values of x.
- For the power function $y = 0$ when $x = 0$.
- The reciprocal equation cannot model $y = 0$.

The following example illustrates the process of fitting a function to a set of data points.

Sample Problem 8-2: Fitting an equation to data points

The following data points are given. Determine a function $w = f(t)$ (t is the independent variable, w is the dependent variable) with a form discussed in this section that best fits the data.

t	0.0	0.5	1.0	1.5	2.0	2.5	3.0	3.5	4.0	4.5	5.0
w	6.00	4.83	3.70	3.15	2.41	1.83	1.49	1.21	0.96	0.73	0.64

Solution

The data is first plotted with linear scales on both axes. The figure indicates that a linear function will not give the best fit since the points do not appear to line up along a straight line. From the other possible functions, the logarithmic function is excluded since for the first point $t = 0$, and the power function is excluded since at $t = 0$, $w \neq 0$. To check if the other two functions (exponential and reciprocal) might give a better fit, two additional plots, shown below, are made. The plot on the left has a log scale on the vertical axis and linear horizontal axis. In the plot on the right both axes have linear scales, and the quantity $1/w$ is plotted on the vertical axis.

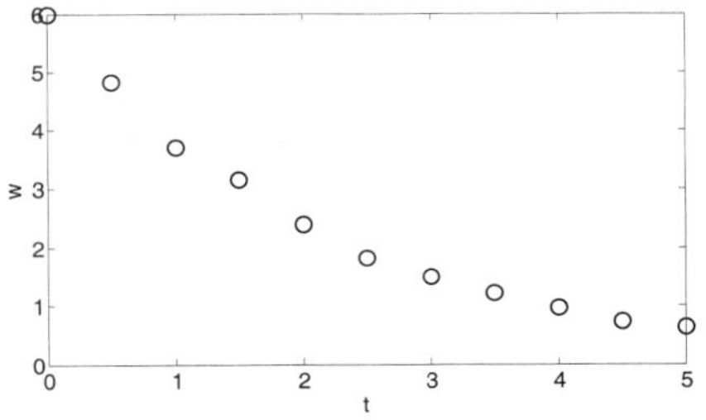

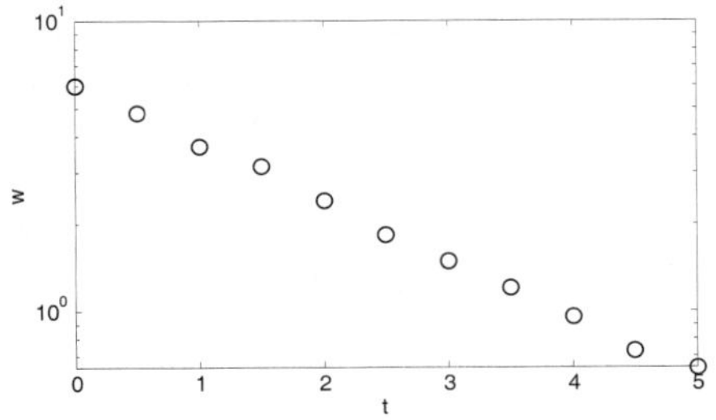

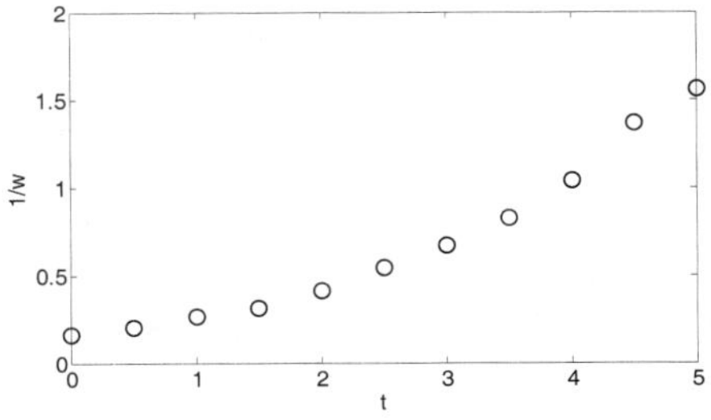

In the left figure the data points appear to line up along a straight line. This indicates that an exponential function of the form $y = be^{mx}$ can give a good fit to the data. A program in a script file that determines the constants b and m, and that plots the data points and the function is given below.

```
t=0:0.5:5;
w=[6 4.83 3.7 3.15 2.41 1.83 1.49 1.21 0.96 0.73 0.64];
p=polyfit(t,log(w),1);
```

Create vectors `t` and `w` with the coordinates of the data points.

Use the `polyfit` function with `t` and `log(w)`.

```
m=p(1)
b=exp(p(2))           Determine the coefficient b.
tm=0:0.1:5;           Create a vector tm to be used for plotting the polynomial.
wm=b*exp(m*tm);       Calculate the function value at each element of tm.
plot(t,w,'o',tm,wm)   Plot the data points and the function.
```

When the program is executed, the values of the constants m and b are displayed in the Command Window.

```
m =
   -0.4580
b =
    5.9889
```

The plot generated by the program, which shows the data points and the function (with axis labels added with the Plot Editor) is

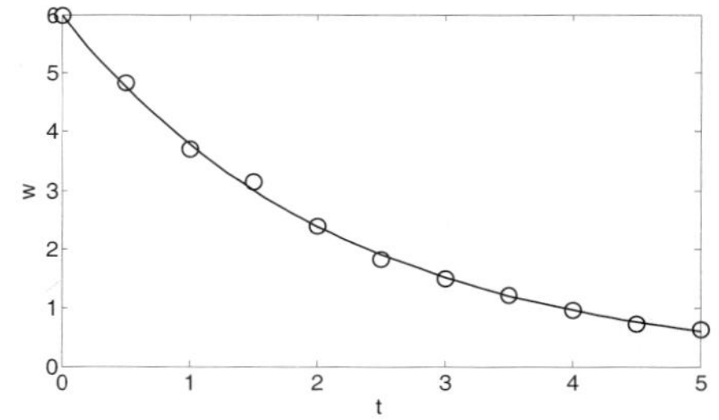

It should be pointed out here that in addition to the power, exponential, logarithmic, and reciprocal functions that are discussed in this section, many other functions can be written in a form suitable for curve fitting with the `polyfit` function. One example where a function of the form $y = e^{(a_2x^2 + a_1x + a_0)}$ is fitted to data points using the `polyfit` function with a third-order polynomial is described in Sample Problem 8-7.

8.3 *INTERPOLATION*

Interpolation is the estimation of values between data points. MATLAB has interpolation functions that are based on polynomials, which are described in this section, and on Fourier transformation, which is outside the scope of this book. In one-dimensional interpolation each point has one independent variable (x) and one dependent variable (y). In two-dimensional interpolation each point has two independent variables (x and y) and one dependent variable (z).

One-dimensional interpolation:

If only two data points exist, the points can be connected with a straight line is and a linear equation (polynomial of first order) can be used to estimate values between the points. As was discussed in the previous section, if three (or four) data points exist, a second- (or a third-) order polynomial that passes through the points can be determined and then be used to estimate values between the points. As the number of points increases, a higher-order polynomial is required for the polynomial to pass through all the points. Such a polynomial, however, will not necessarily give a good approximation of the values between the points. This is illustrated in Figure 8-2 with $n = 6$.

A more accurate interpolation can be obtained if instead of considering all the points in the data set (by using one polynomial that passes through all the points), only a few data points in the neighborhood where the interpolation is needed are considered. In this method, called spline interpolation, many low-order polynomials are used, where each is valid only in a small domain of the data set.

The simplest method of spline interpolation is called linear spline interpolation. In this method, shown on the right, every two adjacent points are connected with a straight line (a polynomial of first degree). The equation of a straight line that passes through two adjacent points (x_i, y_j) and (x_{i+1}, y_{j+1}) and that can be used to calculate the value of y for any x between the points is given by:

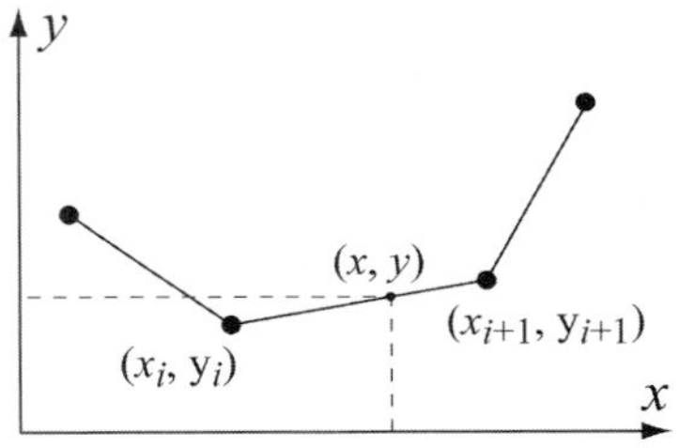

$$y = \frac{y_{i+1} - y_i}{x_{i+1} - x_i}x + \frac{y_i x_{i+1} - y_{i+1} x_i}{x_{i+1} - x_i}$$

In a linear interpolation the line between two data points has a constant slope, and there is a change in the slope at every point. A smoother interpolation curve can be obtained by using quadratic or cubic polynomials. In these methods, called quadratic splines and cubic splines, a second-, or third-order polynomial is used to interpolate between every two points. The coefficients of the polynomial are determined by using data from points that are adjacent to the two data points. The theoretical background for the determination of the constants of the polynomials is beyond the scope of this book and can be found in books on numerical analysis.

One-dimensional interpolation in MATLAB is done with the `interp1` (the last character is the number one) function, which has the form:

```
yi = interp1(x,y,xi,'method')
```

`yi` is the interpolated value.

`x` is a vector with the horizontal coordinates of the input data points (independent variable).
`y` is a vector with the vertical coordinates of the input data points (dependent variable).
`xi` is the horizontal coordinate of the interpolation point (independent variable).

Method of interpolation, typed as a string (optional).

- The vector `x` must be monotonic (with elements in ascending or descending order).
- `xi` can be a scalar (interpolation of one point) or a vector (interpolation of many points). `yi` is a scalar or a vector with the corresponding interpolated values.
- MATLAB can do the interpolation using one of several methods that can be specified. These methods include:

`'nearest'`	returns the value of the data point that is nearest to the interpolated point.
`'linear'`	uses linear spline interpolation.
`'spline'`	uses cubic spline interpolation.
`'pchip'`	uses piecewise cubic Hermite interpolation, also called `'cubic'`

- When the `'nearest'` and the `'linear'` methods are used, the value(s) of `xi` must be within the domain of `x`. If the `'spline'` or the `'pchip'` methods are used, `xi` can have values outside the domain of `x` and the function `interp1` performs extrapolation.
- The `'spline'` method can give large errors if the input data points are nonuniform such that some points are much closer together than others.
- Specification of the method is optional. If no method is specified, the default is `'linear'`.

Sample Problem 8-3: Interpolation

The following data points, which are points of the function $f(x) = 1.5^x\cos(2x)$, are given. Use linear, spline, and pchip interpolation methods to calculate the value of y between the points. Make a figure for each of the interpolation methods. In the figure show the points, a plot of the function, and a curve that corresponds to the interpolation method.

x	0	1	2	3	4	5
y	1.0	–0.6242	–1.4707	3.2406	–0.7366	–6.3717

Solution

The following is a program written in a script file that solves the problem:

```
x=0:1.0:5;                    Create vectors x and y with coordinates of the data points.
y=[1.0 -0.6242 -1.4707 3.2406 -0.7366 -6.3717];
xi=0:0.1:5;                   Create vector xi with points for interpolation.
yilin=interp1(x,y,xi,'linear');   Calculate y points from linear interpolation.
yispl=interp1(x,y,xi,'spline');   Calculate y points from spline interpolation.
yipch=interp1(x,y,xi,'pchip');    Calculate y points from pchip interpolation.
yfun=1.5.^xi.*cos(2*xi);          Calculate y points from the function.
subplot(1,3,1)
plot(x,y,'o',xi,yfun,xi,yilin,'--');
subplot(1,3,2)
plot(x,y,'o',xi,yfun,xi,yispl,'--');
subplot(1,3,3)
plot(x,y,'o',xi,yfun,xi,yipch,'--');
```

The three figures generated by the program are shown below (axes labels were added with the Plot Editor). The data points are marked with circles, the interpolation curves are plotted with dashed lines, and the function is shown with a solid line. The left figure shows the linear interpolation, the middle is the spline, and the figure on the right shows the pchip interpolation.

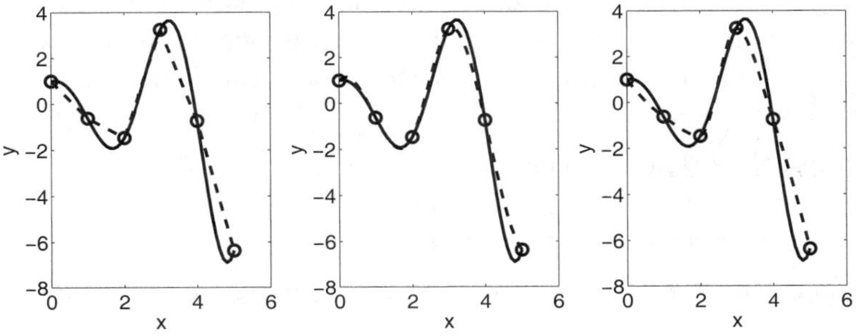

8.4 *THE BASIC FITTING INTERFACE*

The basic fitting interface is a tool that can be used to perform curve fitting and interpolation interactively. By using the interface the user can:

- Curve-fit the data points with polynomials of various degrees up to 10, and with spline and Hermite interpolation methods.
- Plot the various fits on the same graph so that they can be compared.
- Plot the residuals of the various polynomial fits and compare the norms of the residuals.
- Calculate the values of specific points with the various fits.
- Add the equations of the polynomials to the plot.

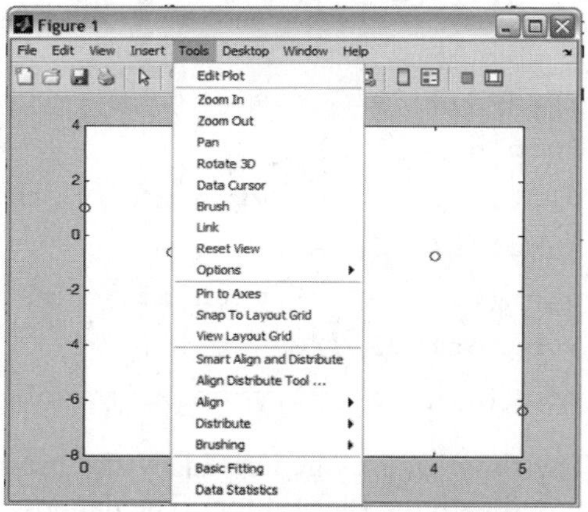

To activate the basic fitting interface, the user first has to make a plot of the data points. Then the interface is activated by selecting **Basic Fitting** in the **Tools** menu, as shown on the right. This opens the Basic Fitting Window, shown in Figure 8-3. When the window first opens, only one panel (the **Plot fits** panel) is visible. The window can be extended to show a second panel (the **Numerical results** panel) by clicking on the → button. One click adds the first section of the panel, and a second click makes the window look as shown in Figure 8-3. The window can be reduced back by clicking on the ← button. The first two items in the Basic Fitting Window are related to the selection of the data points:

Select data: Used to select a specific set of data points for curve fitting in a figure that has more than one set of data points. Only one set of data points can be curve-fitted at a time, but multiple fits can be performed simultaneously on the same set.

Center and scale x data: When this box is checked, the data is centered at zero mean and scaled to unit standard deviation. This might be needed in order to improve the accuracy of numerical computation.

The next four items are in the **Plot fits** panel and are related to the display of the fit.

Check to display fits on figure: The user selects the fits to be displayed in the figure. The selections include interpolation with spline interpolant (interpolation method) that uses the `spline` function, interpolation with Hermite interpolant that uses the `pchip` function, and polynomials of various degrees that use the

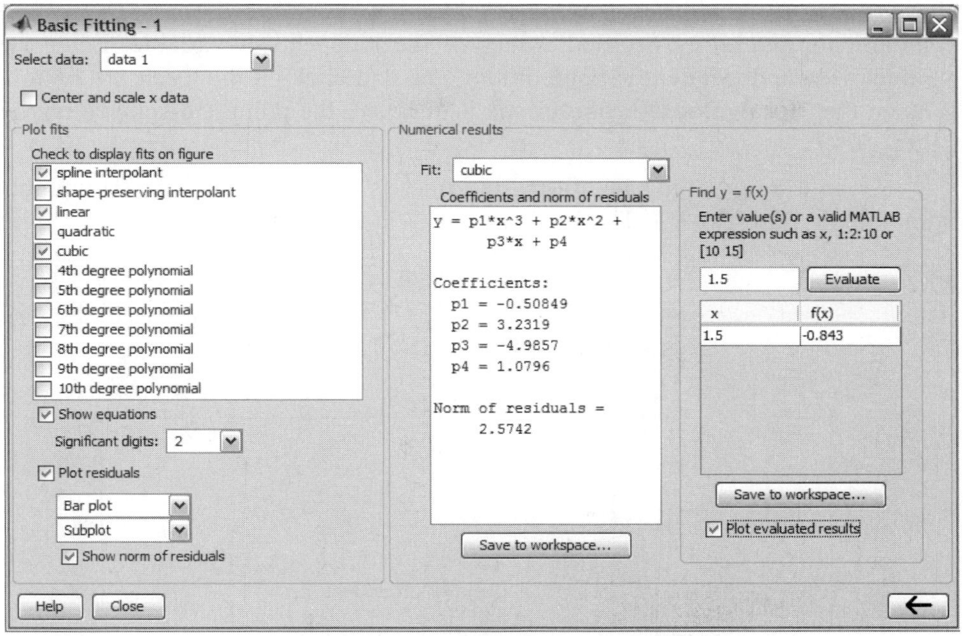

Figure 8-3: The Basic Fitting Window.

`polyfit` function. Several fits can be selected and displayed simultaneously.

Show equations: When this box is checked, the equations of the polynomials that were selected for the fit are displayed in the figure. The equations are displayed with the number of significant digits selected in the adjacent sign menu.

Plot residuals: When this box is checked, a plot that shows the residual at each data point is created (residuals are defined in Section 8.2.1). Choices in the menus include a bar plot, a scatter plot, and a line plot which can be displayed as a subplot in the same Figure Window that has the plot of the data points, or as a separate plot in a different Figure Window.

Show norm of residuals: When this box is checked, the norm of the residuals is displayed in the plot of the residuals. The norm of the residual is a measure of the quality of the fit. A smaller norm corresponds to a better fit.

The next three items are in the **Numerical results** panel. They provide the numerical information for one fit, independently of the fits that are displayed:

Fit: The user selects the fit to be examined numerically. The fit is shown on the plot only if it is selected in the **Plot fit** panel.

Coefficients and norm of residuals: Displays the numerical results for the polynomial fit that is selected in the **Fit** menu. It includes the coefficients of the polynomial and the norm of the residuals. The results can be saved by clicking on the **Save to workspace** button.

Find y = f(x): Provides a means for obtaining interpolated (or extrapolated) numerical values for specified values of the independent variable. Enter the value of the independent variable in the box, and click on the **Evaluate** button. When the **Plot evaluated results** box is checked, the point is displayed on the plot.

As an example, the basic fitting interface is used for fitting the data points from Sample Problem 8-3. The Basic Fitting Window is the one shown in Figure

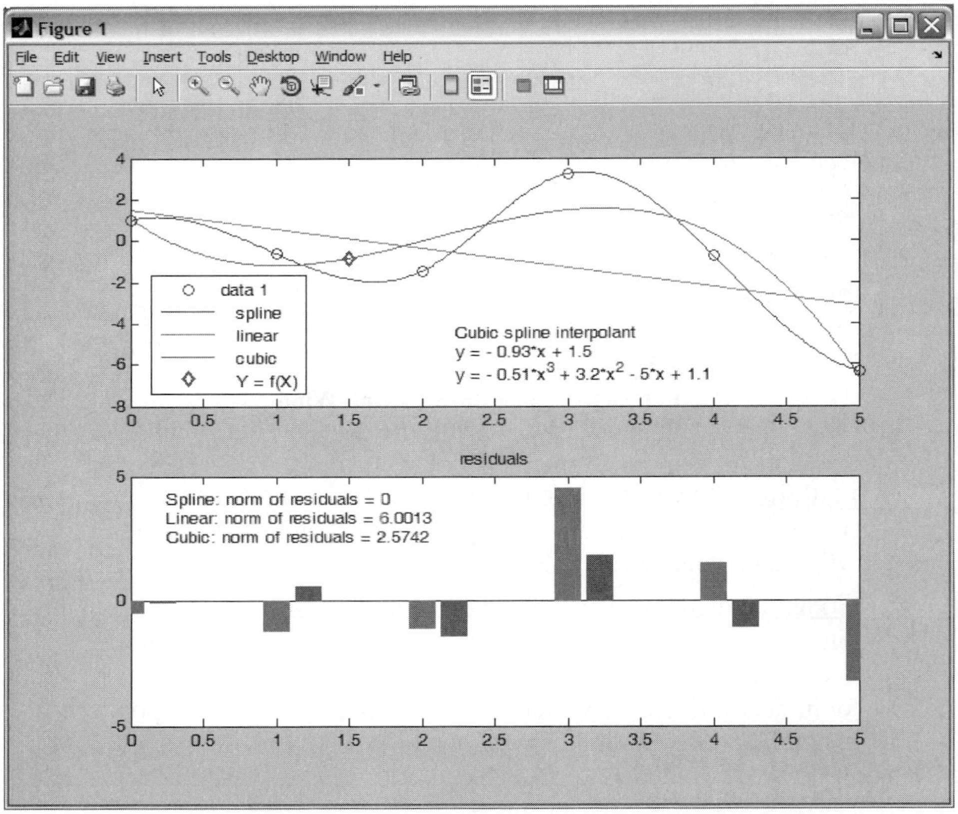

Figure 8-4: A Figure Window modified by the Basic Fitting Interface.

8-3, and the corresponding Figure Window is shown in Figure 8-4. The Figure Window includes a plot of the points, one interpolation fit (spline), two polynomial fits (linear and cubic), a display of the equations of the polynomial fits, and a mark of the point $x = 1.5$ that is entered in the **Find y = f(x)** box of the Basic Fitting Window. The Figure Window also includes a plot of the residuals of the polynomial fits and a display of their norm.

8.5 EXAMPLES OF MATLAB APPLICATIONS

Sample Problem 8-4: Determining wall thickness of a box

The outside dimensions of a rectangular box (bottom and four sides, no top), made of aluminum, are 24 by 12 by 4 inches. The wall thickness of the bottom and the sides is x. Derive an expression that relates the weight of the box and the wall thickness x. Determine the thickness x for a box that weighs 15 lb. The specific weight of aluminum is 0.101 lb/in.^{3}.

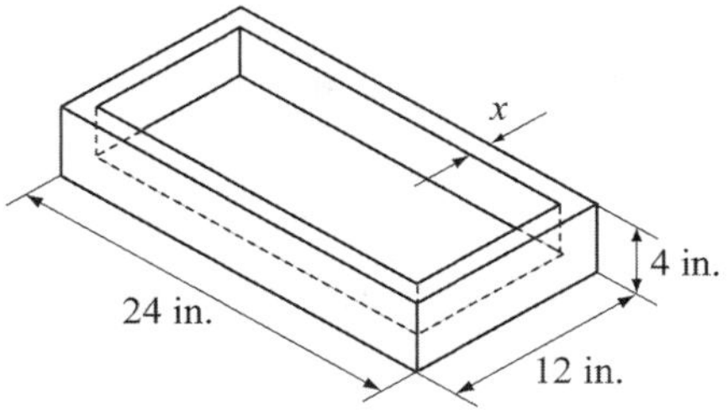

Solution

The volume of the aluminum V_{Al} is calculated from the weight W of the box by:

$$V_{Al} = \frac{W}{\gamma}$$

where γ is the specific weight. The volume of the aluminum based on the dimensions of the box is given by

$$V_{Al} = 24 \cdot 12 \cdot 4 - (24 - 2x)(12 - 2x)(4 - x)$$

where the inside volume of the box is subtracted from the outside volume. This equation can be rewritten as

$$(24 - 2x)(12 - 2x)(4 - x) + V_{Al} - (24 \cdot 12 \cdot 4) = 0$$

which is a third-degree polynomial. A root of this polynomial is the required thickness x. A program in a script file that determines the polynomial and solves for the roots is:

```
W=15; gamma=0.101;                    % Assign W and gamma.
VAlum=W/gamma;                        % Calculate the volume of the aluminum.
a=[-2  24];                           % Assign the polynomial 24 – 2x to a.
b=[-2  12];                           % Assign the polynomial 12 – 2x to b.
c=[-1  4];                            % Assign the polynomial 4 – x to c.
Vin=conv(c, conv(a,b));               % Multiply the three polynomials above.
polyeq=[0 0 0 (VAlum-24*12*4)]+Vin    % Add V_Al – 24*12*4 to Vin.
x=roots(polyeq)                       % Determine the roots of the polynomial.
```

Note in the second to last line that in order to add the quantity $V_{Al} - (24 \cdot 12 \cdot 4)$ to the polynomial `Vin` it has to be written as a polynomial of the same order as `Vin` (`Vin` is a polynomial of third order). When the program (saved as Chap8SamPro4) is executed, the coefficients of the polynomial and the value of x are displayed:

```
>> Chap8SamPro4
polyeq =
  -4.0000   88.0000 -576.0000   148.5149
x =
   10.8656 + 4.4831i
   10.8656 - 4.4831i
   0.2687
```

The polynomial is: $-4x^3 + 88x^2 - 576x + 148.515$.

The polynomial has one real root, $x = 0.2687$ in., which is the thickness of the aluminum wall.

Sample Problem 8-5: Floating height of a buoy

An aluminum thin-walled sphere is used as a marker buoy. The sphere has a radius of 60 cm and a wall thickness of 12 mm. The density of aluminum is ρ_{Al} = 2690 kg/m³. The buoy is placed in the ocean, where the density of the water is 1030 kg/m³. Determine the height h between the top of the buoy and the surface of the water.

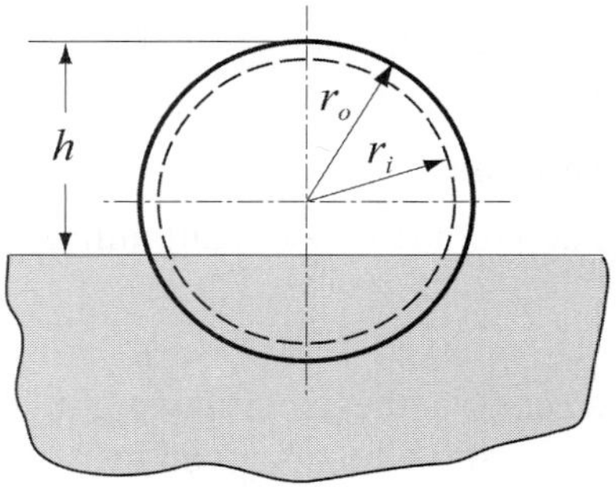

Solution

According to Archimedes' law, the buoyancy force applied to an object that is placed in a fluid is equal to the weight of the fluid that is displaced by the object. Accordingly, the aluminum sphere will be at a depth such that the weight of the sphere is equal to the weight of the fluid displaced by the part of the sphere that is submerged.

The weight of the sphere is given by

$$W_{sph} = \rho_{Al} V_{Al} g = \rho_{Al} \frac{4}{3}\pi(r_o^3 - r_i^3)g$$

where V_{Al} is the volume of the aluminum; r_o and r_i are the outside and inside radii of the sphere, respectively; and g is the gravitational acceleration.

The weight of the water that is displaced by the spherical portion that is submerged is given by:

$$W_{wtr} = \rho_{wtr} V_{wtr} g = \rho_{wtr} \frac{1}{3}\pi(2r_o - h)^2(r_o + h)g$$

Setting the two weights equal to each other gives the following equation:

$$h^3 - 3r_o h^2 + 4r_o^3 - 4\frac{\rho_{Al}}{\rho_{wtr}}(r_o^3 - r_i^3) = 0$$

The last equation is a third-degree polynomial for h. The root of the polynomial is the answer.

A solution with MATLAB is obtained by writing the polynomials and using the `roots` function to determine the value of h. This is done in the following script file:

```
rout=0.60; rin=0.588;                                  Assign the radii to variables.
rhoalum=2690; rhowtr=1030;                             Assign the densities to variables.
a0=4*rout^3-4*rhoalum*(rout^3-rin^3)/rhowtr;           Assign the coefficient a0.
p = [1 -3*rout 0 a0];                                  Assign the coefficient vector of the polynomial.
h = roots(p)                                           Calculate the roots of the polynomial.
```

When the script file is executed in the Command Window, as shown below, the answer is three roots, since the polynomial is of the third degree. The only answer that is physically possible is the second, where $h = 0.9029$ m.

```
>> Chap8SamPro5
h =
    1.4542          The polynomial has three roots. The only one that is
    0.9029          physically possible for the problem is 0.9029 m.
   -0.5570
```

Sample Problem 8-6: Determining the size of a capacitor

An electrical capacitor has an unknown capacitance. In order to determine its capacitance it is connected to the circuit shown. The switch is first connected to *B* and the capacitor is charged. Then, the switch is connected to *A* and the capacitor discharges through the resistor. As the capacitor is discharging, the voltage across the capacitor is measured for 10 s in intervals of 1 s. The recorded measurements are given in the table below. Plot the voltage as a function of time and determine the capacitance of the capacitor by fitting an exponential curve to the data points.

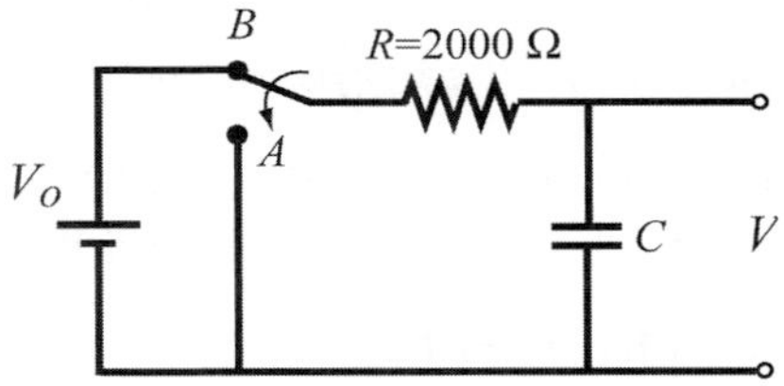

t (s)	1	2	3	4	5	6	7	8	9	10
V (V)	9.4	7.31	5.15	3.55	2.81	2.04	1.26	0.97	0.74	0.58

Solution

When a capacitor discharges through a resistor, the voltage of the capacitor as a function of time is given by

$$V = V_0 e^{(-t)/(RC)}$$

where V_0 is the initial voltage, R the resistance of the resistor, and C the capacitance of the capacitor. As was explained in Section 8.2.2 the exponential function can be written as a linear equation for $\ln(V)$ and t in the form:

$$\ln(V) = \frac{-1}{RC}t + \ln(V_0)$$

This equation, which has the form $y = mx + b$, can be fitted to the data points by using the `polyfit(x,y,1)` function with t as the independent variable `x` and $\ln(V)$ as the dependent variable `y`. The coefficients m and b determined by the `polyfit` function are then used to determine C and V_0 by:

$$C = \frac{-t}{Rm} \quad \text{and} \quad V_0 = e^b$$

The following program written in a script file determines the best-fit exponential function to the data points, determines C and V_0, and plots the points and the fitted function.

```
R=2000;                                    Define R.
t=1:10;                                    Assign the data points to vectors t and v.
v=[9.4 7.31 5.15 3.55 2.81 2.04 1.26 0.97 0.74 0.58];
p=polyfit(t,log(v),1);                     Use the polyfit function with t and log(v).
C=-1/(R*p(1))                              Calculate C from p(1), which is m in the equation.
V0=exp(p(2))                               Calculate V0 from p(2), which is b in the equation.
tplot=0:0.1:10;                            Create vector tplot of time for plotting the function.
vplot=V0*exp(-tplot./(R*C));               Create vector vplot for plotting the function.
plot(t,v,'o',tplot,vplot)
```

When the script file is executed (saved as Chap8SamPro6) the values of C and V_0 are displayed in the Command Window as shown below:

```
>> Chap8SamPro6
C =                                        The capacitance of the capacitor is 1,600 μF.
    0.0016
V0 =
   13.2796
```

The program creates also the following plot (axis labels were added to the plot using the Plot Editor):

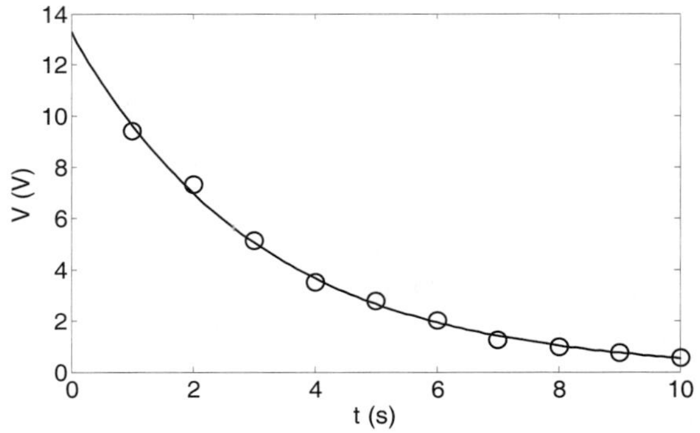

Sample Problem 8-7: Temperature dependence of viscosity

Viscosity, μ, is a property of gases and fluids that characterizes their resistance to flow. For most materials viscosity is highly sensitive to temperature. Below is a table that gives the viscosity of SAE 10W oil at different temperatures (Data from *B.R.* Munson, D.F. Young, and T.H. Okiishi, *Fundamentals of Fluid Mechanics*, 4th ed., John Wiley and Sons, 2002). Determine an equation that can be fitted to the data.

T (°C)	–20	0	20	40	60	80	100	120
μ (N s/m²) ($\times 10^{-5}$)	4	0.38	0.095	0.032	0.015	0.0078	0.0045	0.0032

Solution

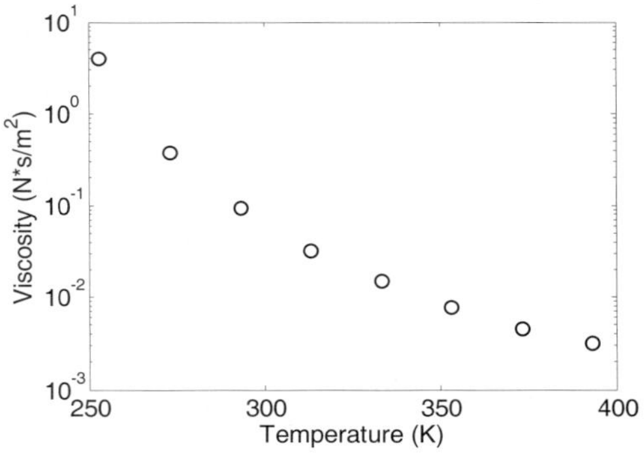

To determine what type of equation might provide a good fit to the data, μ is plotted as a function of T (absolute temperature) with a linear scale for T and a logarithmic scale for μ. The plot, shown on the right, indicates that the data points do not appear to line up along a straight line. This means that a simple exponential function of the form $y = be^{mx}$, which models a straight line with these axes, will not provide the best fit. Since the points in the figure appear to lie along a curved line, a function that can possibly have a good fit to the data is:

$$\ln(\mu) = a_2 T^2 + a_1 T + a_0$$

This function can be fitted to the data by using MATLABs `polyfit(x,y,2)` function (second-degree polynomial), where the independent variable is T and the dependent variable is ln(μ). The equation above can be solved for μ to give the viscosity as a function of temperature:

$$\mu = e^{(a_2 T^2 + a_1 T + a_0)} = e^{a_0} e^{a_1 T} e^{a_2 T^2}$$

The following program determines the best fit to the function and creates a plot that displays the data points and the function.

```
T=[-20:20:120];
mu=[4 0.38 0.095 0.032 0.015 0.0078 0.0045 0.0032];
TK=T+273;
p=polyfit(TK,log(mu),2)
Tplot=273+[-20:120];
```

```
muplot = exp(p(1)*Tplot.^2 + p(2)*Tplot + p(3));
semilogy(TK,mu,'o',Tplot,muplot)
```

When the program executes (saved as Chap8SamPro7), the coefficients that are determined by the `polyfit` function are displayed in the Command Window (shown below) as three elements of the vector `p`.

```
>> Chap8SamPro7
p =
    0.0003   -0.2685   47.1673
```

With these coefficients the viscosity of the oil as a function of temperature is:

$$\mu = e^{(0.0003T^2 - 0.2685T + 47.1673)} = e^{47.1673}e^{(-0.2685)T}e^{0.0003T^2}$$

The plot that is generated shows that the equation correlates well to the data points (axis labels were added with the Plot Editor).

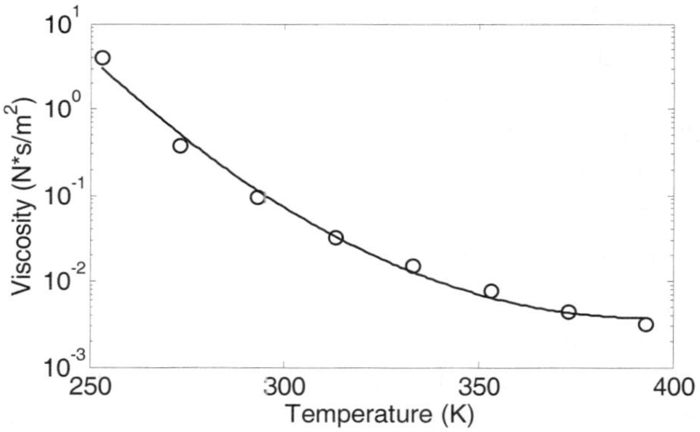

8.6 Problems

1. Plot the polynomial $y = -0.4x^4 + 7x^2 - 20.5x - 28$ in the domain $-5 \le x \le 4$. First create a vector for x, next use the `polyval` function to calculate y, and then use the `plot` function.

2. Plot the polynomial $y = -0.001x^4 + 0.051x^3 - 0.76x^2 + 3.8x - 1.4$ in the domain $1 \le x \le 14$. First create a vector for x, next use the `polyval` function to calculate y, and then use the `plot` function.

3. Use MATLAB to carry out the following multiplication of two polynomials:

$$(2x^2 + 3)(x^3 + 3.5x^2 + 5x - 16)$$

4. Use MATLAB to carry out the following multiplication of polynomials:

$$(x+1.4)(x-0.4)x(x+0.6)(x-1.4)$$

Plot the polynomial for $-1.5 \le x \le 1.5$.

5. Divide the polynomial $-0.6x^5 + 7.7x^3 - 8x^2 - 24.6x + 48$ by the polynomial $-0.6x^3 + 4.1x - 8$.

6. Divide the polynomial $x^4 - 6x^3 + 13x^2 - 12x + 4$ by the polynomial $x^3 - 3x^2 + 2$.

7. The product of three consecutive integers is 1,716. Using MATLAB's built-in function for operations with polynomials, determine the three integers.

8. The product of four consecutive even integers is 13,440. Using MATLAB's built-in function for operations with polynomials, determine the four integers.

9. A cylindrical aluminum fuel tank has an outside diameter of 30 in. and a height of 50 in. The the thickness of the wall is t, and the bottom and top ends are 25% thicker. Determine t if the weight of the tank is 152 lb. The specific weight of aluminum is 165 lb/ft³.

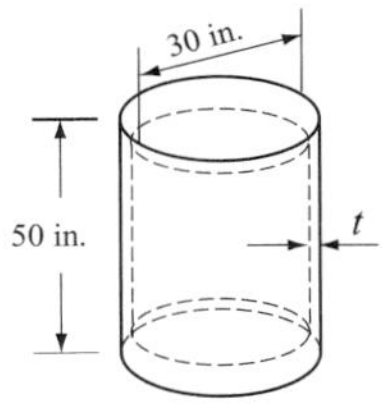

10. A cylindrical aluminum fuel tank has a flat bottom and a semi-spherical top. The outside diameter is 25 cm, and the height of the cylindrical section is 40 cm. The thickness of the side and the semi-spherical top walls is t, and the thickness of the flat bottom is $1.5t$. Determine t if the mass of the tank is 27.5 kg. The density of aluminum is 2.7 g/cm³.

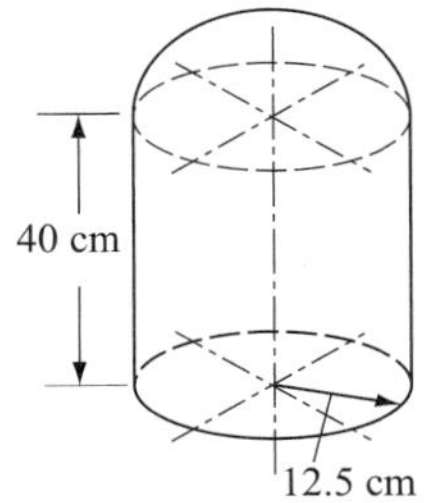

11. A 24 ft–long rod is cut into 12 pieces, which are welded together to form the frame of a rectangular box. The length of the box's base is three times its width.
   (*a*) Create a polynomial expression for the volume V in terms of x.
   (*b*) Make a plot of V versus x.
   (*c*) Determine the x that maximizes the volume and determine that volume.

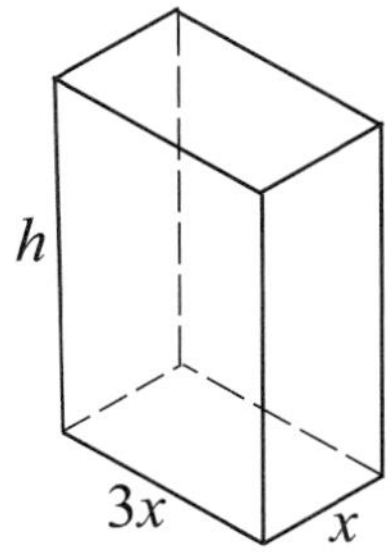

12. A rectangular piece of cardboard, 40 inches long by 22 inches wide, is used for making a rectangular box (open top) by cutting out squares of x by x from the corners and folding up the sides.

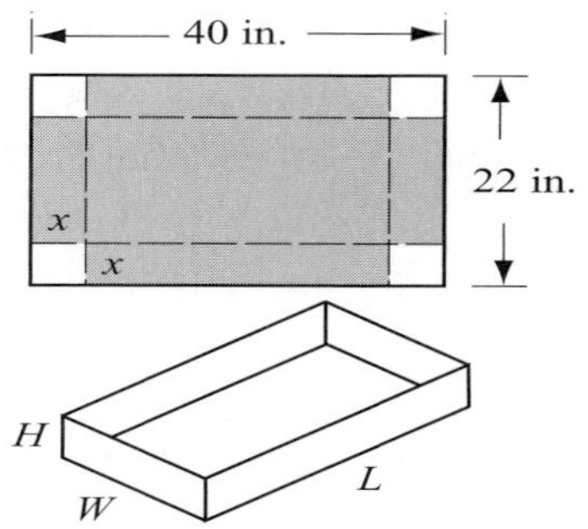

(*a*) Create a polynomial expression for the volume V in terms of x.
(*b*) Make a plot of V versus x.
(*c*) Determine x if the volume of the box is 1,000 in.^{3}.
(*d*) Determine the value of x that corresponds to the box with the largest possible volume, and determine that volume.

13. Write a user-defined function that adds or subtracts two polynomials of any order. Name the function `p=polyadd(p1,p2,operation)`. The first two input arguments `p1` and `p2` are the vectors of the coefficients of the two polynomials. (If the two polynomials are not of the same order, the function adds the necessary zero elements to the shorter vector.) The third input argument `operation` is a string that can be either `'add'` or `'sub'`, for adding or subtracting the polynomials, respectively, and the output argument is the resulting polynomial.

Use the function to add and subtract the following polynomials:
$f_1(x) = x^5 - 7x^4 + 11x^3 - 4x^2 - 5x - 2$ and $f_2(x) = 9x^2 - 10x + 6$.

14. Write a user-defined function that multiplies two polynomials. Name the function `p=polymult(p1,p2)`. The two input arguments `p1` and `p2` are vectors of the coefficients of the two polynomials. The output argument `p` is the resulting polynomial.

Use the function to multiply the following polynomials:
$f_1(x) = x^5 - 7x^4 + 11x^3 - 4x^2 - 5x - 2$ and $f_2(x) = 9x^2 - 10x + 6$.
Check the answer with MATLAB's built-in function `conv`.

15. Write a user-defined function that calculates the maximum (or minimum) of a quadratic equation of the form:

$$f(x) = ax^2 + bx + c$$

Name the function `[x,y,w] = maxormin(a,b,c)`. The input arguments are the coefficients `a`, `b`, and `c`. The output arguments are `x`, the coordinate of the maximum (or minimum); `y`, the maximum (or minimum) value; and `w`, which is equal to 1 if `y` is a maximum and equal to 2 if `y` is a minimum.

Use the function to determine the maximum or minimum of the following functions:

(*a*) $f(x) = 3x^2 - 7x + 14$ (*b*) $f(x) = -5x^2 - 11x + 15$

16. A cylinder of radius r and height h is constructed inside a cone with base radius $R = 10$ in. and height $H = 30$ in., as shown in the figure.
   (*a*) Create a polynomial expression for the volume V of the cylinder in terms of r.
   (*b*) Make a plot of V versus r.
   (*c*) Determine r if the volume of the cylinder is 800 in.^{3}.
   (*d*) Determine the value of r that corresponds to the cylinder with the largest possible volume, and determine that volume.

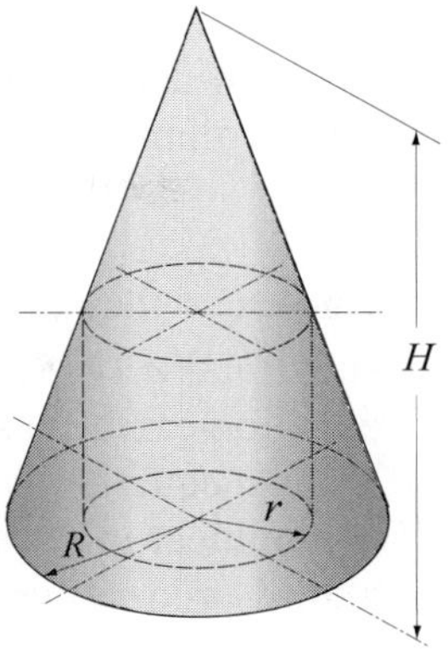

17. Consider the parabola $y = 1.5(x-5)^2 + 1$ and the point $P(3, 5.5)$.
   (*a*) Create a polynomial expression for the distance d from point P to an arbitrary point Q on the parabola.
   (*b*) Make a plot of d versus x for $3 \leq x \leq 6$.
   (*c*) Determine the coordinates of Q if $d = 28$.
   (*d*) Determine the coordinates of Q that correspond to the smallest d, and calculate the corresponding value of d.

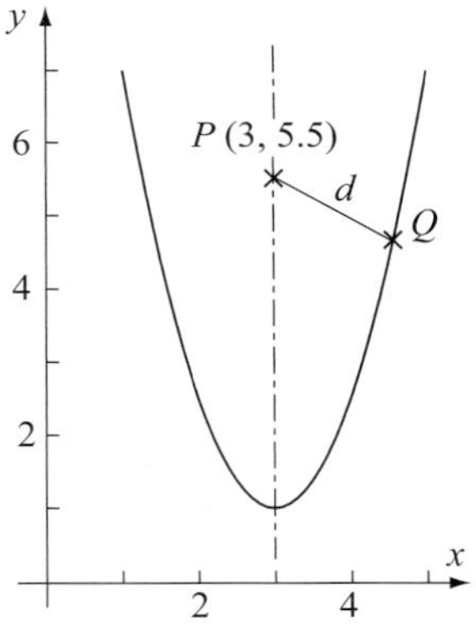

18. The boiling temperature of water T_B at various altitudes h is given in the following table. Determine a linear equation in the form $T_B = mh + b$ that best fits the data. Use the equation for calculating the boiling temperature at 16,000 ft. Make a plot of the points and the equation.

h (ft)	0	2000	5000	7500	10000	20000	26000
T (°F)	212	210	203	198	194	178	168

19. The number of bacteria N_B measured at different times t is given in the following table. Determine an exponential function in the form $N_B = Ne^{\alpha t}$ that best fits the data. Use the equation to estimate the number of bacteria after 60 min. Make a plot of the points and the equation.

t (min)	10	20	30	40	50
N_B	15,000	215,000	335,000	480,000	770,000

20. The van der Waals equation gives a relationship between the pressure p (in atm), volume V (in L), and temperature T (in K) for a real gas:

$$p = \frac{nRT}{V-nb} - \frac{n^2 a}{V^2}$$

where n is the number of moles, $R = 0.08206$ (L atm)/(mol K) is the gas constant, and a (in L^2 atm/mol^2), and b (in L/mol) are material constants. The equation can be easily used for calculating p (given T and V) or T (given p and V). The equation is not as readily solved for V when p and T are given, since it is nonlinear in V. One useful way to solve for V is by rewriting the equation as a third-order polynomial

$$V^3 - \left(nb + \frac{nRT}{p}\right)V^2 + \frac{n^2 a}{p}V - \frac{n^3 ab}{p} = 0$$

and calculating the root of the polynomial.

Write a user-defined function that calculates V for given p, T, n, a, and b. For function name and arguments use `V=waals(p,T,n,a,b)`. The function calculates V by using MATLAB's built-in function `roots`. Note that the solution of the polynomial can have non-real (complex) roots. The output argument `V` in `waals` should be the physically realistic solution (positive and real). (MATLAB's built-in function `imag(x)` can be used for determining which root is real.)

Use the user-defined function to calculate V for $p = 30$ atm, $T = 300$ K, $n = 1.5$, $a = 1.345$ L^2 atm/mol^2, $b = 0.0322$ L/mol.

21. The population of the world for selected years from 1750 to 2009 is given in the following table:

Year	1750	1800	1850	1900	1950	1990	2000	2009
Population (millions)	791	980	1,260	1,650	2,520	5,270	6,060	6,800

(*a*) Determine the exponential function that best fits the data. Use the function to estimate the population in 1980. Make a plot of the points and the function.

(*b*) Curve-fit the data with a third-order polynomial. Use the polynomial to estimate the population in 1980. Make a plot of the points and the polynomial.

(*c*) Fit the data with linear and spline interpolations. Estimate the population in 1975 with linear and spline interpolations. Make a plot of the data points and curves made of the interpolated points.

In each part make a plot of the data points (circle markers) and the fit curve or the interpolation curves. Note that part (*c*) has two interpolation curves.
The actual population of the world in 1980 was 4453.8 million.

22. The following points are given:

x	−5	−3.4	−2.0	−0.8	0	1.2	2.5	4	5.0	7	8.5
y	4.4	4.5	4	3.6	3.9	3.8	3.5	2.5	1.2	0.5	-0.2

(*a*) Fit the data with a first-order polynomial. Make a plot of the points and the polynomial.
(*b*) Fit the data with a second-order polynomial. Make a plot of the points and the polynomial.
(*c*) Fit the data with a fourth-order polynomial. Make a plot of the points and the polynomial.
(*d*) Fit the data with an eight-order polynomial. Make a plot of the points and the polynomial.

23. The standard air density, D (average of measurements made), at different heights, h, from sea level up to a height of 33 km is given below.

h (km)	0	3	6	9	12	15
D (kg/m^3)	1.2	0.91	0.66	0.47	0.31	0.19
h (km)	18	21	24	27	30	33
D (kg/m^3)	0.12	0.075	0.046	0.029	0.018	0.011

(*a*) Make the following four plots of the data points (density as a function of height): (1) both axes with linear scale; (2) h with log axis, D with linear axis; (3) h with linear axis, D with log axis; (4) both log axes. According to the plots choose a function (linear, power, exponential, or logarithmic) that best fits the data points and determine the coefficients of the function.
(*b*) Plot the function and the points using linear axes.

24. Write a user-defined function that fits data points to a power function of the form $y = bx^m$. Name the function `[b,m] = powerfit(x,y)`, where the input arguments `x` and `y` are vectors with the coordinates of the data points, and the output arguments `b` and `m` are the constants of the fitted exponential equation. Use `powerfit` to fit the data below. Make a plot that shows the data points and the function.

x	0.5	2.4	3.2	4.9	6.5	7.8
y	0.8	9.3	37.9	68.2	155	198

25. The aerodynamic drag force F_D that is applied to a car is given by:

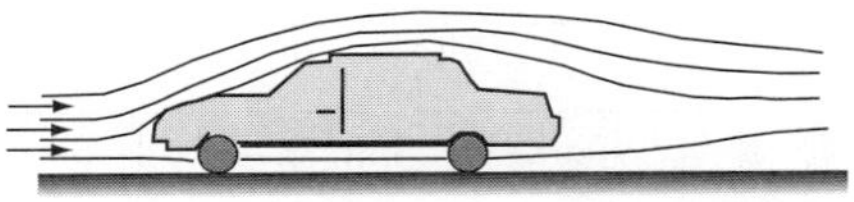

$$F_D = \frac{1}{2}\rho C_D A v^2$$

where $\rho = 1.2$ kg/m^3 is the air density, C_D is the drag coefficient, A is the projected front area of the car, and v is the speed of the car (in units of m/s) relative to the wind. The product $C_D A$ characterizes the air resistance of a car. (At speeds above 70 km/h the aerodynamic drag force is typically more than half of the total resistance to motion.) Data obtained in a wind tunnel test is displayed in the table. Use the data to determine the product $C_D A$ for the tested car using curve fitting. Make a plot of the data points and the curve-fitted equation.

v (km/h)	20	40	60	80	100	120	140	160
F_D (N)	10	50	109	180	300	420	565	771

26. Viscosity is a property of gases and fluids that characterizes their resistance to flow. For most materials viscosity is highly sensitive to temperature. For gases, the variation of viscosity with temperature is frequently modeled by an equation of the form

$$\mu = \frac{CT^{3/2}}{T+S}$$

where μ is the viscosity, T is the absolute temperature, and C and S are empirical constants. Below is a table that gives the viscosity of air at different temperatures (data from B.R. Munson, D.F. Young, and T.H. Okiishi, *Fundamentals of Fluid Mechanics*, 4th ed., John Wiley and Sons, 2002).

T (°C)	−20	0	40	100	200	300	400	500	1,000
μ (N s/m^2) ($\times 10^{-5}$)	1.63	1.71	1.87	2.17	2.53	2.98	3.32	3.64	5.04

Determine the constants C and S by curve-fitting the equation to the data points. Make a plot of viscosity versus temperature (in °C). In the plot show the data points with markers and the curve-fitted equation with a solid line.

The curve fitting can be done by rewriting the equation in the form

$$\frac{T^{3/2}}{\mu} = \frac{1}{C}T + \frac{S}{C}$$

and using a first-order polynomial.

27. Measurements of the fuel efficiency of a car F_E at various speeds v are shown in the table.

v (mi/h)	5	15	25	35	45	55	65	75
F_E (mpg)	11	22	28	29.5	30	30	27	23

(*a*) Curve-fit the data with a second-order polynomial. Use the polynomial to estimate the fuel efficiency at 60 mi/h. Make a plot of the points and the polynomial.
(*b*) Curve-fit the data with a third-order polynomial. Use the polynomial to estimate the fuel efficiency at 60 mi/h. Make a plot of the points and the polynomial.
(*c*) Fit the data with linear and spline interpolations. Estimate the fuel efficiency at 60 mi/h with linear and spline interpolations. Make a plot that shows the data points and curves made of interpolated points.

28. The relationship between two variables P and t is known to be:

$$P = \frac{mt}{b+t}$$

The following data points are given

t	1	3	4	7	8	10
P	2.1	4.6	5.4	6.1	6.4	6.6

Determine the constants m and b by curve-fitting the equation to the data points. Make a plot of P versus t. In the plot show the data points with markers and the curve-fitted equation with a solid line. (The curve fitting can be done by writing the reciprocal of the equation and using a first-order polynomial.)

29. The yield strength, σ_y, of many metals depends on the size of the grains. For these metals the relationship between the yield stress and the average grain diameter d can be modeled by the Hall-Petch equation:

$$\sigma_y = \sigma_0 + kd^{\left(\frac{-1}{2}\right)}$$

The following are results from measurements of average grain diameter and yield stress.

d (mm)	0.005	0.009	0.016	0.025	0.040	0.062	0.085	0.110
σ_y (MPa)	205	150	135	97	89	80	70	67

(*a*) Using curve fitting, determine the constants σ_0 and k in the Hall-Petch equation for this material. Using the constants determine with the equation the yield stress of material with a grain size of 0.05 mm. Make a plot that shows the data points with circle markers and the curve derived from the Hall-Petch equation with a solid line.

(*b*) Use linear interpolation to determine the yield stress of material with a grain size of 0.05 mm. Make a plot that shows the data points with circle markers and the linear interpolation with a solid line.

(*c*) Use cubic interpolation to determine the yield stress of material with a grain size of 0.05 mm. Make a plot that shows the data points with circle markers and cubic interpolation with a solid line.

30. The stress concentration factor k is the ratio between the maximum stress τ_{max} and the average stress τ_{ave}, $k = \tau_{max}/\tau_{ave}$. For a stepped shaft loaded in torsion, with dimensions as shown in the figure, k is a function of r/d and the maximum stress is at the rounded corner. The average stress is given by $\tau_{ave} = (16T)/(\pi d^3)$, where T is the applied torque. The stress concentration factors measured in tests using shafts with $d/D = 2$ and various ratios of r/d are given in the table.

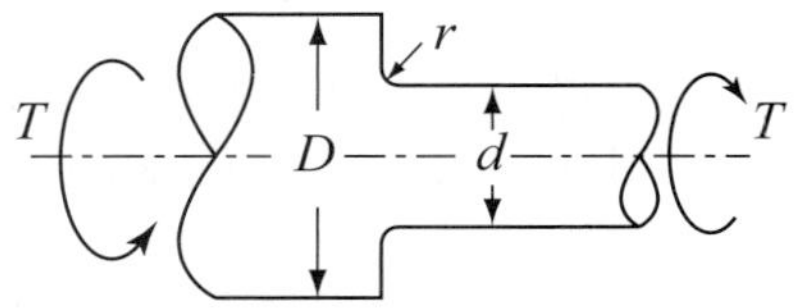

r/d	0.3	0.26	0.22	0.18	0.14	0.1	0.06	0.02
k	1.18	1.19	1.21	1.26	1.32	1.43	1.6	1.98

(*a*) Use an power function $k = b(r/d)^m$ to model the relationship between k and r/d. Determine the values of b and m that best-fit the data.
(*b*) Plot the data points and the curve-fitted model.
(*c*) Use the model to predict the stress concentration factor for $r/d = 0.04$.

31. The ideal gas equation relates the volume, pressure, temperature, and the quantity of a gas by:

$$V = \frac{nRT}{P}$$

where V is the volume in liters, P is the pressure in atm, T is the temperature in kelvins, n is the number of moles, and R is the gas constant.

An experiment is conducted for determining the value of the gas constant R. In the experiment 0.05 mol of gas is compressed to different volumes by applying pressure to the gas. At each volume the pressure and temperature of the gas are recorded. Using the data given below, determine R by plotting V versus T/P and fitting the data points with a linear equation.

V (L)	0.75	0.65	0.55	0.45	0.35
T (°C)	25	37	45	56	65
P (atm)	1.63	1.96	2.37	3.00	3.96

Chapter 9
Applications in Numerical Analysis

Numerical methods are commonly used for solving mathematical problems that are formulated in science and engineering where it is difficult or impossible to obtain exact solutions. MATLAB has a large library of functions for numerically solving a wide variety of mathematical problems. This chapter explains a number of the most frequently used of these functions. It should be pointed out here that the purpose of this book is to show users how to use MATLAB. Some general information on the numerical methods is given, but the details, which can be found in books on numerical analysis, are not included.

The following topics are presented in this chapter: solving an equation with one unknown, finding a minimum or a maximum of a function, numerical integration, and solving a first-order ordinary differential equation.

9.1 Solving an Equation with One Variable

An equation with one variable can be written in the form $f(x) = 0$. A solution to the equation (also called a root) is a numerical value of x that satisfies the equation. Graphically, a solution is a point where the function $f(x)$ crosses or touches the x axis. An exact solution is a value of x for which the value of the function is exactly zero. If such a value does not exist or is difficult to determine, a numerical solution can be determined by finding an x that is very close to the solution. This is done by the iterative process, where in each iteration the computer determines a value of x that is closer to the solution. The iterations stop when the difference in x between two iterations is smaller than some measure. In general, a function can have zero, one, several, or an infinite number of solutions.

In MATLAB a zero of a function can be determined with the command (built-in function) `fzero` with the form:

```
x = fzero(function,x0)
```

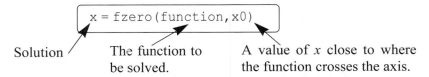

Solution — The function to be solved. — A value of x close to where the function crosses the axis.

The built-in function `fzero` is a MATLAB function function (see Section 7.9), which means that it accepts another function (the function to be solved) as an input argument.

Additional details on the arguments of `fzero`:

- `x` is the solution, which is a scalar.
- `function` is the function to be solved. It can be entered in several different ways:
  1. The simplest way is to enter the mathematical expression as a string.
  2. The function is created as a user-defined function in a function file and then the function handle is entered (see Section 7.9.1).
  3. The function is created as an anonymous function (see Section 7.8.1) and then the name of the anonymous function (which is the name of the handle) is entered (see Section 7.9.1).

  (As explained in Section 7.9.2, it is also possible to pass a user-defined function and an inline function into a function function by using its name. However, function handles are more efficient and easier to use, and should be the preferred method.)

- The function has to be written in a standard form. For example, if the function to be solved is $xe^{-x} = 0.2$, it has to be written as $f(x) = xe^{-x} - 0.2 = 0$. If this function is entered into the `fzero` command as a string, it is typed as: `'x*exp(-x)-0.2'`.
- When a function is entered as an expression (string), it cannot include predefined variables. For example, if the function to be entered is $f(x) = xe^{-x} - 0.2$, it is not possible to define `b=0.2` and then enter `'x*exp(-x)-b'`.
- `x0` can be a scalar or a two-element vector. If it is entered as a scalar, it has to be a value of x near the point where the function crosses (or touches) the x axis. If `x0` is entered as a vector, the two elements have to be points on opposite sides of the solution. If $f(x)$ crosses the x axis, then $f(x0(1))$ has a different sign than $f(x0(2))$. When a function has more than one solution, each solution can be determined separately by using the `fzero` function and entering values for `x0` that are near each of the solutions.

- A good way to find approximately where a function has a solution is to make a plot of the function. In many applications in science and engineering the domain of the solution can be estimated. Often when a function has more than one solution only one of the solutions will have a physical meaning.

Sample Problem 9-1: Solving a nonlinear equation

Determine the solution of the equation $xe^{-x} = 0.2$.

Solution

The equation is first written in the form of a function: $f(x) = xe^{-x} - 0.2$. A plot of the function, shown on the right, shows that the function has one solution between 0 and 1 and another solution between 2 and 3. The plot is obtained by typing

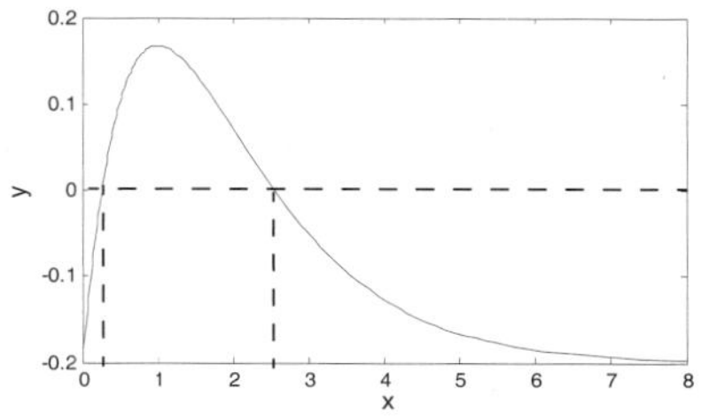


```
>> fplot('x*exp(-x)-0.2',[0 8])
```

in the Command Window. The solutions of the function are found by using the `fzero` command twice. First the equation is entered as a string expression, and a value of `x0` between 0 and 1 (`x0` = 0.7) is used. Second, the equation to be solved is written as an anonymous function, which is then used in `fzero` with `x0` between 2 and 3 (`x0` = 2.8). This is shown below:

```
>> x1=fzero('x*exp(-x)-0.2',0.7)      The function is entered as a string expression.
x1 =
    0.2592                             The first solution is 0.2592.
>> F=@(x)x*exp(-x)-0.2
F =                                    Creating an anonymous function.
    @(x)x*exp(-x)-0.2
>> fzero(F,2.8)                        Using the name of the anonymous function in fzero.
ans =
    2.5426                             The second solution is 2.5426.
```

Additional comments:

- The `fzero` command finds zeros of a function only where the function crosses the x axis. The command does not find a zero at points where the function touches but does not cross the x axis.
- If a solution cannot be determined, `NaN` is assigned to `x`.

- The `fzero` command has additional options (see the Help Window). Two of the more important options are:
  `[x fval]=fzero(function, x0)` assigns the value of the function at `x` to the variable `fval`.
  `x=fzero(function, x0, optimset('display','iter'))` displays the output of each iteration during the process of finding the solution.
- When the function can be written in the form of a polynomial, the solution, or the roots, can be found with the `roots` command, as explained in Chapter 8 (Section 8.1.2).
- The `fzero` command can also be used to find the value of x where the function has a specific value. This is done by translating the function up or down. For example, in the function of Sample Problem 9-1 the first value of x where the function is equal to 0.1 can be determined by solving the equation $xe^{-x} - 0.3 = 0$. This is shown below:

```
>> x=fzero('x*exp(-x)-0.3',0.5)
x =
     0.4894
```

9.2 Finding a Minimum or a Maximum of a Function

In many applications there is a need to determine the local minimum or maximum of a function of the form $y = f(x)$. In calculus the value of x that corresponds to a local minimum or maximum is determined by finding the zero of the derivative of the function. The value of y is determined by substituting the x into the function. In MATLAB the value of x where a one-variable function $f(x)$ within the interval $x_1 \le x \le x_2$ has a minimum can be determined with the `fminbnd` command which has the form:

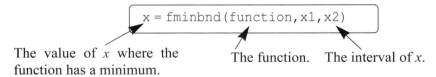

```
x = fminbnd(function,x1,x2)
```

The value of x where the function has a minimum. The function. The interval of x.

- The function can be entered as a string expression, or as a function handle, in the same way as with the `fzero` command. See Section 9.1 for details.
- The value of the function at the minimum can be added to the output by using the option

  `[x fval]=fminbnd(function,x1,x2)`

  where the value of the function at `x` is assigned to the variable `fval`.
- Within a given interval, the minimum of a function can either be at one of the end points of the interval or at a point within the interval where the slope of the

function is zero (local minimum). When the fminbnd command is executed, MATLAB looks for a local minimum. If a local minimum is found, its value is compared to the value of the function at the end points of the interval. MATLAB returns the point with the actual minimum value for the interval.

For example, consider the function $f(x) = x^3 - 12x^2 + 40.25x - 36.5$, which is plotted in the interval $0 \le x \le 8$ in the figure on the right. It can be observed that there is a local minimum between 5 and 6, and that the absolute minimum is at $x = 0$. Using the fminbnd command with the interval $3 \le x \le 8$ to find the location of the local minimum and the value of the function at this point gives:

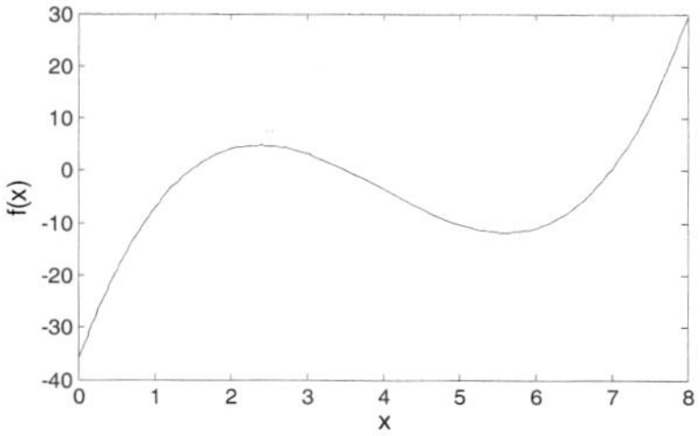


```
>> [x fval]=fminbnd('x^3-12*x^2+40.25*x-36.5',3,8)
x =
    5.6073
fval =
  -11.8043
```

The local minimum is at $x = 5.6073$. The value of the function at this point is –11.8043.

Notice that the fminbnd command gives the local minimum. If the interval is changed to $0 \le x \le 8$, fminbnd gives:

```
>> [x fval]=fminbnd('x^3-12*x^2+40.25*x-36.5',0,8)
x =
     0
fval =
  -36.5000
```

The minimum is at $x = 0$. The value of the function at this point is –36.5.

For this interval the fminbnd command gives the absolute minimum which is at the end point $x = 0$.

- The fminbnd command can also be used to find the maximum of a function. This is done by multiplying the function by –1 and finding the minimum. For example, the maximum of the function $f(x) = xe^{-x} - 0.2$ (from Sample Problem 9-1) in the interval $0 \le x \le 8$ can be determined by finding the minimum of the function $f(x) = -xe^{-x} + 0.2$ as shown below:

```
>> [x fval]=fminbnd('-x*exp(-x)+0.2',0,8)
x =
    1.0000
fval =
   -0.1679
```

The maximum is at $x = 1.0$. The value of the function at this point is 0.1679.

9.3 NUMERICAL INTEGRATION

Integration is a common mathematical operation in science and engineering. Calculating area and volume, velocity from acceleration, and work from force and displacement are just a few examples where integrals are used. Integration of simple functions can be done analytically, but more involved functions are frequently difficult or impossible to integrate analytically. In calculus courses the integrand (the quantity to be integrated) is usually a function. In applications of science and engineering the integrand can be a function or a set of data points. For example, data points from discrete measurements of flow velocity can be used to calculate volume.

It is assumed in the presentation below that the reader has knowledge of integrals and integration. A definite integral of a function $f(x)$ from a to b has the form:

$$q = \int_a^b f(x)dx$$

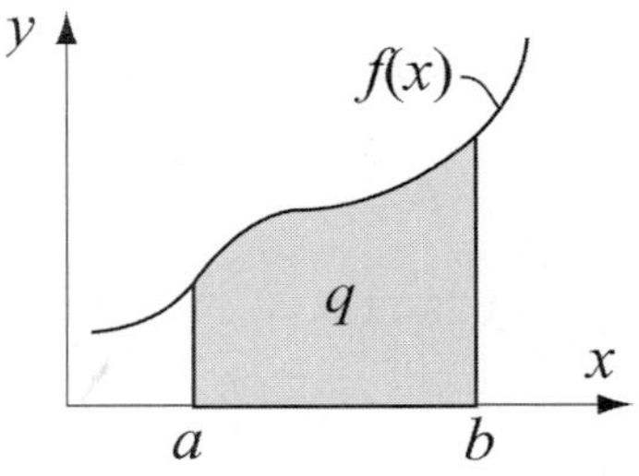

The function $f(x)$ is called the integrand, and the numbers a and b are the limits of integration. Graphically, the value of the integral q is the area between the graph of the function, the x axis, and the limits a and b (the shaded area in the figure). When a definite integral is calculated analytically $f(x)$ is always a function. When the integral is calculated numerically $f(x)$ can be a function or a set of points. In numerical integration the total area is obtained by dividing the area into small sections, calculating the area of each section, and adding them up. Various numerical methods have been developed for this purpose. The difference between the methods is in the way that the area is divided into sections and the method by which the area of each section is calculated. Books on numerical analysis include details of the numerical techniques.

The following discussion describes how to use the three MATLAB built-in integration functions `quad`, `quadl`, and `trapz`. The `quad` and `quadl` commands are used for integration when $f(x)$ is a function, and `trapz` is used when $f(x)$ is given by data points.

The `quad` command:

The form of the `quad` command, which uses the adaptive Simpson method of integration, is:

```
q = quad(function,a,b)
```

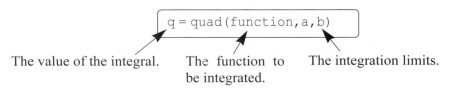

The value of the integral. — The function to be integrated. — The integration limits.

- The function can be entered as a string expression or as a function handle, in the same way as with the `fzero` command. See Section 9.1 for details. The first two methods are demonstrated in Sample Problem 9-2.
- The function $f(x)$ must be written for an argument x that is a vector (use element-by-element operations) such that it calculates the value of the function for each element of x.
- The user has to make sure that the function does not have a vertical asymptote between a and b.
- `quad` calculates the integral with an absolute error that is smaller than 1.0e–6. This number can be changed by adding an optional `tol` argument to the command:

  ```
  q = quad('function',a,b,tol)
  ```

  `tol` is a number that defines the maximum error. With larger `tol` the integral is calculated less accurately but faster.

The `quadl` command:

The form of the `quadl` (the last letter is a lowercase L) command is exactly the same as that of the `quad` command:

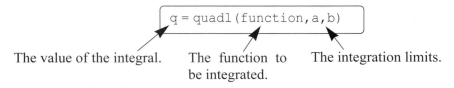

```
q = quadl(function,a,b)
```

The value of the integral. The function to be integrated. The integration limits.

All of the comments that are listed for the `quad` command are valid for the `quadl` command. The difference between the two commands is the numerical method used for calculating the integration. The `quadl` command uses the adaptive Lobatto method, which can be more efficient for high accuracies and smooth integrals.

Sample Problem 9-2: Numerical integration of a function

Use numerical integration to calculate the following integral:

$$\int_0^8 (xe^{-x^{0.8}} + 0.2)dx$$

Solution

For illustration, a plot of the function for the interval $0 \le x \le 8$ is shown on the right. The solution uses the quad command and shows how to enter the function in the command in two ways. In the first, it is entered directly by typing the expression as an argument. In the second, an anonymous function is created and its name is subsequently entered in the command.

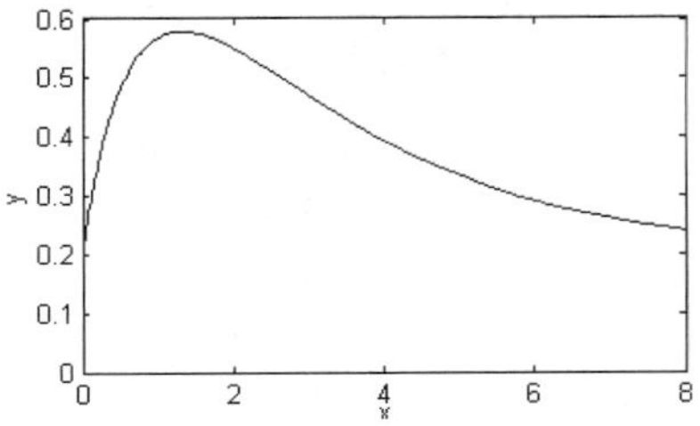

The use of the quad command in the Command Window, with the function to be integrated typed in as a string, is shown below. Note that the function is typed with element-by-element operations.

```
>> quad('x.*exp(-x.^0.8)+0.2',0,8)
ans =
    3.1604
```

The second method is to first create a user-defined function that calculates the function to be integrated. The function file (named y=Chap9Sam2(x)) is:

```
function y=Chap9Sam2(x)
y=x.*exp(-x.^0.8)+0.2;
```

Note again that the function is written with element-by-element operations such that the argument x can be a vector. The integration is then done in the Command Window by typing the handle @Chap9Sam2 for the argument function in the quad command as shown below:

```
>> q=quad(@Chap9Sam2,0,8)
q =
    3.1604
```

The trapz command:

The trapz command can be used for integrating a function that is given as data points. It uses the numerical trapezoidal method of integration. The form of the command is

```
q = trapz(x,y)
```

where x and y are vectors with the x and y coordinates of the points, respectively. The two vectors must be of the same length.

9.4 *Ordinary Differential Equations*

Differential equations play a crucial role in science and engineering since they are in the foundation of virtually every physical phenomenon that is involved in engineering applications. Only a limited number of differential equations can be solved analytically. Numerical methods, on the other hand, can result in an approximate solution to almost any equation. Obtaining a numerical solution might not be simple task however. This is because a numerical method that can solve any equation does not exist. Instead, there are many methods that are suitable for solving different types of equations. MATLAB has a large library of tools that can be used for solving differential equations. To fully utilize the power of MATLAB, however, requires that the user have knowledge of differential equations and the various numerical methods that can be used for solving them.

This section describes in detail how to use MATLAB to solve a first-order ordinary differential equation. The possible numerical methods that can be used for solving such an equation are described in general terms, but are not explained from a mathematical point of view. This section provides information for solving simple, "nonproblematic" first-order equations. This solution provides the basis for solving higher-order equations and systems of equations.

An ordinary differential equation (ODE) is an equation that contains an independent variable, a dependent variable, and derivatives of the dependent variable. The equations that are considered here are of first order with the form

$$\frac{dy}{dx} = f(x, y)$$

where x and y are the independent and dependent variables, respectively. A solution is a function $y = f(x)$ that satisfies the equation. In general, many functions can satisfy a given ODE, and more information is required for determining the solution of a specific problem. The additional information is the value of the function (the dependent variable) at some value of the independent variable.

Steps for solving a single first-order ODE:

For the remainder of this section the independent variable is taken as t (time). This is done because in many applications time is the independent variable, and also to be consistent with the information in the **Help** menu of MATLAB.

Step 1: Write the problem in a standard form.

Write the equation in the form:

$$\frac{dy}{dt} = f(t, y) \quad \text{for } t_0 \le t \le t_f, \text{ with } \quad y = y_0 \text{ at } t = t_0.$$

As shown above, three pieces of information are needed for solving a first order ODE: An equation that gives an expression for the derivative of y with respect to t, the interval of the independent variable, and the initial value of y. The solution is the value of y as a function of t between t_0 and t_f.

An example of a problem to solve is:

$$\frac{dy}{dt} = \frac{t^3 - 2y}{t} \quad \text{for } 1 \le t \le 3 \quad \text{with } y = 4.2 \text{ at } t = 1\,.$$

Step 2: Create a user-defined function (in a function file) or an anonymous function.

The ODE to be solved has to be written as a user-defined function (in a function file) or as an anonymous function. Both calculate $\frac{dy}{dt}$ for given values of t and y. For the example problem above, the user-defined function (which is saved as a separate file) is:

```
function dydt=ODEexp1(t,y)
dydt=(t^3-2*y)/t;
```

When an anonymous function is used, it can be defined in the Command Window, or be within a script file. For the example problem here the anonymous function (named ode1) is:

```
>> ode1=@(t,y)(t^3-2*y)/t
ode1 =
    @(t,y)(t^3-2*y)/t
```

Step 3: Select a method of solution.

Select the numerical method that you would like MATLAB to use in the solution. Many numerical methods have been developed to solve first-order ODEs, and several of the methods are available as built-in functions in MATLAB. In a typical numerical method, the time interval is divided into small time steps. The solution starts at the known point y_0, and then by using one of the integration methods the value of y is calculated at each time step. Table 9-1 lists seven ODE solver commands, which are MATLAB built-in functions that can be used for solving a first-order ODE. A short description of each solver is included in the table.

Table 9-1: MATLAB ODE Solvers

ODE Solver Name	Description
`ode45`	For nonstiff problems, one-step solver, best to apply as a first try for most problems. Based on explicit Runge-Kutta method.
`ode23`	For nonstiff problems, one-step solver. Based on explicit Runge-Kutta method. Often quicker but less accurate than `ode45`.
`ode113`	For nonstiff problems, multistep solver.

Table 9-1: MATLAB ODE Solvers (Continued)

ODE Solver Name	Description
`ode15s`	For stiff problems, multistep solver. Use if `ode45` failed. Uses a variable order method.
`ode23s`	For stiff problems, one-step solver. Can solve some problems that `ode15s` cannot.
`ode23t`	For moderately stiff problems.
`ode23tb`	For stiff problems. Often more efficient than `ode15s`.

In general, the solvers can be divided into two groups according to their ability to solve stiff problems and according to whether they use on-step or multi-step methods. Stiff problems are ones that include fast and slowly changing components and require small time steps in their solution. One-step solvers use information from one point to obtain a solution at the next point. Multistep solvers use information from several previous points to find the solution at the next point. The details of the different methods are beyond the scope of this book.

It is impossible to know ahead of time which solver is the most appropriate for a specific problem. A suggestion is to first try `ode45`, which gives good results for many problems. If a solution is not obtained because the problem is stiff, trying the solver `ode15s` is suggested.

Step 4: Solve the ODE.

The form of the command that is used to solve an initial value ODE problem is the same for all the solvers and for all the equations that are solved. The form is:

```
[t,y] = solver_name(ODEfun,tspan,y0)
```

Additional information:

`solver_name` Is the name of the solver (numerical method) that is used (e.g. `ode45` or `ode23s`)

`ODEfun` The function from Step 2 that calculates $\frac{dy}{dt}$ for given values of t and y. If it was written as a user-defined function, the function handle is entered. If it was written as an anonymous function, the name of the anonymous function is entered. (See the example that follows.)

`tspan` A vector that specifies the interval of the solution. The vector must have at least two elements but can have more. If the vector has only two elements, the elements must be `[t0 tf]`, which are the initial and final points of the solution interval. The vector `tspan` can have, however, additional points between the first and last points. The number of elements in `tspan` affects the output from the command. See `[t,y]` below.

`y0` The initial value of y (the value of y at the first point of the interval).

`[t,y]` The output, which is the solution of the ODE. `t` and `y` are column vectors. The first and the last points are the beginning and end points of the interval. The spacing and number of points in between depends on the input vector `tspan`. If `tspan` has two elements (the beginning and end points), the vectors `t` and `y` contain the solution at every integration step calculated by the solver. If `tspan` has more than two points (additional points between the first and the last), the vectors `t` and `y` contain the solution only at these points. The number of points in `tspan` does not affect the time steps used for the solution by the program.

For example, consider the solution to the problem stated in Step 1:

$$\frac{dy}{dt} = \frac{t^3 - 2y}{t} \quad \text{for } 1 \le t \le 3 \quad \text{with} \quad y = 4.2 \quad \text{at } t = 1 \text{,}$$

If the ODE function is written as a user-defined function (see Step 2), then the solution with MATLAB's built-in function `ode45` is obtained by:

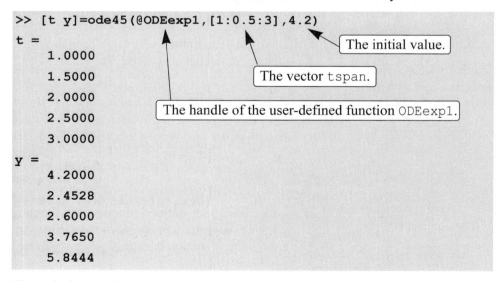

```
>> [t y]=ode45(@ODEexp1,[1:0.5:3],4.2)
t =
    1.0000
    1.5000
    2.0000
    2.5000
    3.0000
y =
    4.2000
    2.4528
    2.6000
    3.7650
    5.8444
```

The initial value.

The vector `tspan`.

The handle of the user-defined function `ODEexp1`.

The solution is obtained with the solver `ode45`. The name of the user-defined function from Step 2 is `ODEexp1`. The solution starts at $t = 1$ and ends at $t = 3$ with increments of 0.5 (according to the vector `tspan`). To show the solution, the problem is solved again below using `tspan` with smaller spacing, and the solution is plotted with the `plot` command.

```
>> [t y]=ode45(@ODEexp1,[1:0.01:3],4.2);
>> plot(t,y)
>> xlabel('t'), ylabel('y')
```

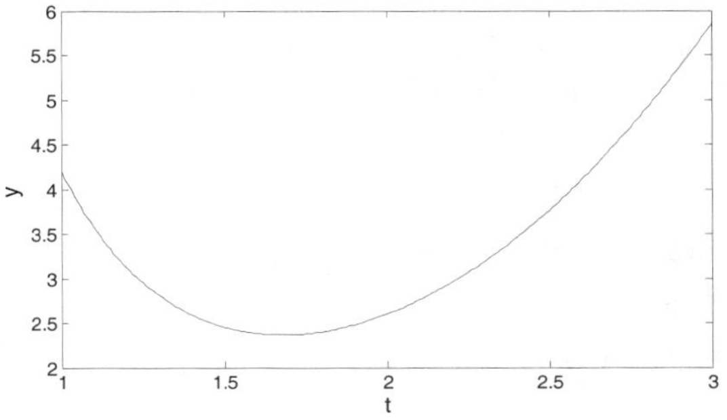

If the ODE function is written as an anonymous function called `ode1` (see Step 2), then the solution (same as shown above) is obtained by typing: `[t y]=ode45(ode1,[1:0.5:3],4.2)`.

9.5 EXAMPLES OF MATLAB APPLICATIONS

Sample Problem 9-3: The gas equation

The ideal gas equation relates the volume (V in L), temperature (T in K), pressure (P in atm), and the amount of gas (number of moles n) by:

$$p = \frac{nRT}{V}$$

where $R = 0.08206$ (L atm)/(mol K) is the gas constant.

The van der Waals equation gives the relationship between these quantities for a real gas by

$$\left(P + \frac{n^2 a}{V^2}\right)(V - nb) = nRT$$

where a and b are constants that are specific for each gas.

Use the `fzero` function to calculate the volume of 2 mol CO_2 at temperature of 50° C, and pressure of 6 atm. For CO_2, $a = 3.59$ (L^2 atm)/mol^2, and $b = 0.0427$ L/mol.

Solution

The solution written in a script file is shown below.

```
global P T n a b R
```

```
R=0.08206;
P=6; T=323.2; n=2; a=3.59; b=0.047;
Vest=n*R*T/P;
V=fzero(@Waals,Vest)
```

Calculating an estimated value for V.

Function handle `@waals` is used to pass the user-defined function `waals` into `fzero`.

The program first calculates an estimated value of the volume using the ideal gas equation. This value is then used in the `fzero` command for the estimate of the solution. The van der Waals equation is written as a user-defined function named `Waals`, which is shown below:

```
function fofx=Waals(x)
global P T n a b R
fofx=(P+n^2*a/x^2)*(x-n*b)-n*R*T;
```

In order for the script and function files to work correctly, the variables `P`, `T`, `n`, `a`, `b`, and `R` are declared global. When the script file (saved as Chap9SamPro3) is executed in the Command Window, the value of V is displayed, as shown next:

```
>> Chap9SamPro3
V =
    8.6613
```

The volume of the gas is 8.6613 L.

Sample Problem 9-4: Maximum viewing angle

To get the best view of a movie, a person has to sit at a distance x from the screen such that the viewing angle θ is maximum. Determine the distance x for which θ is maximum for the configuration shown in the figure.

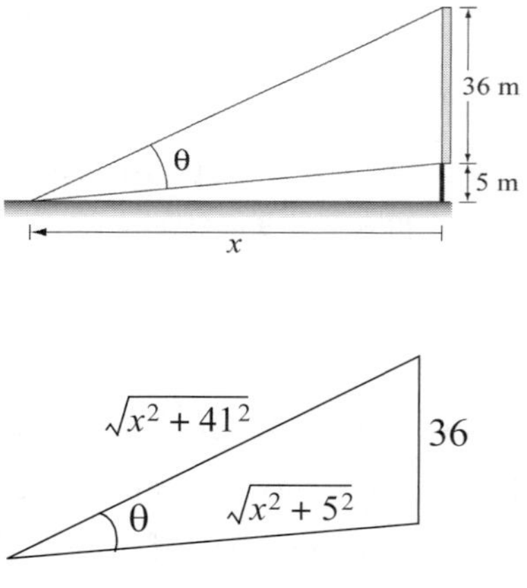

Solution

The problem is solved by writing a function for the angle θ in terms of x, and then finding the x for which the angle is maximum. In the triangle that includes θ, one side is given (the height of the screen), and the other two sides can be written in terms of x, as shown in the figure. One way in which θ can be written in terms of x is by using the Law of Cosines:

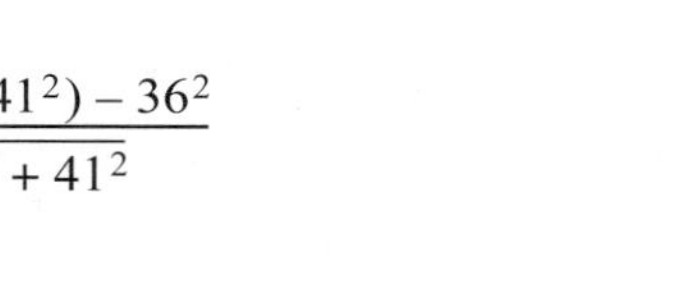

$$\cos(\theta) = \frac{(x^2+5^2)+(x^2+41^2)-36^2}{2\sqrt{x^2+5^2}\sqrt{x^2+41^2}}$$

The angle θ is expected to be between 0 and π/2. Since cos(0) = 1 and the cosine is decreasing with increasing θ, the maximum angle corresponds to the smallest cos(θ). A plot of cos(θ) as a function of x shows that the function has a minimum between 10 and 20. The commands for the plot are:

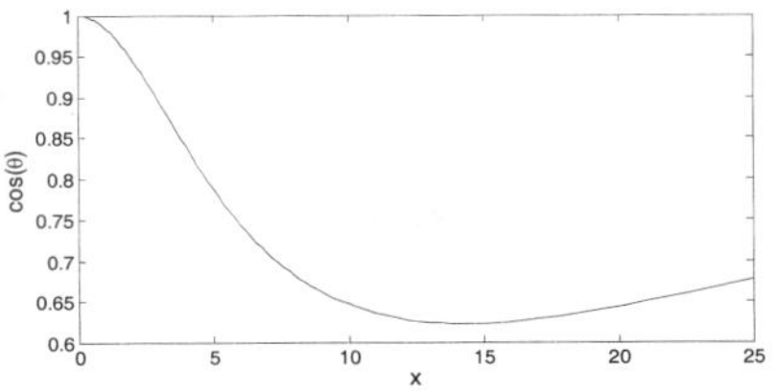


```
>>fplot('((x^2+5^2)+(x^2+41^2)-36^2)/(2*sqrt(x^2+ 5^2)*sqrt(x^2+
                                                   41^2))',[0 25])
>> xlabel('x'); ylabel('cos(\theta)')
```

The minimum can be determined with the `fminbnd` command:

```
>>[x anglecos]=fminbnd('((x^2+5^2)+(x^2+41^2)-36^2)/
                    (2*sqrt(x^2+5^2)*sqrt(x^2+41^2))',10,20)
x =
   14.3178
anglecos =
    0.6225
>> angle=anglecos*180/pi
angle =
   35.6674
```

The minimum is at x = 14.3178 m. At this point cos(θ) = 0.6225.

In degrees the angle is 35.6674°.

Sample Problem 9-5: Water flow in a river

To estimate the amount of water that flows in a river during a year, a section of the river is made to have a rectangular cross section as shown. In the beginning of every month (starting at January 1st) the height h of the water and the speed v of the water flow are measured. The first day of measurement is taken as 1, and the last day—which is January 1st of the next year—is day 366. The following data was measured:

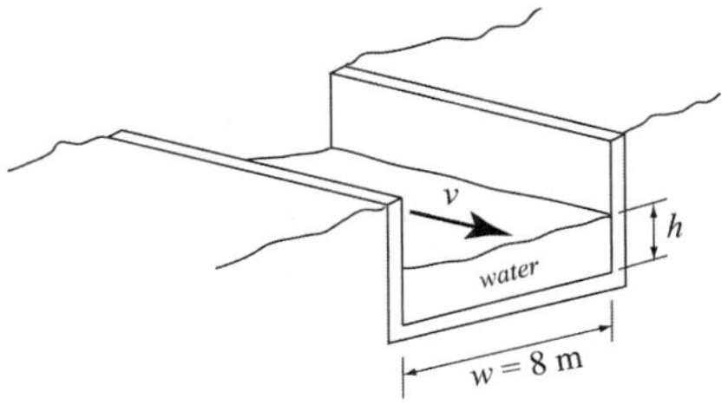

Day	1	32	60	91	121	152	182	213	244	274	305	335	366
h (m)	2.0	2.1	2.3	2.4	3.0	2.9	2.7	2.6	2.5	2.3	2.2	2.1	2.0
v (m/s)	2.0	2.2	2.5	2.7	5	4.7	4.1	3.8	3.7	2.8	2.5	2.3	2.0

Use the data to calculate the flow rate, and then integrate the flow rate to obtain an estimate of the total amount of water that flows in the river during a year.

Solution

The flow rate, Q (volume of water per second), at each data point is obtained by multiplying the water speed by the width and height of the cross-sectional area of the water that flows in the channel:

$$Q = vwh \quad (\text{m}^3/\text{s})$$

The total amount of water that flows is estimated by the integral:

$$V = (60 \cdot 60 \cdot 24) \int_{t_1}^{t_2} Q dt$$

The flow rate is given in cubic meters per second, which means that time must have units of seconds. Since the data is given in terms of days, the integral is multiplied by $(60 \cdot 60 \cdot 24)$ s/day.

The following is a program written in a script file that first calculates Q and then carries out the integration using the `trapz` command. The program also generates a plot of the flow rate versus time.

```
w=8;
d=[1 32 60 91 121 152 182 213 244 274 305 335 366];
h=[2 2.1 2.3 2.4 3.0 2.9 2.7 2.6 2.5 2.3 2.2 2.1 2.0];
speed=[2 2.2 2.5 2.7 5 4.7 4.1 3.8 3.7 2.8 2.5 2.3 2];
Q=speed.*w.*h;
Vol=60*60*24*trapz(d,Q);
fprintf('The estimated amount of water that flows in the
river in a year is %g cubic meters.',Vol)
plot(d,Q)
xlabel('Day'), ylabel('Flow Rate (m^3/s)')
```

When the file (saved as Chap9SamPro5) is executed in the Command Window, the estimated amount of water is displayed and the plot is generated. Both are shown below:.

```
>> Chap9SamPro5
The estimated amount of water that flows in the river in a
year is 2.03095e+009 cubic meters.
```

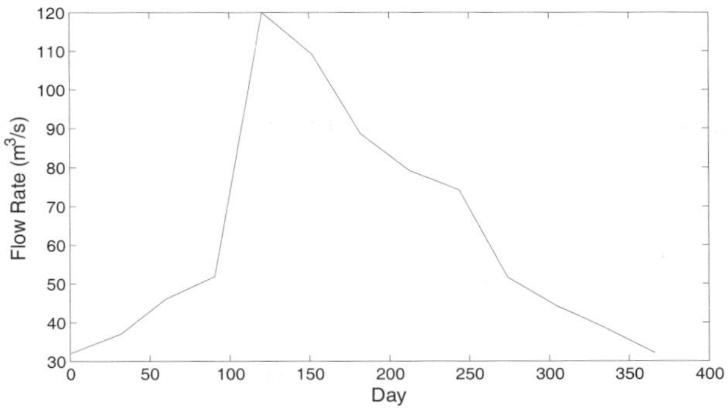

Sample Problem 9-6: Car crash into a safety bumper

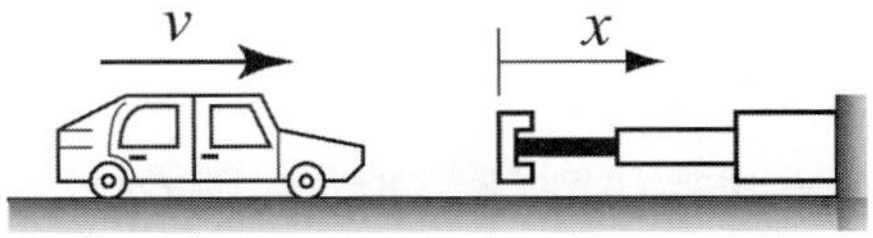

A safety bumper is placed at the end of a racetrack to stop out-of-control cars. The bumper is designed such that the force that the bumper applies to the car is a function of the velocity v and the displacement x of the front edge of the bumper according to the equation:

$$F = Kv^3(x+1)^3$$

where $K = 30$ (s kg)/m^5 is a constant.

A car with a mass m of 1,500 kg hits the bumper at a speed of 90 km/h. Determine and plot the velocity of the car as a function of its position for $0 \le x \le 3$ m.

Solution

The deceleration of the car once it hits the bumper can be calculated from Newton's second law of motion,

$$ma = -Kv^3(x+1)^3$$

which can be solved for the acceleration a as a function of v and x:

$$a = \frac{-Kv^3(x+1)^3}{m}$$

The velocity as a function of x can be calculated by substituting the acceleration in the equation

$$vdv = adx$$

which gives

$$\frac{dv}{dx} = \frac{-Kv^2(x+1)^3}{m}$$

The last equation is a first-order ODE that needs to be solved for the interval $0 \le x \le 3$ with the initial condition $v = 90$ km/h at $x = 0$.

A numerical solution of the differential equation with MATLAB is shown in the following program, which is written in a script file:

```
global k m
k=30; m=1500; v0=90;
xspan=[0:0.2:3];
v0mps=v0*1000/3600;
[x v]=ode45(@bumper,xspan,v0mps)
plot(x,v)
xlabel('x (m)'); ylabel('velocity (m/s)')
```

A vector that specifies the interval of the solution.

Changing the units of v_0 to m/s.

Solving the ODE.

Note that the function handle `@bumper` is used for passing the user-defined function `bumper` into `ode45`. The listing of the user-defined function with the differential equation, named `bumper`, is:

```
function dvdx=bumper(x,v)
global k m
dvdx=-(k*v^2*(x+1)^3)/m;
```

When the script file executes (saved as Chap9SamPro6) the vectors `x` and `v` are displayed in the Command Window (actually, they are displayed on the screen one after the other, but to save room they are displayed below next to each other).

```
>> Chap9SamPro6
x =              v =
         0          25.0000
    0.2000          22.0420
    0.4000          18.4478
    0.6000          14.7561
    0.8000          11.4302
    1.0000           8.6954
    1.2000           6.5733
    1.4000           4.9793
    1.6000           3.7960
    1.8000           2.9220
    2.0000           2.2737
    2.2000           1.7886
    2.4000           1.4226
    2.6000           1.1435
    2.8000           0.9283
```

```
3.0000          0.7607
```

The plot generated by the program of the velocity as a function of distance is:

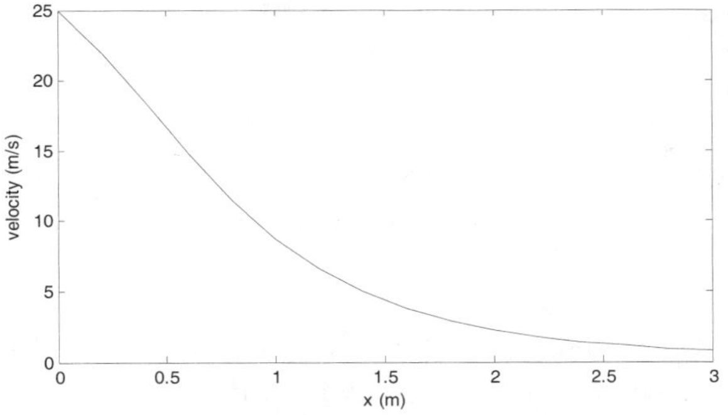

9.6 PROBLEMS

1. Determine the solution of the equation $e^{0.5x} - \sqrt{x} = 3$.

2. Determine the solution of the equation $3 + 3\sin x = 0.5x^3$.

3. Determine the three positive roots of the equation $x^3 - 8x^2 + 17x + \sqrt{x} = 10$.

4. Determine the positive roots of the equation $x^2 - 5x\sin(3x) + 3 = 0$.

5. A block of mass $m = 20$ kg is being pulled by a cable as shown. The force that is required to move the box is given by:

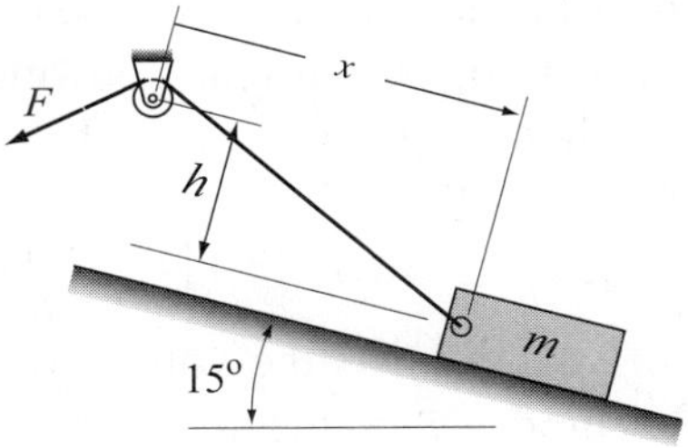

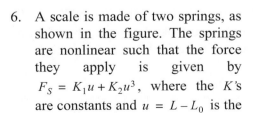

$$F = \frac{(\mu mg\cos 15° + mg\sin 15°)\sqrt{x^2 + h^2}}{x + \mu h}$$

where $h = 8$ m, $\mu = 0.45$ is the friction coefficient, and $g = 9.81$ m/s². Determine the distance x when the pulling force is equal to 230 N.

6. A scale is made of two springs, as shown in the figure. The springs are nonlinear such that the force they apply is given by $F_S = K_1u + K_2u^3$, where the K's are constants and $u = L - L_0$ is the

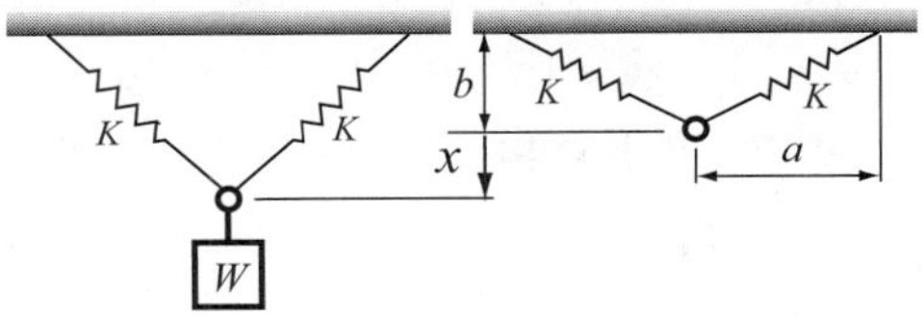

elongation of the spring ($L = \sqrt{a^2 + (b + x)^2}$ and $L_0 = \sqrt{a^2 + b^2}$ are the current and initial lengths of the springs, respectively). Initially, the springs are not stretched. When an object is attached to the ring, the springs stretch and the ring is displaced downward a distance x. The weight of the object can be expressed in terms of the distance x by:

$$W = 2F_S\frac{(b+x)}{L}$$

For the given scale $a = 0.22$ m, $b = 0.08$ m, and the springs' constants are $K_1 = 1600$ N/m and $K_2 = 100000$ N/m^3. Plot W as a function of x for $0 \le x \le 0.25$. Determine the distance x when a 400 N object is attached to the scale.

7. An estimate of the minimum velocity required for a round flat stone to skip when it hits the water is given by (Lyderic Bocquet, "The Physics of Stone Skipping," Am. J. Phys., vol. 71, no. 2, February 2003)

$$V = \frac{\sqrt{\dfrac{16Mg}{\pi C\rho_w d^2}}}{\sqrt{1-\dfrac{8M\tan^2\beta}{\pi d^3 C\rho_w \sin\theta}}}$$

where M and d are the stone mass and diameter, ρ_w is the water density, C is a coefficient, θ is the tilt angle of the stone, β is the incidence angle, and $g = 9.81$ m/s^2. Determine d if $V = 0.8$ m/s. (Assume that $M = 0.1$ kg, $C = 1$, $\rho_w = 1000$ kg/m^3, and $\beta = \theta = 10°$.)

8. The diode in the circuit shown is forward biased. The current I flowing through the diode is given by:

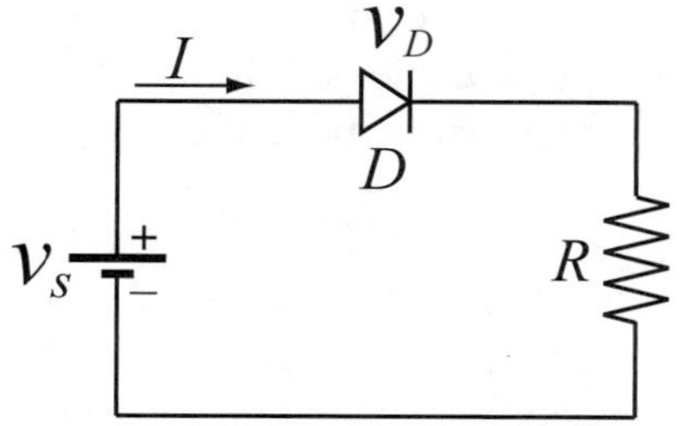

$$I = I_S\left(e^{\frac{qv_D}{kT}} - 1\right)$$

where v_D is the voltage drop across the diode, T is the temperature in kelvins, $I_S = 10^{-12}$ A is the saturation current, $q = 1.6 \times 10^{-19}$ coulombs is the elementary charge value, and $k = 1.38 \times 10^{-23}$ joule/K is Boltzmann's constant. The current I flowing through the circuit (the same as the current in the diode) is given also by:

$$I = \frac{v_S - v_D}{R}$$

Determine v_D if $v_S = 2$ V, $T = 297$ K, and $R = 1000\ \Omega$. (Substitute I from one equation into the other equation and solve the resulting nonlinear equation.)

9. Determine the minimum and the maximum of the function

$$f(x) = \frac{x-2}{[(x-2)^4+2]^{1.8}}.$$

10. A paper cup shaped as a cone is designed to have a volume of 250 cm^3. Determine the radius R and height h such that the least amount of paper will be used for making the cup.

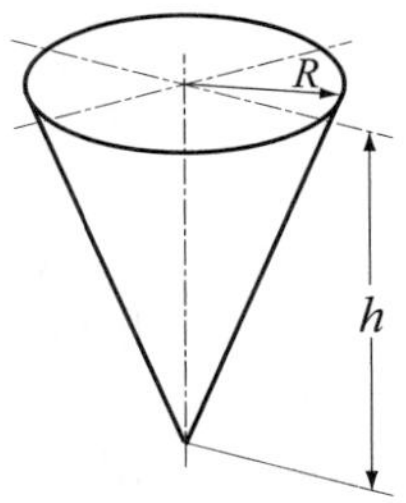

11. Consider again the block that is being pulled in Problem 5. Determine the distance x at which the force that is necessary to pull the box is the smallest. What is the magnitude of this force?

12. Determine the dimensions (radius r and height h) and the volume of the cylinder with the largest volume that can be made inside of a sphere with a radius R of 14 in.

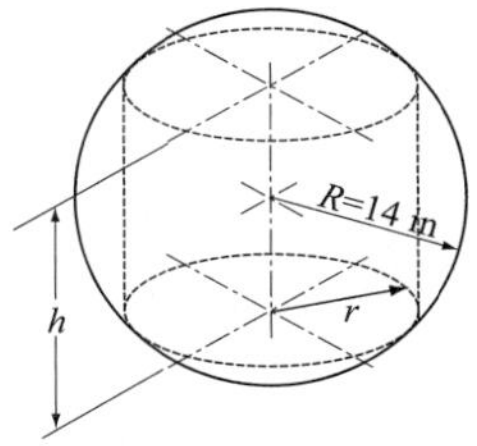

13. Consider the ellipse $\frac{x^2}{19^2}+\frac{y^2}{5^2} = 1$. Determine the sides a and b of the rectangle with the largest area that can be enclosed by the ellipse.

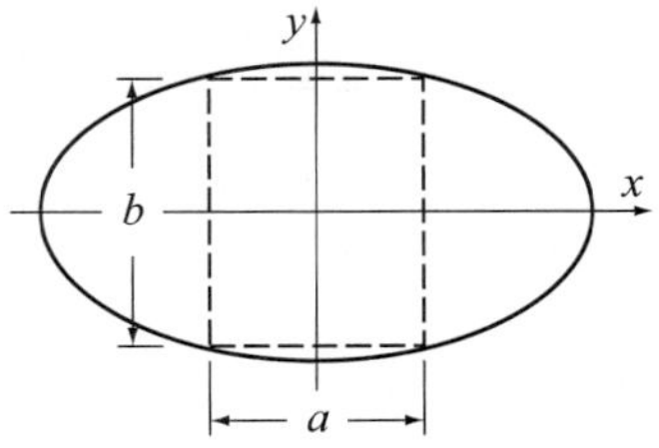

14. Planck's radiation law gives the spectral radiancy R as a function of the wave length λ and temperature T (in kelvins):

$$R = \frac{2\pi c^2 h}{\lambda^5}\frac{1}{e^{(hc)/(\lambda kT)}-1}$$

where $c = 3.0 \times 10^8$ m/s is the speed of light, $h = 6.63 \times 10^{-34}$ J s is Planck's constant, and $k = 1.38 \times 10^{-23}$ J/K is the Boltzmann's constant.

Plot R as a function of λ for $0.2 \times 10^{-6} \le \lambda \le 6.0 \times 10^{-6}$ m at $T = 1500$ K, and determine the wavelength that gives the maximum R at this temperature.

15. A 108 in.–long beam AB is attached to the wall with a pin at point A and to a 68 in.–long cable CD. A load $W = 250$ lb is attached to the beam at point B. The tension in the cable T is given by

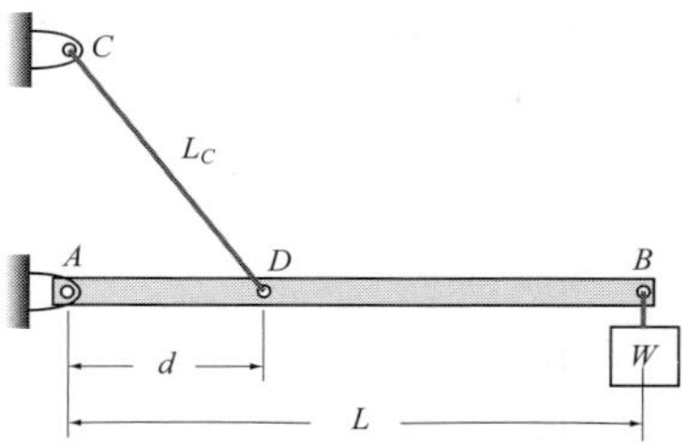

$$T = \frac{W L L_C}{d\sqrt{L_C^2 - d^2}}$$

where L and L_C are the lengths of the beam and the cable, respectively, and d is the distance from point A to point D, where the cable is attached. Make a plot of T versus d. Determine the distance d where the tension in the cable is the smallest.

16. Use MATLAB to calculate the following integral:

(a) $\displaystyle\int_1^6 \frac{2x^2}{\sqrt{1+x}}\,dx$ (b) $\displaystyle\int_1^2 \frac{\cos 2x}{x}\,dx$

17. Use MATLAB to calculate the following integrals:

(a) $\displaystyle\int_1^2 \frac{e^{2x}}{x}\,dx$ (b) $\displaystyle\int_{-1}^1 e^{-x^2}\,dx$

18. The speed of a race car during the first seven seconds of a race is given by:

t (s)	0	1	2	3	4	5	6	7
v (mi/h)	0	14	39	69	95	114	129	139

Determine the distance the car traveled during the first six seconds.

19. The length L of the main supporting cable of a suspension bridge can be calculated by

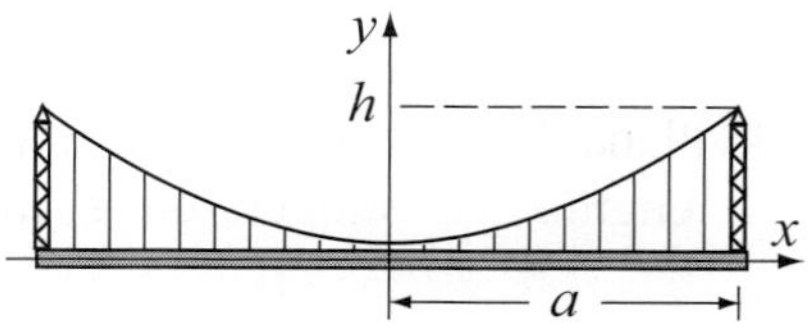

$$L = 2\int_0^a \left(1 + \frac{4h^2}{a^4}x^2\right)^{1/2} dx$$

where a is half the length of the bridge and h is the distance from the deck to the top of the tower where the cable is attached. Determine the length of a bridge with $a = 80$ m and $h = 18$ m.

20. The flow rate Q (volume of fluid per second) in a round pipe can be calculated by:

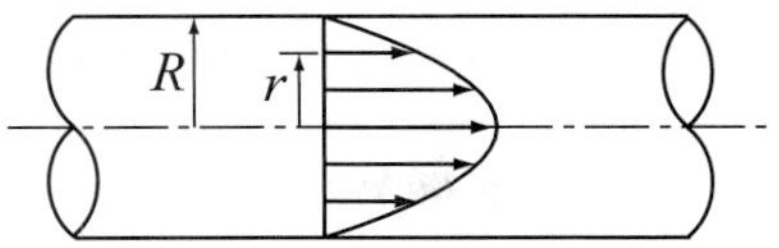

$$Q = \int_0^r 2\pi v r dr$$

For turbulent flow the velocity profile can be estimated by: $v = v_{max}\left(1 - \frac{r}{R}\right)^{1/n}$. Determine Q for $R = 0.25$ in., $n = 7$, $v_{max} = 80$ in./s.

21. The electric field E due to a charged circular disk at a point at a distance z along the axis of the disk is given by

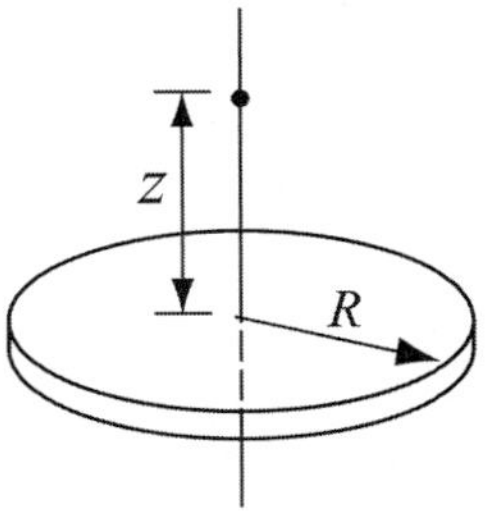

$$E = \frac{\sigma z}{4\varepsilon_0}\int_0^R (z^2 + r^2)^{-3/2}(2r)dr$$

where σ is the charge density, ε_0 is the permittivity constant, $\varepsilon_0 = 8.85 \times 10^{-12}$ C²/(N m²), and R is the radius of the disk. Determine the electric field at a point located 5 cm from a disk with a radius of 6 cm, charged with $\sigma = 300$ μC/m².

22. The length of a curve given by a parametric equation $x(t)$, $y(t)$ is given by:

$$\int_a^b \sqrt{[x'(t)]^2 + [y'(t)]^2}\ dt$$

The cycloid curve is given by $x = R(t - \sin t)$, and $y = R(1 - \cos t)$. Determine the length of a cycloid with $R = 8$ in. for $0 \le t \le 2\pi$.

23. The variation of gravitational acceleration g with altitude y is given by

$$g = \frac{R^2}{(R+y)^2}g_0$$

where $R = 6371$ km is the radius of the earth, and $g_0 = 9.81$ m/s² is the gravitational acceleration at sea level. The change in the gravitational potential energy, ΔU, of an object that is raised from the earth is given by:

$$\Delta U = \int_0^h mgdy$$

Determine the change in the potential energy of a satellite with a mass of 500 kg that is raised from the surface of the earth to a height of 800 km.

24. A cross section of a river with measurements of its depth at intervals of 40 ft is shown in the figure. Use numerical integration to estimate the cross-sectional area of the river.

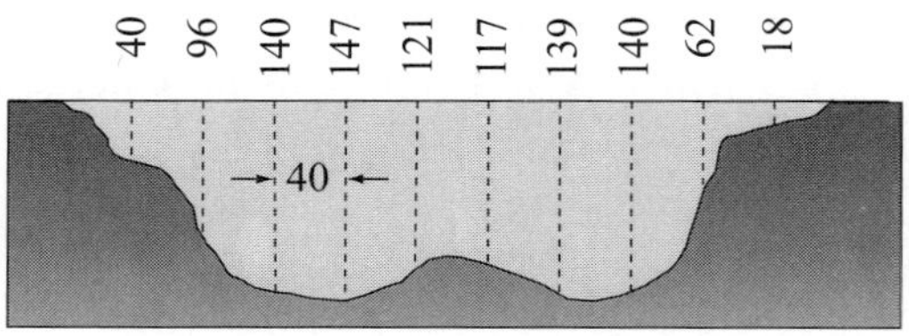

25. An approximate map of the state of Ohio is shown in the figure. Measurements of the width of the state are marked at intervals of 30 miles. Use numerical integration to estimate the area of the state. Compare the result with the actual area of Ohio, which is 44,825 square miles.

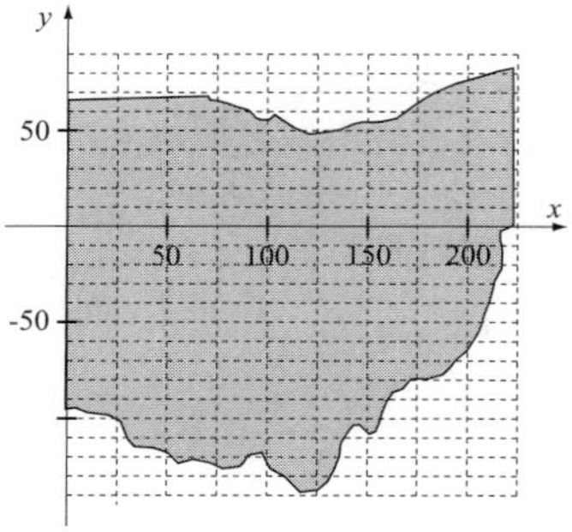

26. The time-dependent relaxation modulus $G(t)$ of many biological materials can be described by Fung's reduced relaxation function:

$$G(t) = G_\infty \left(1 + c\int_{\tau_1}^{\tau_2} \frac{e^{(-t)/x}}{x}\, dx\right)$$

Use numerical integration to find the relaxation modulus at 10 s, 100 s, and 1,000 s. Assume $G_\infty = 5\,\text{ksi}$, $c = 0.05$, $\tau_1 = 0.05\,\text{s}$, and $\tau_2 = 500\,\text{s}$.

27. The orbit of Pluto is elliptical in shape, with $a = 5.9065 \times 10^9\,\text{km}$ and $b = 5.7208 \times 10^9\,\text{km}$. The perimeter of an ellipse can be calculated by

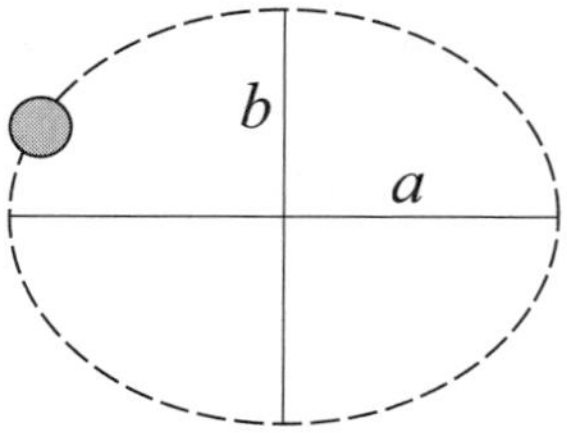

$$P = 4a\int_0^{\pi/2} \sqrt{1 - k^2\sin^2\theta}\, d\theta$$

where $k = \frac{\sqrt{a^2 - b^2}}{a}$. Determine the distance Pluto travels in one orbit. Calculate the average speed at which Pluto travels (in km/h) if one orbit takes about 248 years.

28. The Fresnel integrals are:

$$S(x) = \int_0^x \sin(t^2)dt \text{ and } C(x) = \int_0^x \cos(t^2)dt$$

Calculate $S(x)$ and $C(x)$ for $0 \le x \le 4$ (use spacing of 0.05). In one figure plot two graphs—one of $S(x)$ versus x and the other of $C(x)$ versus x. In a second figure plot $S(x)$ versus $C(x)$.

29. Solve:

$$\frac{dy}{dx} = \sqrt{x} + \frac{x^2\sqrt{y}}{4} \quad \text{for } 1 \le x \le 5 \quad \text{with } y(1) = 1$$

Plot the solution.

30. Solve:

$$\frac{dy}{dx} = \sqrt{xy} - 0.5ye^{-0.1x} \quad \text{for } 0 \le x \le 4 \quad \text{with } y(0) = 6.5$$

Plot the solution.

31. Solve:

$$\frac{dy}{dt} = 80e^{-1.6t}\cos(4t) - 0.4y \quad \text{for } 0 \le t \le 4 \quad \text{with } y(0) = 0$$

Plot the solution.

32. A water tank shaped as an ellipsoid ($a = 1.5$ m, $b = 4.0$ m, $c = 3$m) has a circular hole at the bottom, as shown. According to Torricelli's law, the speed v of the water that is discharging from the hole is given by

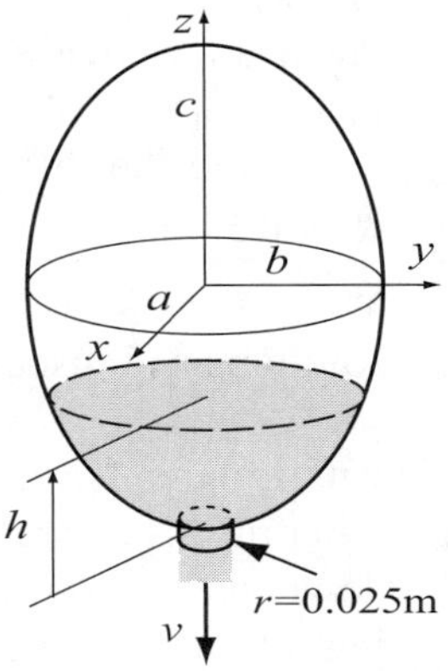

$$v = \sqrt{2gh}$$

where h is the height of the water and $g = 9.81\text{m/s}^2$. The rate at which the height, h, of the water in the tank changes as the water flows out through the hole is given by

$$\frac{dy}{dt} = \frac{\sqrt{2gy}\, r^2}{ac\left[-1 + \frac{(h-c)^2}{c^2}\right]}$$

where r_h is the radius of the hole.

Solve the differential equation for y. The initial height of the water is $h = 5.9$ m. Solve the problem for different times and find an estimate for the time when $h = 0.1$ m. Make a plot of y as a function of time.

33. The growth of a fish is often modeled by the von Bertalanffy growth model:

$$\frac{dw}{dt} = aw^{2/3} - bw$$

where w is the weight and a and b are constants. Solve the equation for w for the case $a = 5\ \text{lb}^{1/3}$, $b = 2\ \text{day}^{-1}$, and $w(0) = 0.5\ \text{lb}$. Make sure that the selected time span is just long enough so that the maximum weight is approached. What is the maximum weight for this case? Make a plot of w as a function of time.

34. The sudden outbreak of an insect population can be modeled by the equation

$$\frac{dN}{dt} = RN\left(1 - \frac{N}{C}\right) - \frac{rN^2}{N_c^2 + N^2}$$

The first term relates to the well-known logistic population growth model where N is the number of insects, R is an intrinsic growth rate, and C is the carrying capacity of the local environment. The second term represents the effects of bird predation. Its effect becomes significant when the population reaches a critical size N_c. r is the maximum value that the second term can reach at large values of N.

Solve the differential equation for $0 \le t \le 50$ days and two growth rates, $R = 0.55$ and $R = 0.58\ \text{day}^{-1}$, and with $N(0) = 10000$. The other parameters are $C = 10^4$, $N_c = 10^4$, $r = 10^4\ \text{day}^{-1}$. Make one plot comparing the two solutions and discuss why this model is called an "outbreak" model.

35. An airplane uses a parachute and other means of braking as it slows down on the runway after landing. Its acceleration is given by $a = -0.0035v^2 - 3\ \text{m/s}^2$. Since $a = \frac{dv}{dt}$, the rate of change of the velocity is given by:

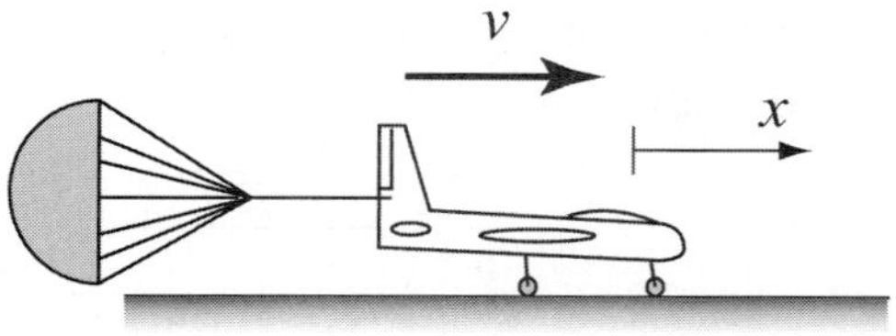

$$\frac{dv}{dt} = -0.0035v^2 - 3$$

Consider an airplane with a velocity of 300 km/h that opens its parachute and starts decelerating at $t = 0$ s.

(*a*) By solving the differential equation, determine and plot the velocity as a function of time from $t = 0$ s until the airplane stops.

(*b*) Use numerical integration to determine the distance x the airplane travels as a function of time. Make a plot of x versus time.

36. An RC circuit includes a voltage source v_s, a resistor $R = 48\ \Omega$, and a capacitor $C = 2.4 \times 10^{-6}$ F, as shown in the figure. The differential equation that describes the response of the circuit is:

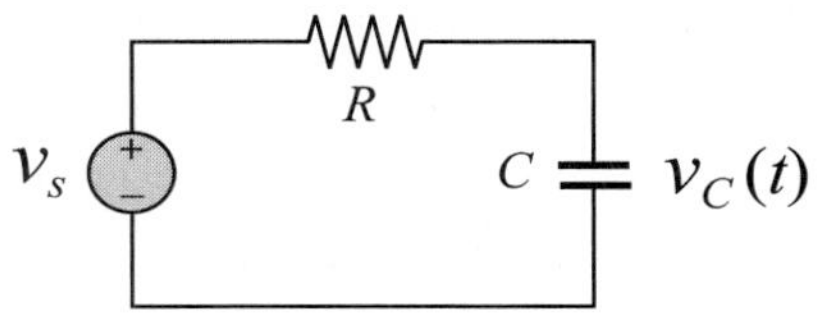

$$\frac{dv_c}{dt} + \frac{1}{RC}v_c = \frac{1}{RC}v_s$$

where v_c is the voltage of the capacitor. Initially, $v_s = 0$, and then at $t = 0$ the voltage source is changed. Determine the response of the circuit for the following three cases:

(*a*) $v_s = 5\sin(20\pi t)$ V for $t \geq 0$.

(*b*) $v_s = 5e^{-t/0.08}\sin(20\pi t)$ V for $t \geq 0$.

(*c*) $v_s = 12$ V for $0 \leq t \leq 0.1$ s, and then $v_s = 0$ for $t \geq 0.1$ s (rectangular pulse).

Each case corresponds to a different differential equation. The solution is the voltage of the capacitor as a function of time. Solve each case for $0 \leq t \leq 0.4$ s. For each case plot v_s and v_c versus time (make two separate plots on the same page).

37. An RL circuit includes a voltage source v_s, a resistor $R = 1.8\ \Omega$, and an inductor $L = 0.4$ H, as shown in the figure. The differential equation that describes the response of the circuit is

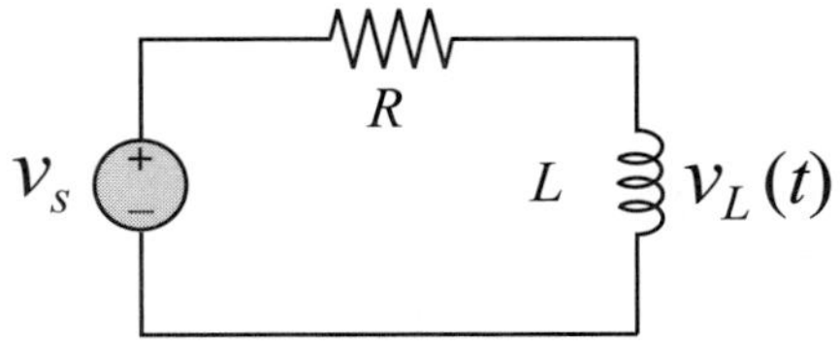

$$\frac{L}{R}\frac{di_L}{dt} + i_L = \frac{v_s}{R}$$

where i_L is the current in the inductor. Initially $i_L = 0$, and then at $t = 0$ the voltage source is changed. Determine the response of the circuit for the following three cases:

(*a*) $v_s = 10\sin(30\pi t)$ V for $t \geq 0$.

(*b*) $v_s = 10e^{-t/0.06}\sin(30\pi t)$ V for $t \geq 0$.

Each case corresponds to a different differential equation. The solution is the current in the inductor as a function of time. Solve each case for $0 \leq t \leq 0.4$ s. For each case plot v_s and i_L versus time (make two separate plots on the same page).

38. Tumor growth can be modeled with the equation

$$\frac{dA}{dt} = \alpha A\left[1 - \left(\frac{A}{k}\right)^{\upsilon}\right]$$

where $A(t)$ is the area of the tumor and α, k, and υ are constants. Solve the equation for $0 \le t \le 30$ days, given $\alpha = 0.8$, $k = 60$, $\upsilon = 0.25$, and $A(0) = 1$ mm^2. Make a plot of A as a function of time.

Chapter 10
Three-Dimensional Plots

Three-dimensional (3-D) plots can be a useful way to present data that consists of more than two variables. MATLAB provides various options for displaying three-dimensional data. They include line and wire, surface, mesh plots, and many others. The plots can also be formatted to have a specific appearance and special effects. Many of the three-dimensional plotting features are described in this chapter. Additional information can be found in the Help Window under **Plotting and Data Visualization**.

In many ways this chapter is a continuation of Chapter 5, where two-dimensional plots were introduced. The 3-D plots are presented in a separate chapter because not all MATLAB users use them. In addition, new users of MATLAB will probably find it easier to practice 2-D plotting first and learn the material in Chapters 6–9 before attempting 3-D plotting. It is assumed throughout the rest of this chapter that the reader is familiar with 2-D plotting.

10.1 Line Plots

A three-dimensional line plot is a line that is obtained by connecting points in three-dimensional space. A basic 3-D plot 'is created with the `plot3` command, which is very similar to the `plot` command and has the form:

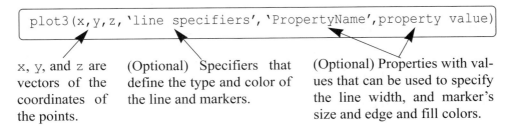

```
plot3(x,y,z,'line specifiers','PropertyName',property value)
```

`x`, `y`, and `z` are vectors of the coordinates of the points.

(Optional) Specifiers that define the type and color of the line and markers.

(Optional) Properties with values that can be used to specify the line width, and marker's size and edge and fill colors.

- The three vectors with the coordinates of the data points must have the same number of elements.
- The line specifiers, properties, and property values are the same as in 2-D plots (see Section 5.1).

For example, if the coordinates x, y, and z are given as a function of the parameter t by

$$x = \sqrt{t}\sin(2t)$$
$$y = \sqrt{t}\cos(2t)$$
$$z = 0.5t$$

a plot of the points for $0 \le t \le 6\pi$ can be produced by the following script file:

```
t=0:0.1:6*pi;
x=sqrt(t).*sin(2*t);
y=sqrt(t).*cos(2*t);
z=0.5*t;
plot3(x,y,z,'k','linewidth',1)
grid on
xlabel('x'); ylabel('y'); zlabel('z')
```

The plot shown in Figure 10-1 is created when the script is executed.

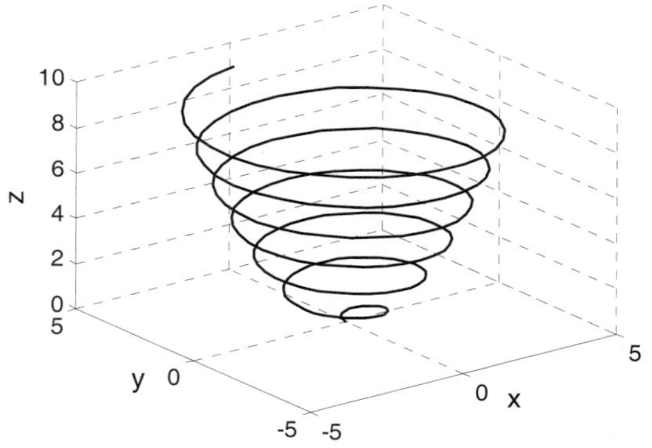

Figure 10-1: A plot of the function $x = \sqrt{t}\sin(2t)$, $y = \sqrt{t}\cos(2t)$, $z = 0.5t$ for $0 \le t \le 6\pi$.

10.2 Mesh and Surface Plots

Mesh and surface plots are three-dimensional plots used for plotting functions of the form $z = f(x, y)$ where x and y are the independent variables and z is the dependent variable. It means that within a given domain the value of z can be calculated for any combination of x and y. Mesh and surface plots are created in three steps. The first step is to create a grid in the $x\,y$ plane that covers the domain of the function. The second step is to calculate the value of z at each point of the grid. The third step is to create the plot. The three steps are explained next.

Creating a grid in the $x\,y$ plane (Cartesian coordinates):

The grid is a set of points in the $x\,y$ plane in the domain of the function. The density of the grid (number of points used to define the domain) is defined by the user. Figure 10-2 shows a grid in the domain $-1 \le x \le 3$ and $1 \le y \le 4$. In this grid

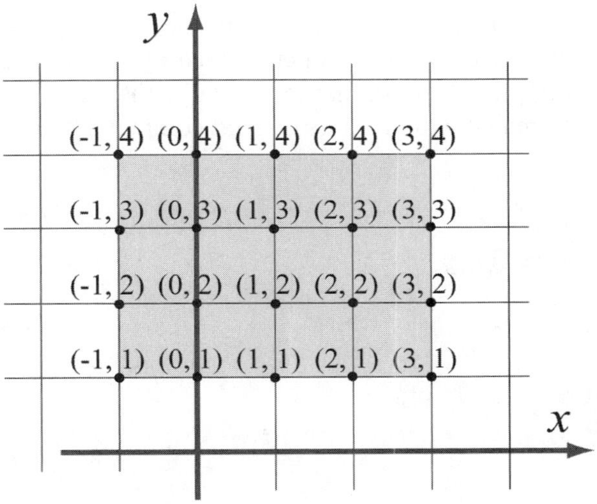

Figure 10-2: A grid in the $x\,y$ plane for the domain $-1 \le x \le 3$ and $1 \le y \le 4$ with spacing of 1.

the distance between the points is one unit. The points of the grid can be defined by two matrices, X and Y. Matrix X has the x coordinates of all the points, and matrix Y has the y coordinates of all the points:

$$X = \begin{bmatrix} -1 & 0 & 1 & 2 & 3 \\ -1 & 0 & 1 & 2 & 3 \\ -1 & 0 & 1 & 2 & 3 \\ -1 & 0 & 1 & 2 & 3 \end{bmatrix} \quad \text{and} \quad Y = \begin{bmatrix} 4 & 4 & 4 & 4 & 4 \\ 3 & 3 & 3 & 3 & 3 \\ 2 & 2 & 2 & 2 & 2 \\ 1 & 1 & 1 & 1 & 1 \end{bmatrix}$$

The X matrix is made of identical rows since in each row of the grid the points have the same x coordinate. In the same way the Y matrix is made of identical columns since in each column of the grid the y coordinate of the points is the same.

MATLAB has a built-in function, called `meshgrid`, that can be used for creating the X and Y matrices. The form of the `meshgrid` function is:

```
[X,Y] = meshgrid(x,y)
```

`X` is the matrix of the x coordinates of the grid points.
`Y` is the matrix of the y coordinates of the grid points.

`x` is a vector that divides the domain of x.
`y` is a vector that divides the domain of y.

In the vectors `x` and `y` the first and last elements are the respective boundaries of the domain. The density of the grid is determined by the number of elements in the vectors. For example, the mesh matrices X and Y that correspond to the grid in Figure 10-2 can be created with the `meshgrid` command by:

```
>> x=-1:3;
>> y=1:4;
>> [X,Y]=meshgrid(x,y)
X =
    -1     0     1     2     3
    -1     0     1     2     3
    -1     0     1     2     3
    -1     0     1     2     3
Y =
     1     1     1     1     1
     2     2     2     2     2
     3     3     3     3     3
     4     4     4     4     4
```

Once the grid matrices exist, they can be used for calculating the value of z at each grid point.

Calculating the value of z at each point of the grid:

The value of z at each point is calculated by using element-by-element calculations in the same way it is used with vectors. When the independent variables x and y are matrices (they must be of the same size), the calculated dependent variable is also a matrix of the same size. The value of z at each address is calculated from the corresponding values of x and y. For example, if z is given by

$$z = \frac{xy^2}{x^2 + y^2}$$

the value of z at each point of the grid above is calculated by:

```
>> Z = X.*Y.^2./(X.^2 + Y.^2)
```

```
Z =
   -0.5000         0    0.5000    0.4000    0.3000
   -0.8000         0    0.8000    1.0000    0.9231
   -0.9000         0    0.9000    1.3846    1.5000
   -0.9412         0    0.9412    1.6000    1.9200
```

Once the three matrices have been created, they can be used to plot mesh or surface plots.

Making mesh and surface plots:

A mesh or surface plot is created with the `mesh` or `surf` command, which has the form:

`mesh(X,Y,Z)` `surf(X,Y,Z)`

where `X` and `Y` are matrices with the coordinates of the grid and `Z` is a matrix with the value of z at the grid points. The mesh plot is made of lines that connect the points. In the surface plot, areas within the mesh lines are colored.

As an example, the following script file contains a complete program that creates the grid and then makes a mesh (or surface) plot of the function $z = \frac{xy^2}{x^2+y^2}$ over the domain $-1 \le x \le 3$ and $1 \le y \le 4$.

```
x=-1:0.1:3;
y=1:0.1:4;
[X,Y]=meshgrid(x,y);
Z=X.*Y.^2./(X.^2+Y.^2);
mesh(X,Y,Z)
xlabel('x'); ylabel('y'); zlabel('z')
```

Type `surf(X,Y,Z)` for surface plot.

Note that in the program above the vectors `x` and `y` have a much smaller spacing than the spacing earlier in the section. The smaller spacing creates a denser grid. The figures created by the program are:

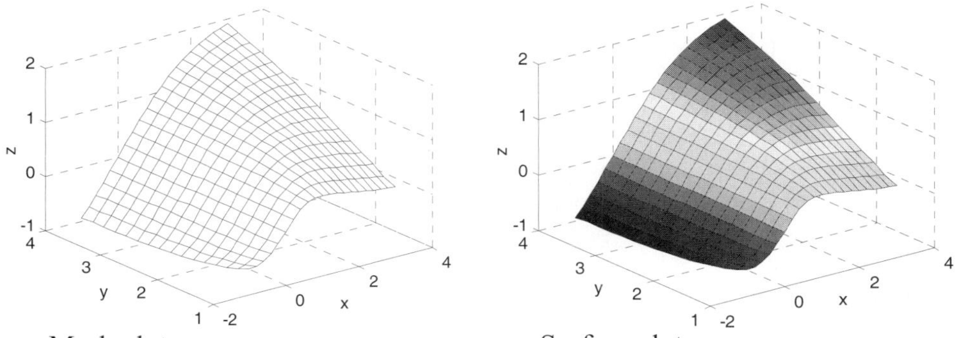

Mesh plot Surface plot

Additional comments on the mesh command:

- The plots that are created have colors that vary according to the magnitude of z. The variation in color adds to the three-dimensional visualization of the plots. The color can be changed to be a constant either by using the Plot Editor in the Figure Window (select the edit arrow, click on the figure to open the Property Editor Window, then change the color in the Mesh Properties list), or by using the `colormap(C)` command. In this command `C` is a three-element vector in which the first, second, and third elements specify the intensity of Red, Green, and Blue (RGB) colors, respectively. Each element can be a number between 0 (minimum intensity) and 1 (maximum intensity). Some typical colors are:

  C = [0 0 0] black   C = [1 0 0] red   C = [0 1 0] green
  C = [0 0 1] blue   C = [1 1 0] yellow   C = [1 0 1] magenta
  C = [0.5 0.5 0.5] gray

- When the mesh command executes, the grid is on by default. The grid can be turned off with the `grid off` command.
- A box can be drawn around the plot with the `box on` command.
- The `mesh` and `surf` commands can also be used with the form `mesh(Z)` and `surf(Z)`. In this case the values of `Z` are plotted as a function of their addresses in the matrix. The row number is on the x axis and the column number is on the y axis.

There are several additional plotting commands that are similar to the `mesh` and `surf` commands that create plots with different features. Table 10-1 shows a summary of the mesh and surface plotting commands. All the examples in the table are plots of the function $z = 1.8^{-1.5\sqrt{x^2+y^2}}\sin(x)\cos(0.5y)$ over the domain $-3 \le x \le 3$ and $-3 \le y \le 3$.

Table 10-1: Mesh and surface plots

Plot type	Example of plot	Program

Mesh Plot
 	
Function format:
`mesh(X,Y,Z)`
	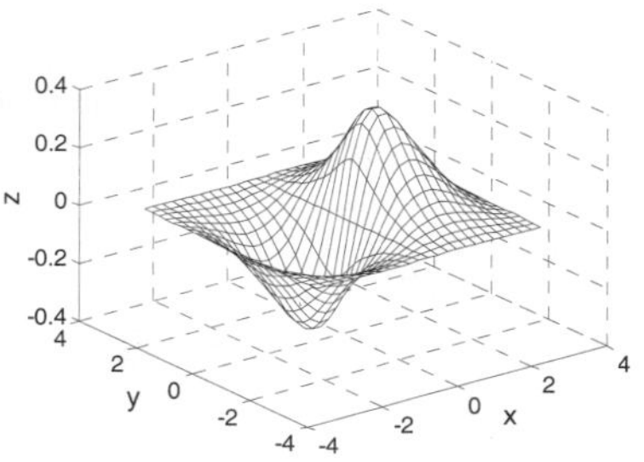	`x=-3:0.25:3;`
		`y=-3:0.25:3;`
		`[X,Y] = meshgrid(x,y);`
		`Z=1.8.^(-1.5*sqrt(X.^2+Y.^2)).*cos(0.5*Y).*sin(X);`
		`mesh(X,Y,Z)`
		`xlabel('x'); ylabel('y')`
		`zlabel('z')`

Table 10-1: Mesh and surface plots (Continued)

Plot type	Example of plot	Program

<u>Surface Plot</u>
 	
Function format:
`surf(X,Y,Z)`
	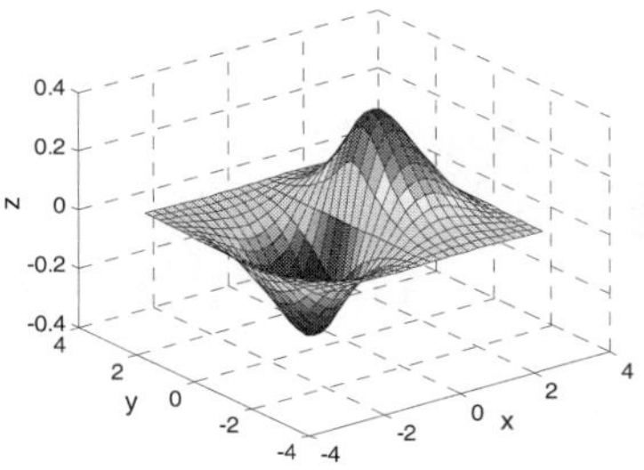	`x=-3:0.25:3;`
		`y=-3:0.25:3;`
		`[X,Y]=meshgrid(x,y);`
		`Z=1.8.^(-1.5*sqrt(X.^2+Y.^2)).*cos(0.5*Y).*sin(X);`
		`surf(X,Y,Z)`
		`xlabel('x'); ylabel('y')`
		`zlabel('z')`

<u>Mesh Curtain Plot</u> (draws a curtain around the mesh)
 	
Function format:
`meshz(X,Y,Z)`
	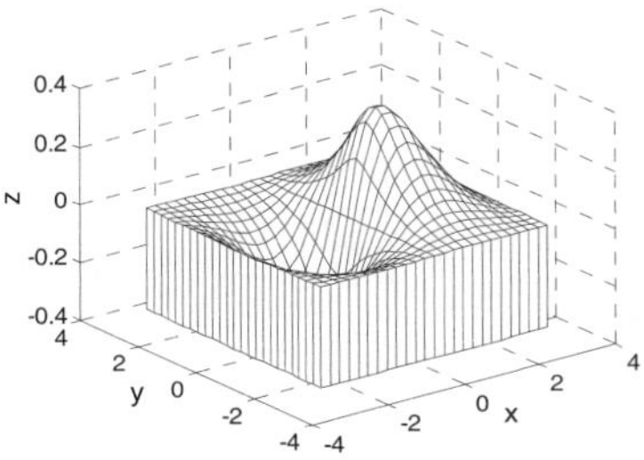	`x=-3:0.25:3;`
		`y=-3:0.25:3;`
		`[X,Y]=meshgrid(x,y);`
		`Z=1.8.^(-1.5*sqrt(X.^2+Y.^2)).*cos(0.5*Y).*sin(X);`
		`meshz(X,Y,Z)`
		`xlabel('x'); ylabel('y')`
		`zlabel('z')`

<u>Mesh and Contour Plot</u> (draws a contour plot beneath the mesh)
 	
Function format:
`meshc(X,Y,Z)`
	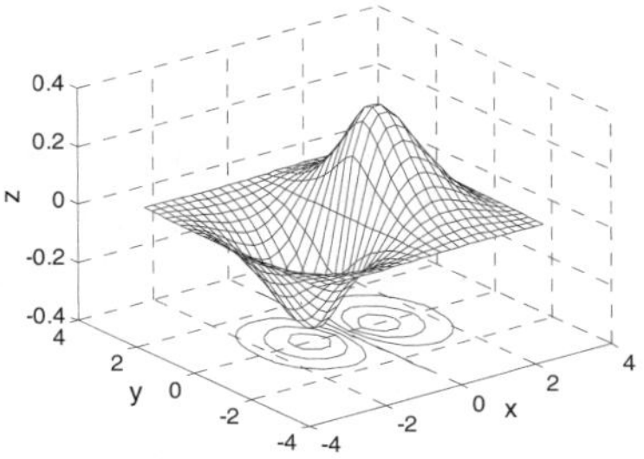	`x=-3:0.25:3;`
		`y=-3:0.25:3;`
		`[X,Y]=meshgrid(x,y);`
		`Z=1.8.^(-1.5*sqrt(X.^2+Y.^2)).*cos(0.5*Y).*sin(X);`
		`meshc(X,Y,Z)`
		`xlabel('x'); ylabel('y')`
		`zlabel('z')`

<u>Surface and Contour Plot</u> (draws a contour plot beneath the surface)
 	
Function format:
`surfc(X,Y,Z)`
	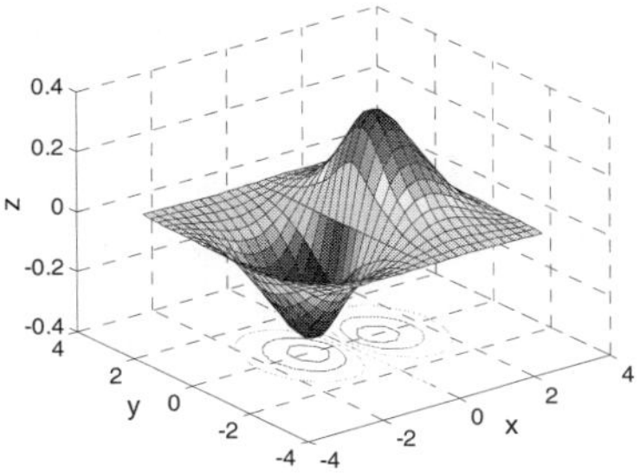	`x=-3:0.25:3;`
		`y=-3:0.25:3;`
		`[X,Y]=meshgrid(x,y);`
		`Z=1.8.^(-1.5*sqrt(X.^2+Y.^2)).*cos(0.5*Y).*sin(X);`
		`surfc(X,Y,Z)`
		`xlabel('x'); ylabel('y')`
		`zlabel('z')`

Table 10-1: Mesh and surface plots (Continued)

Plot type	Example of plot	Program

<u>Surface Plot with Lighting</u>
 	
Function format:
`surfl(X,Y,Z)`
	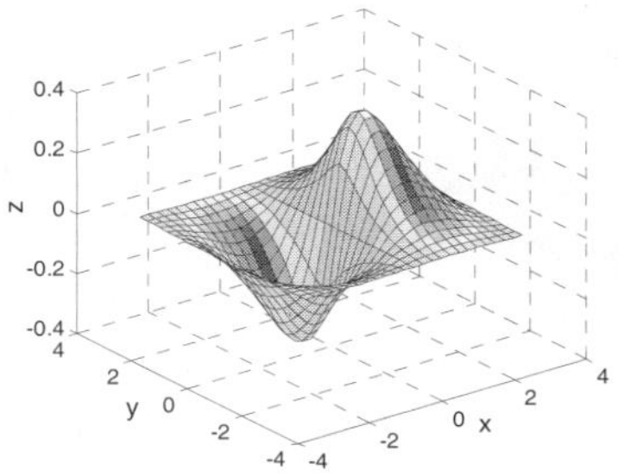	`x=-3:0.25:3;`
		`y=-3:0.25:3;`
		`[X,Y]=meshgrid(x,y);`
		`Z=1.8.^(-1.5*sqrt(X.^2+ Y.^2)).*cos(0.5*Y).*sin(X);`
		`surfl(X,Y,Z)`
		`xlabel('x'); ylabel('y')`
		`zlabel('z')`

<u>Waterfall Plot</u> (draws a mesh in one direction only)
 	
Function format:
`water-fall(X,Y,Z)`
	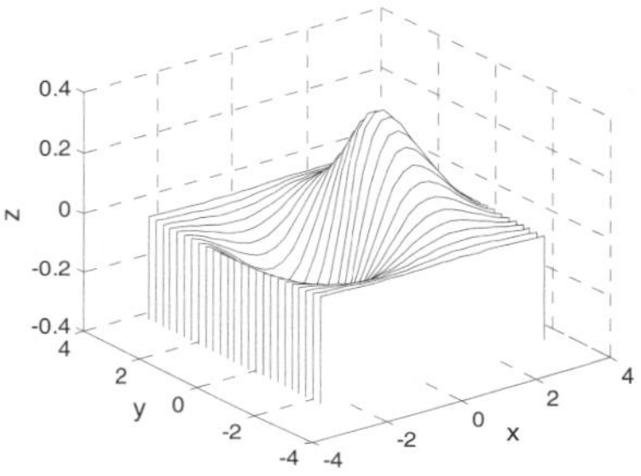	`x=-3:0.25:3;`
		`y=-3:0.25:3;`
		`[X,Y] = meshgrid(x,y);`
		`Z=1.8.^(-1.5*sqrt(X.^2+ Y.^2)).*cos(0.5*Y).*sin(X);`
		`waterfall(X,Y,Z)`
		`xlabel('x'); ylabel('y')`
		`zlabel('z')`

<u>3-D Contour Plot</u>
 	
Function format:
`contour3(X, Y,Z,n)`
 	
`n` is the number of contour levels (optional)
	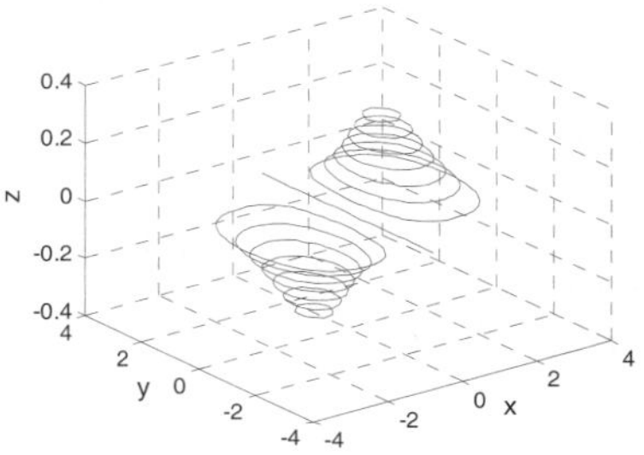	`x=-3:0.25:3;`
		`y=-3:0.25:3;`
		`[X,Y]=meshgrid(x,y);`
		`Z=1.8.^(-1.5*sqrt(X.^2+ Y.^2)).*cos(0.5*Y).*sin(X);`
		`contour3(X,Y,Z,15)`
		`xlabel('x'); ylabel('y')`
		`zlabel('z')`

<u>2-D Contour Plot</u> (draws projections of contour levels on the *x y* plane)
Function format:
`contour (X,Y,Z,n)`
`n` is the number of contour levels (optional)
	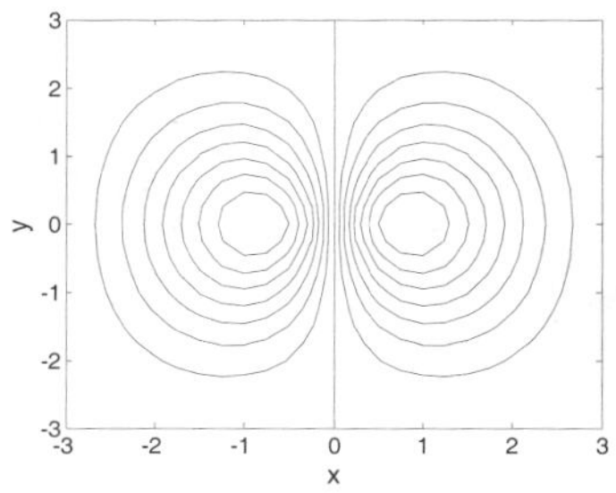	`x=-3:0.25:3;`
		`y=-3:0.25:3;`
		`[X,Y ]=meshgrid(x,y);`
		`Z=1.8.^(-1.5*sqrt(X.^2+ Y.^2)).*cos(0.5*Y).*sin(X);`
		`contour(X,Y,Z,15)`
		`xlabel('x'); ylabel('y')`
		`zlabel('z')`

10.3 Plots with Special Graphics

MATLAB has additional functions for creating various types of special three-dimensional plots. A complete list can be found in the Help Window under Plotting and Data Visualization. Several of these 3-D plots are presented in Table 10-2. The examples in the table do not show all the options available with each

Table 10-2: Specialized 3-D plots

Plot type	Example of plot	Program

<u>Plot a Sphere</u>
 	
Function format:
`sphere`
Returns the *x*, *y*, and *z* coordinates of a unit sphere with 20 faces.
`sphere(n)`
Same as above with *n* faces.
	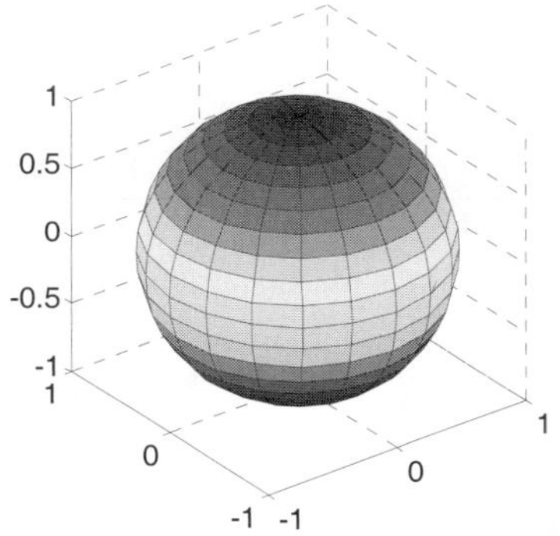	`sphere`
		
		or:
		
		`[X,Y,Z]=sphere(20);`
		`surf(X,Y,Z)`

<u>Plot a Cylinder</u>
 	
Function format:
`[X,Y,Z]=`
`cylinder(r)`
Returns the *x*, *y*, and *z* coordinates of cylinder with profile *r*.
	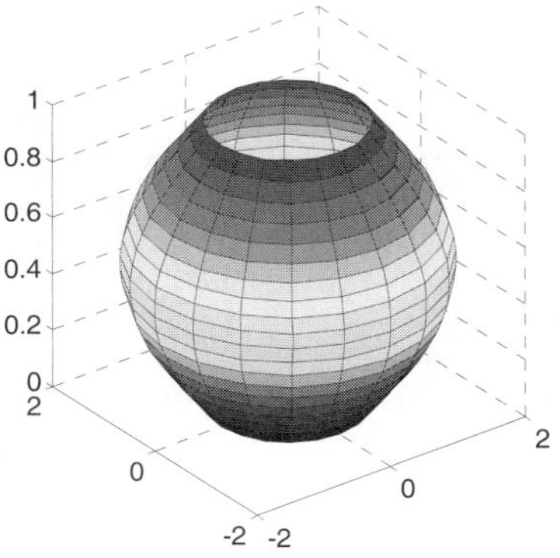	`t=linspace(0,pi,20);`
		`r=1+sin(t);`
		`[X,Y,Z]=cylinder(r);`
		`surf(X,Y,Z)`
		`axis square`

<u>3-D Bar Plot</u> Function format: `bar3(Y)` Each element in `Y` is one bar. Columns are grouped together.	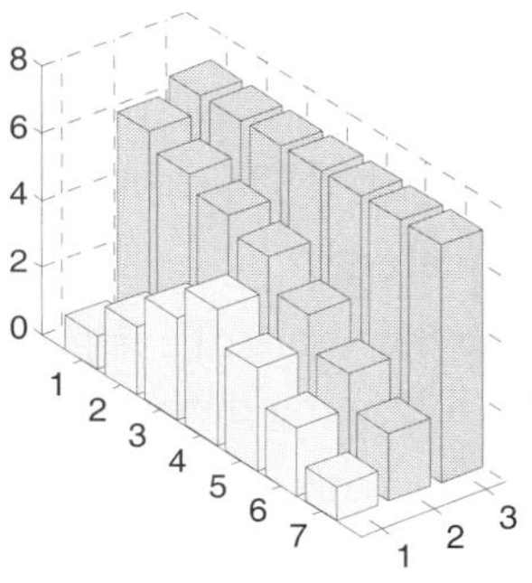	`Y=[1 6.5 7; 2 6 7; 3 5.5 7; 4 5 7; 3 4 7; 2 3 7; 1 2 7];` `bar3(Y)`

Table 10-2: Specialized 3-D plots (Continued)

Plot type	Example of plot	Program

<u>3-D Stem Plot</u>
(draws sequential points with markers and vertical lines from the *x y* plane)
 	
Function format:
`stem3(X,Y,Z)`
	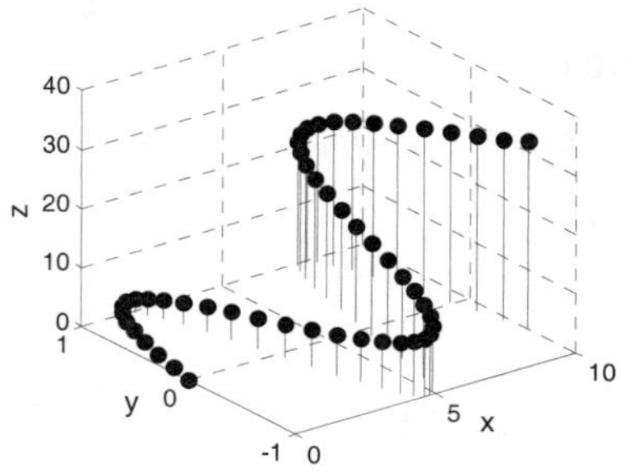	`t=0:0.2:10;`
		`x=t;`
		`y=sin(t);`
		`z=t.^1.5;`
		`stem3(x,y,z,'fill')`
		`grid on`
		`xlabel('x');`
		`ylabel('y')`
		`zlabel('z')`

<u>3-D Scatter Plot</u>
 	
Function format:
`scatter3(X,Y,Z)`
	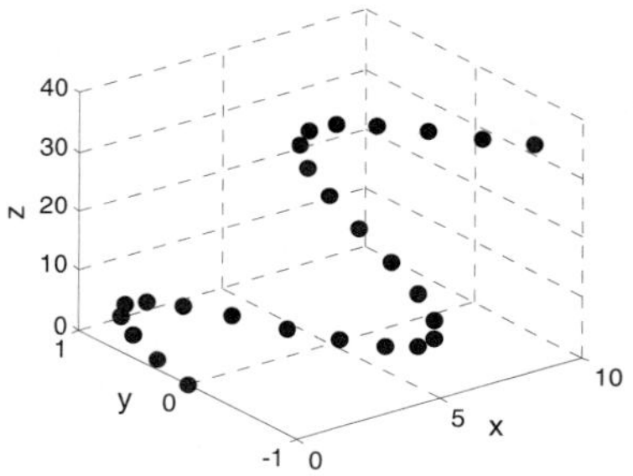	`t=0:0.4:10;`
		`x=t;`
		`y=sin(t);`
		`z=t.^1.5;`
		`scatter3(x,y,z,'filled')`
		`grid on`
		`colormap([0.1 0.1 0.1])`
		`xlabel('x');`
		`ylabel('y')`
		`zlabel('z')`

<u>3-D Pie Plot</u>
 	
Function format:
`pie3(X, explode)`
	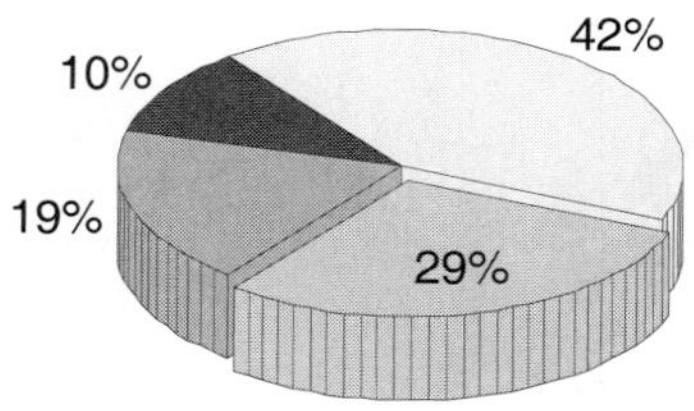	`X=[5 9 14 20];`
		`explode=[0 0 1 0];`
		`pie3(X,explode)`
		
		`explode` is a vector (same length as `X`) of 0's and 1's. 1 offsets the slice from the center.

plot type. More details on each type of plot can be obtained in the Help Window, or by typing `help command_name` in the Command Window.

Polar coordinates grid in the *x y* plane:

A 3-D plot of a function in which the value of z is given in polar coordinates (for example $z = r\theta$) can be done by following these steps:

- Create a grid of values of θ and r with the `meshgrid` function.

- Calculate the value of z at each point of the grid.
- Convert the polar coordinates grid to a grid in Cartesian coordinates. This can be done with MATLAB's built-in function pol2cart (see example below).
- Make a 3-D plot using the values of z and the Cartesian coordinates.

For example, the following script creates a plot of the function $z = r\theta$ over the domain $0 \le \theta \le 360°$ and $0 \le r \le 2$.

```
[th,r]=meshgrid((0:5:360)*pi/180,0:.1:2);
Z=r.*th;
[X,Y] = pol2cart(th,r);
mesh(X,Y,Z)
```

Type surf(X,Y,Z) for surface plot.

The figures created by the program are:

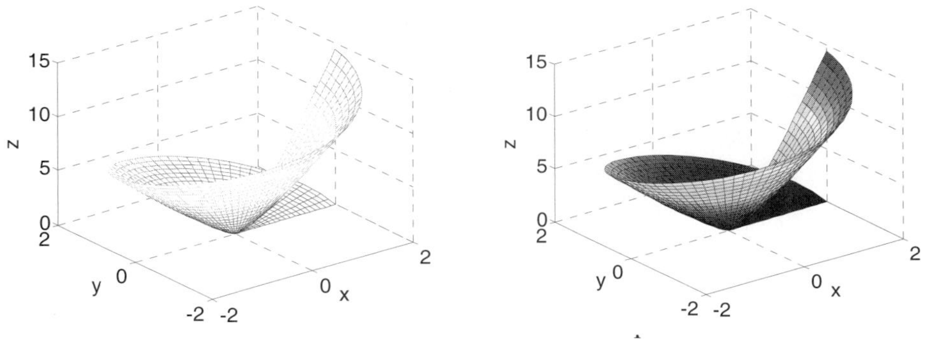

10.4 THE view *COMMAND*

The view command controls the direction from which the plot is viewed. This is done by specifying a direction in terms of azimuth and elevation angles, as seen in Figure 10-3, or by defining a point in space from which the plot is viewed. To set the viewing angle of the plot, the view command has the form:

`view(az,el)` or `view([az,el])`

- az is the azimuth, which is an angle (in degrees) in the $x\,y$ plane measured relative to the negative y axis direction and defined as positive in the counterclockwise direction.
- el is the angle of elevation (in degrees) from the $x\,y$ plane. A positive value corresponds to opening an angle in the direction of the z axis.
- The default view angles are $az = -37.5°$, and $el = 30°$.

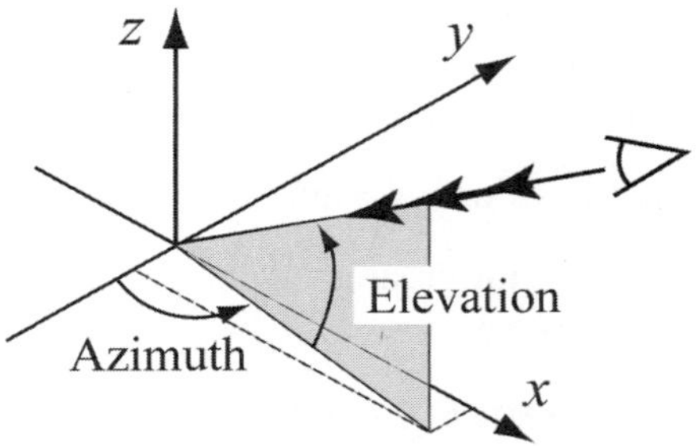

Figure 10-3: Azimuth and elevation angles.

As an example, the surface plot from Table 10-1 is plotted again in Figure 10-4, with viewing angles $az = 20°$ and $el = 35°$.

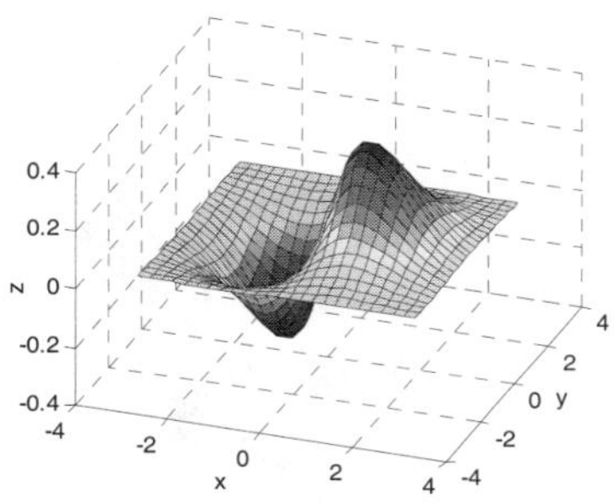

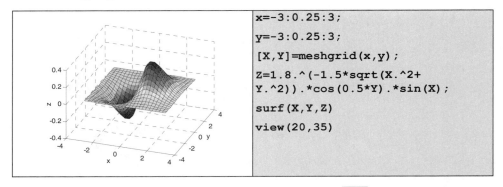

```
x=-3:0.25:3;
y=-3:0.25:3;
[X,Y]=meshgrid(x,y);
Z=1.8.^(-1.5*sqrt(X.^2+
Y.^2)).*cos(0.5*Y).*sin(X);
surf(X,Y,Z)
view(20,35)
```

Figure 10-4: A surface plot of the function $z = 1.8^{-1.5\sqrt{x^2+y^2}}\sin(x)\cos(0.5y)$ with viewing angles of $az = 20°$ and $el = 35°$.

- With the choice of appropriate azimuth and elevation angles, the `view` command can be used to plot projections of 3-D plots on various planes according to the following table:

Projection plane	*az* value	*el* value
x y (top view)	0	90
x z (side view)	0	0
y z (side view)	90	0

An example of a top view is shown next. Figure 10-5 shows the top view of the function that is plotted in Figure 10-1. Examples of projections onto the *x z* and *y z* planes are shown next, in Figures 10-6 and 10-7, respectively. The figures show mesh plot projections of the function plotted in Table 10-1.

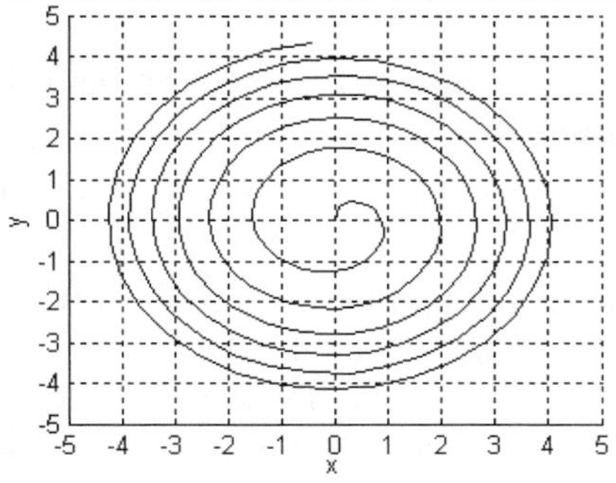


```
t=0:0.1:6*pi;
x=sqrt(t).*sin(2*t);
y=sqrt(t).*cos(2*t);
z=0.5*t;
plot3(x,y,z,'k','linewidth',1)
view(0,90)
grid on
xlabel('x'); ylabel('y')
zlabel('z')
```

Figure 10-5: A top view plot of the function $x = \sqrt{t}\sin(2t)$, $y = \sqrt{t}\cos(2t)$, $z = 0.5t$ for $0 \le t \le 6\pi$.

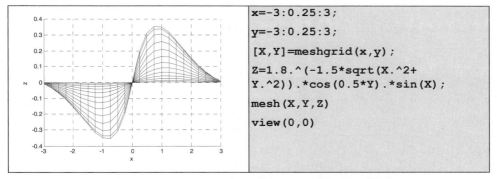

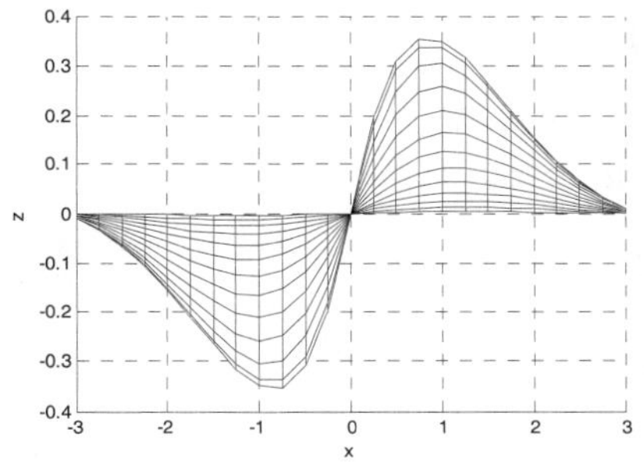


```
x=-3:0.25:3;
y=-3:0.25:3;
[X,Y]=meshgrid(x,y);
Z=1.8.^(-1.5*sqrt(X.^2+
Y.^2)).*cos(0.5*Y).*sin(X);
mesh(X,Y,Z)
view(0,0)
```

Figure 10-6: Projections onto the x z plane of the function $z = 1.8^{-1.5\sqrt{x^2+y^2}}\sin(x)\cos(0.5y)$.

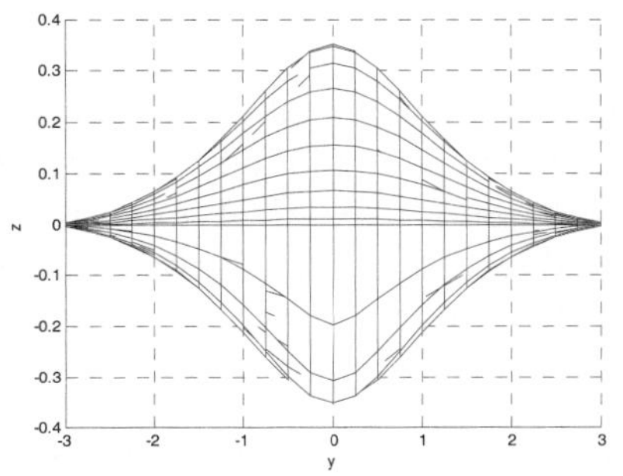


```
x=-3:0.25:3;
y=-3:0.25:3;
[X,Y]=meshgrid(x,y);
Z=1.8.^(-1.5*sqrt(X.^2+
Y.^2)).*cos(0.5*Y).*sin(X);
surf(X,Y,Z)
view(90,0)
```

Figure 10-7: Projections onto the y-z plane of the function $z = 1.8^{-1.5\sqrt{x^2+y^2}}\sin(x)\cos(0.5y)$.

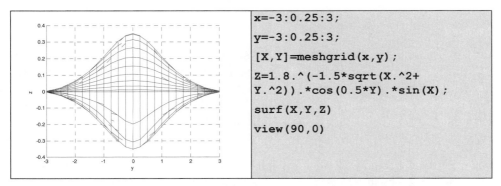

- The view command can also set a default view:

  `view(2)` sets the default to the top view, which is a projection onto the *x*-*y* plane with $az = 0°$, and $el = 90°$.

  `view(3)` sets the default to the standard 3-D view with $az = -37.5°$ and $el = 30°$.

- The viewing direction can also be set by selecting a point in space from which the plot is viewed. In this case the `view` command has the form `view([x,y,z])`, where `x`, `y`, and `z` are the coordinates of the point. The direction is determined by the direction from the specified point to the origin of the coordinate system and is independent of the distance. This means that the view is the same with point [6, 6, 6] as with point [10, 10, 10]. Top view can be set up with [0, 0, 1]. A side view of the *x z* plane from the negative *y* direction can be set with [0, –1, 0], and so on.

10.5 EXAMPLES OF MATLAB APPLICATIONS

Sample Problem 10-1: 3-D projectile trajectory

A projectile is fired with an initial velocity of 250 m/s at an angle of $\theta = 65°$ relative to the ground. The projectile is aimed directly north. Because of a strong wind blowing to the west, the projectile also moves in this direction at a constant speed of 30 m/s. Determine and plot the trajectory of the projectile until it hits the ground. For comparison, plot also (in the same figure) the trajectory that the projectile would have had if there was no wind.

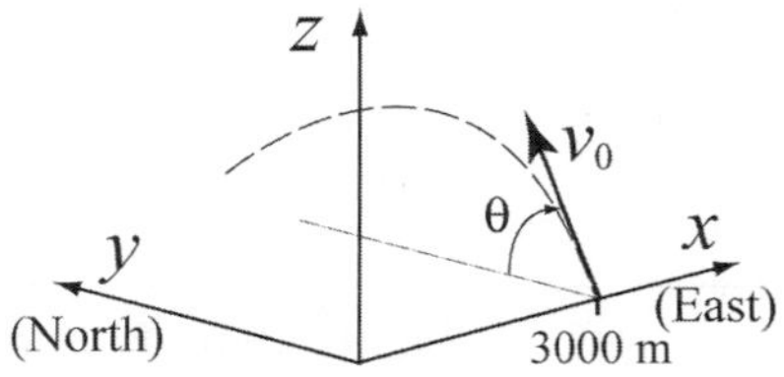

Solution

As shown in the figure, the coordinate system is set up such that the *x* and *y* axes point in the east and north directions, respectively. Then the motion of the projectile can be analyzed by considering the vertical direction *z* and the two horizontal components *x* and *y*. Since the projectile is fired directly north, the initial velocity v_0 can be resolved into a horizontal *y* component and a vertical *z* component:

$$v_{0y} = v_0\cos(\theta) \quad \text{and} \quad v_{0z} = v_0\sin(\theta)$$

In addition, due to the wind the projectile has a constant velocity in the negative *x* direction, $v_x = -30$ m/s.

The initial position of the projectile (x_0, y_0, z_0) is at point (3000, 0, 0). In the vertical direction the velocity and position of the projectile are given by:

$$v_z = v_{0z} - gt \quad \text{and} \quad z = z_0 + v_{0z}t - \frac{1}{2}gt^2$$

The time it takes the projectile to reach the highest point ($v_z = 0$) is $t_{hmax} = \frac{v_{0z}}{g}$. The total flying time is twice this time, $t_{tot} = 2t_{hmax}$. In the horizontal direction the velocity is constant (both in the x and y directions), and the position of the projectile is given by:

$$x = x_0 + v_x t \quad \text{and} \quad y = y_0 + v_{0y}t$$

The following MATLAB program written in a script file solves the problem by following the equations above.

```
v0=250; g=9.81; theta=65;
x0=3000; vx=-30;
v0z=v0*sin(theta*pi/180);
v0y=v0*cos(theta*pi/180);
t=2*v0z/g;
tplot=linspace(0,t,100);          Creating a time vector with 100 elements.
z=v0z*tplot-0.5*g*tplot.^2;
y=v0y*tplot;                      Calculating the x, y, and z coordinates
x=x0+vx*tplot;                    of the projectile at each time.
xnowind(1:length(y))=x0;          Constant x coordinate when no wind.
plot3(x,y,z,'k-',xnowind,y,z,'k--')   Two 3-D line plots.
grid on
axis([0 6000 0 6000 0 2500])
xlabel('x (m)'); ylabel('y (m)'); zlabel('z (m)')
```

The figure generated by the program is shown below.

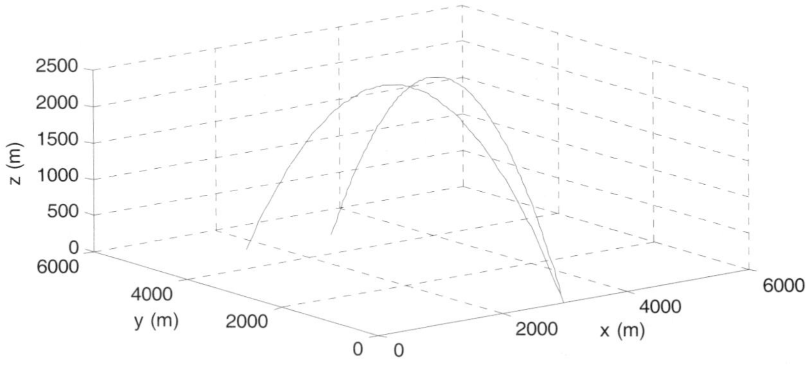

Sample Problem 10-2: Electric potential of two point charges

The electric potential V around a charged particle is given by

$$V = \frac{1}{4\pi\varepsilon_0}\frac{q}{r}$$

where $\varepsilon_0 = 8.8541878 \times 10^{-12}\frac{\text{C}}{\text{N m}^2}$ is the permittivity constant, q is the magnitude of the charge in coulombs, and r is the distance from the particle in meters. The electric field of two or more particles is calculated by using superposition. For example, the electric potential at a point due to two particles is given by

$$V = \frac{1}{4\pi\varepsilon_0}\left(\frac{q_1}{r_1} + \frac{q_2}{r_2}\right)$$

where q_1, q_2, r_1, and r_2 are the charges of the particles and the distance from the point to the corresponding particle, respectively.

Two particles with a charge of $q_1 = 2 \times 10^{-10}$ C and $q_2 = 3 \times 10^{-10}$ C are positioned in the $x\,y$ plane at points (0.25, 0, 0) and (–0.25, 0, 0), respectively, as shown. Calculate and plot the electric potential due to the two particles at points in the $x\,y$ plane that are located in the domain $-0.2 \le x \le 0.2$ and $-0.2 \le y \le 0.2$ (the units in the $x\,y$ plane are meters). Make the plot such that the $x\,y$ plane is the plane of the points, and the z axis is the magnitude of the electric potential.

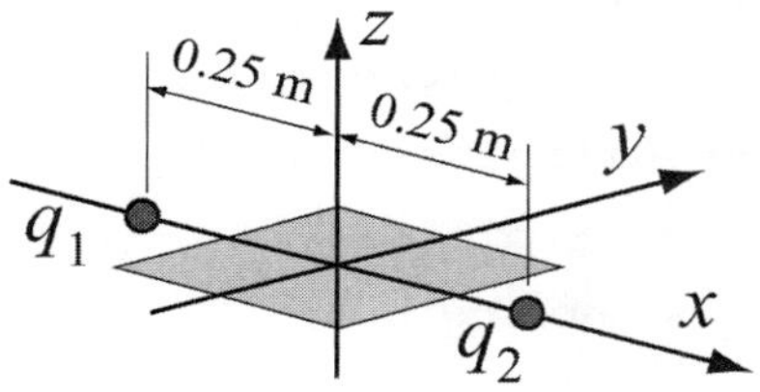

Solution

The problem is solved by following these steps:

(*a*) A grid is created in the $x\,y$ plane with the domain $-0.2 \le x \le 0.2$ and $-0.2 \le y \le 0.2$.

(*b*) The distance from each grid point to each of the charges is calculated.

(*c*) The electric potential at each point is calculated.

(*d*) The electric potential is plotted.

The following is a program in a script file that solves the problem.

```
eps0=8.85e-12; q1=2e-10; q2=3e-10;
k=1/(4*pi*eps0);
x=-0.2:0.01:0.2;
y=-0.2:0.01:0.2;
[X,Y]=meshgrid(x,y);          Creating a grid in the x y plane.
```

```
r1=sqrt((X+0.25).^2+Y.^2);       Calculating the distance r1 for each grid point.
r2=sqrt((X-0.25).^2+Y.^2);       Calculating the distance r2 for each grid point.
V=k*(q1./r1+q2./r2);             Calculating the electric potential V at each grid point.
mesh(X,Y,V)
xlabel('x (m)'); ylabel('y (m)'); zlabel('V (V)')
```

The plot generated when the program runs is:

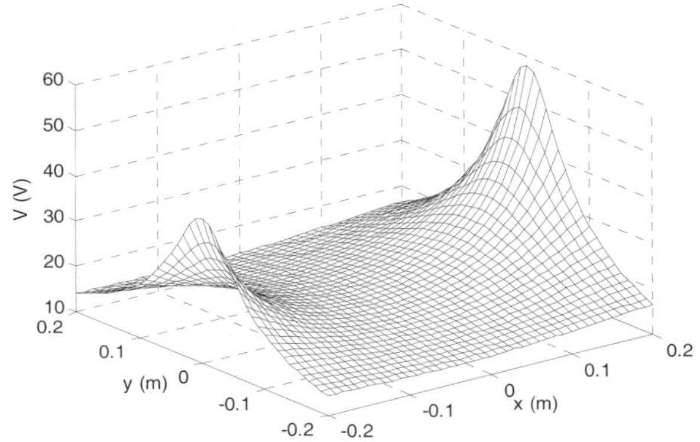

Sample Problem 10-3: Heat conduction in a square plate

Three sides of a rectangular plate ($a = 5$ m, $b = 4$ m) are kept at a temperature of 0 °C and one side is kept at a temperature T_1 = 80 °C, as shown in the figure. Determine and plot the temperature distribution $T(x, y)$ in the plate.

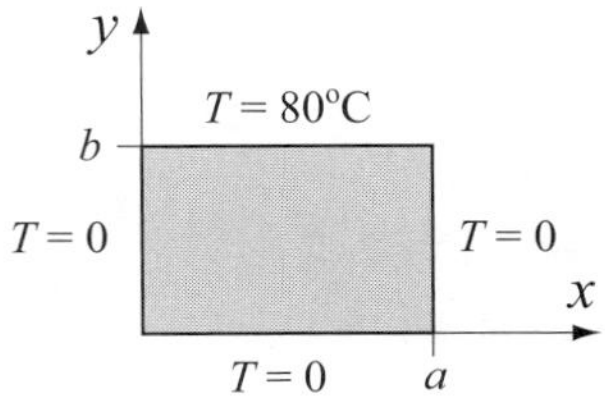

Solution

The temperature distribution, $T(x, y)$ in the plate can be determined by solving the two-dimensional heat equation. For the given boundary conditions $T(x, y)$ can be expressed analytically by a Fourier series (Erwin Kreyszig, *Advanced Engineering Mathematics*, John Wiley and Sons, 1993):

$$T(x,y) = \frac{4T_1}{\pi}\sum_{n=1}^{\infty}\frac{\sin\left[(2n-1)\frac{\pi x}{a}\right]}{(2n-1)}\frac{\sinh\left[(2n-1)\frac{\pi y}{a}\right]}{\sinh\left[(2n-1)\frac{\pi b}{a}\right]}$$

A program in a script file that solves the problem is listed below. The program follows these steps:

(*a*) Create an *X*, *Y* grid in the domain $0 \le x \le a$ and $0 \le y \le b$. The length of the plate, *a*, is divided into 20 segments, and the width of the plate, *b*, is divided into 16 segments.

(*b*) Calculate the temperature at each point of the mesh. The calculations are done point by point using a double loop. At each point the temperature is determined by adding *k* terms of the Fourier series.

(*c*) Make a surface plot of *T*.

```
a=5; b=4; na=20; nb=16; k=5; T0=80;
clear T
x=linspace(0,a,na);
y=linspace(0,b,nb);
[X,Y]=meshgrid(x,y);                 Creating a grid in the x y plane.
for i=1:nb                           First loop, i, is the index of the grid's row.
    for j=1:na                       Second loop, j, is the index of the grid's column.
        T(i,j)=0;
        for n=1:k                    Third loop, n, is the nth term of the Fourier
            ns=2*n-1;                series, k is the number of terms.
       T(i,j)=T(i,j)+sin(ns*pi*X(i,j)/a).*sinh(ns*pi*Y(i,j)/
a)/(sinh(ns*pi*b/a)*ns);
        end
        T(i,j) = T(i,j)*4*T0/pi;
    end
end
mesh(X,Y,T)
xlabel('x (m)'); ylabel('y (m)'); zlabel('T ( ^oC)')
```

The program was executed twice, first using five terms ($k = 5$) in the Fourier series to calculate the temperature at each point, and then with $k = 50$. The mesh plots created in each execution are shown in the figures below. The temperature should be uniformly 80°C at $y = 4$ m. Note the effect of the number of terms (k) on the accuracy at $y = 4$ m.

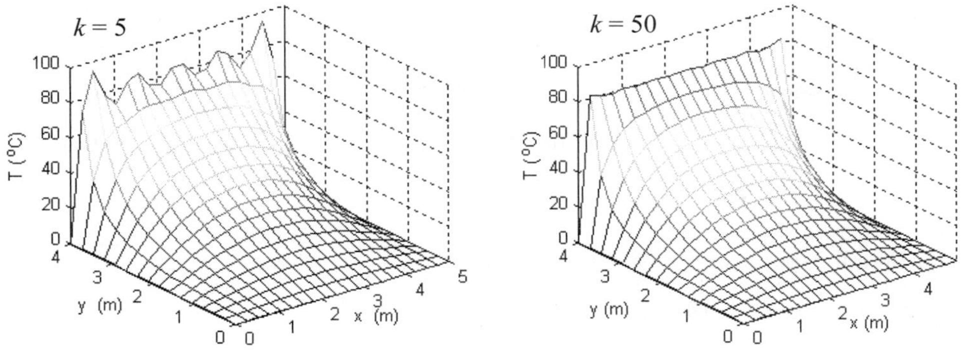

10.6 Problems

1. The position of a moving particle as a function of time is given by:

$$x = (4-0.1t)\sin(0.8t) \qquad y = (4-0.1t)\cos(0.8t) \qquad z = 0.4t^{(3/2)}$$

Plot the position of the particle for $0 \le t \le 30$.

2. An elliptical staircase that decreases in size with height can be modeled by the parametric equations

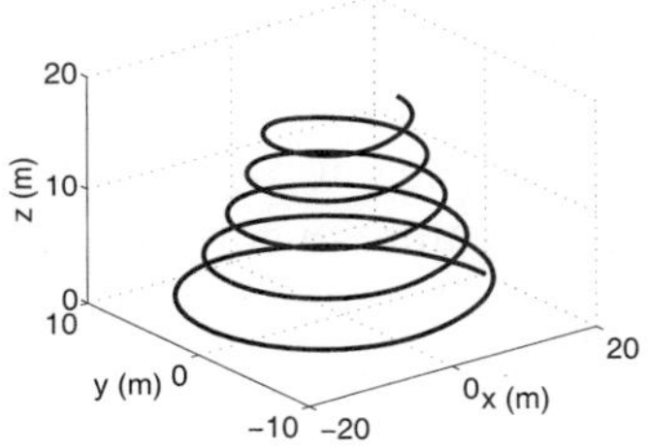

$$x = r\cos(t) \qquad y = r\sin(t) \qquad z = \frac{ht}{2\pi n}$$

where $r = \dfrac{ab}{\sqrt{[b\cos(t)]^2 + [a\sin(t)]^2}}e^{-0.04t}$,

a and b are the semimajor and semiminor axes of the ellipse, h is the staircase height, and n is the number of revolutions that the staircase makes. Make a 3-D plot of the staircase with $a = 20$ m, $b = 10$ m, $h = 18$ m, and $n = 5$. (Create a vector t for the domain 0 to $2\pi n$, and use the `plot3` command.)

3. The ladder of a fire truck can be elevated (increase of angle ϕ), rotated about the z axis (increase of angle θ), and extended (increase of r). Initially the ladder rests on the truck ($\phi = 0$, $\theta = 0$, and $r = 8$ m). Then the ladder is moved to a new position by raising the ladder at a rate of 5 deg/s, rotating at a rate of 8 deg/s, and extending the ladder at a rate of 0.6 m/s. Determine and plot the position of the tip of the ladder for 10 seconds.

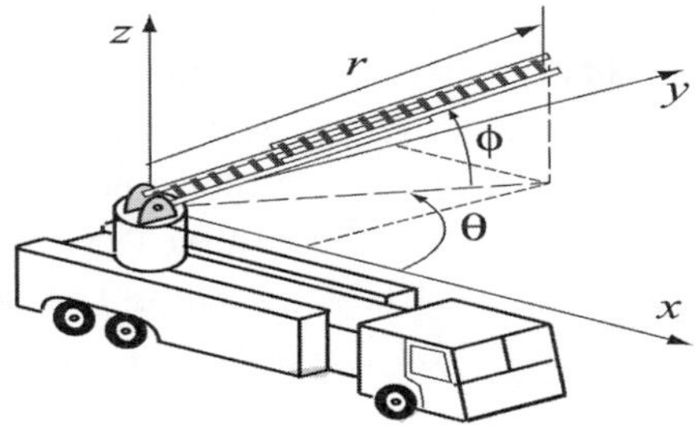

4. Make a 3-D surface plot of the function $z = \frac{x^2}{3} + 2\sin(3y)$ in the domain $-3 \le x \le 3$ and $-3 \le y \le 3$.

5. Make a 3-D surface plot of the function $z = 0.5|x| + |y|0.5$ in the domain $-2 \le x \le 2$ and $-2 \le y \le 2$.

6. Make a 3-D mesh plot of the function $z = \frac{\sin R}{R}$, where $R = \sqrt{x^2 + y^2}$ in the domain $-10 \le x \le 10$ and $-10 \le y \le 10$.

7. Make a 3-D surface plot of the function $z = \cos(xy)\cos(\sqrt{x^2 + y^2})$ in the domain $-\pi \le x \le \pi$ and $-\pi \le y \le \pi$.

8. An anti-symmetric cross-ply composite laminate has two layers in which the fibers are aligned perpendicular to one another. A laminate of this type will deform into a saddle shape due to residual thermal stresses as described by the equation

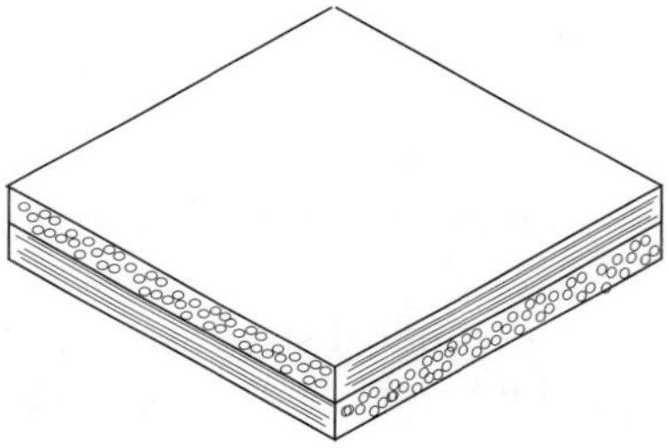

$$w = k(x^2 - y^2)$$

where x and y are the in-plane coordinates, w is the out-of-plane deflection, and k is the curvature (a complicated function of material properties and geometry). Make a surface plot showing the deflection of a six-inch square plate ($-3 \le x \le 3$ in., $-3 \le y \le 3$ in.), assuming $k = 0.01$ in^{-1}.

9. The van der Waals equation gives a relationship between the pressure p (atm), volume V, (L), and temperature T (K) for a real gas:

$$P = \frac{nRT}{V - b} - \frac{n^2 a}{V^2}$$

where n is the number of moles, $R = 0.08206$ (L atm)/(mol K) is the gas constant, and a (L^2 atm/mol^2), and b (L/mol) are material constants.

Consider 1.5 moles of nitrogen ($a = 1.39$ L^2 atm/mol^2, $b = 0.03913$ L/mol). Make a 3-D plot that shows the variation of pressure (dependent variable, z axis) with volume (independent variable, x axis) and temperature (independent variable, y axis). The domains for the volume and temperature are $0.3 \le V \le 1.2$ L and $273 \le T \le 473$ K.

10. Molecules of a gas in a container are moving around at different speeds. Maxwell's speed distribution law gives the probability distribution $P(v)$ as a function of temperature and speed:

$$P(v) = 4\pi\left(\frac{M}{2\pi RT}\right)^{3/2} v^2 e^{(-Mv^2)/(2RT)}$$

where M is the molar mass of the gas in kg/mol, $R = 8.31$ J/(mol K), is the gas constant, T is the temperature in kelvins, and v is the molecule's speed in m/s.

Make a 3-D plot of $P(v)$ as a function of v and T for $0 \le v \le 1000$ m/s and $70 \le T \le 320$ K for oxygen (molar mass 0.032 kg/mol).

11. The wind chill temperature, T_{wc}, is the air temperature felt on exposed skin due to wind. In U.S. customary units it is calculated by:

$$T_{wc} = 35.74 + 0.6215T - 35.75v^{0.16} + 0.4275T\,v^{0.16}$$

where T is the temperature in degrees F, and v is the wind speed in mi/h. Make a 3-D plot of T_{wc} as a function of v and T for $0 \le v \le 70$ mi/h and $0 \le T \le 50$ F.

12. The flow Q (m^3/s) in a rectangular channel is given by the Manning's equation:

$$Q = \frac{kdw}{n}\left(\frac{wd}{w+2d}\right)^{2/3}\sqrt{S}$$

where d is the depth of water (m), w is the width of the channel (m), S is the slope of the channel (m/m), n is the roughness coefficient of the channel walls, and k is a conversion constant (equal to 1 when the units above are used). Make a 3-D plot of Q (z axis) as a function of w (x axis) for $0 \le w \le 8$ m, and a function of d (y-axis) for $0 \le d \le 4$ m. Assume $n = 0.05$ and $S = 0.001$ m/m.

13. An RLC circuit with an alternating voltage source is shown. The source voltage v_s is given by $v_s = v_m \sin(\omega_d t)$, where $\omega_d = 2\pi f_d$, in which f_d is the driving frequency. The amplitude of the current, I, in this circuit is given by

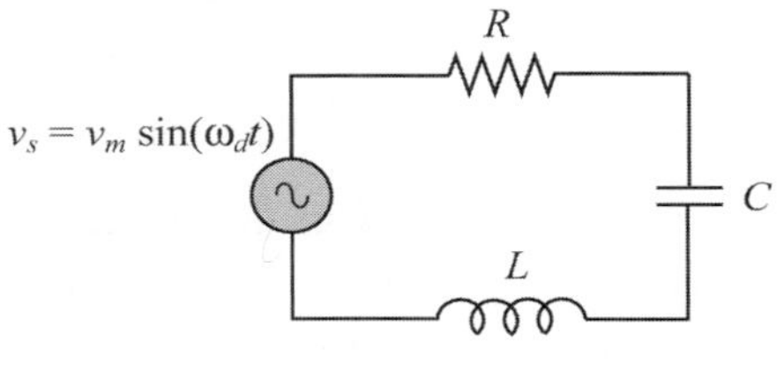

$$I = \frac{v_m}{\sqrt{R^2 + (\omega_d L - 1/(\omega_d C))^2}}$$

where R and C are the resistance of the resistor and capacitance of the capacitor, respectively. For the circuit in the figure $C = 15 \times 10^{-6}$ F,

$L = 240 \times 10^{-3}$ H, and $v_m = 24$ V.

a) Make a 3-D plot of I (z axis) as a function of ω_d (x axis) for $60 \le f \le 110$ Hz, and as a function of R (y axis) for $10 \le R \le 40\ \Omega$.

b) Make a plot that is a projection on the $x\,z$ plane. Estimate from this plot the natural frequency of the circuit (the frequency at which I is maximum). Compare the estimate with the calculated value of $1/(2\pi\sqrt{LC})$.

14. A defect in a crystal lattice where a row of atoms is missing is called an edge dislocation. The stress field around an edge dislocation is given by:

$$\sigma_{xx} = \frac{-Gb}{2\pi(1-\nu)}\frac{y(3x^2+y^2)}{(x^2+y^2)^2}$$

$$\sigma_{yy} = \frac{Gb}{2\pi(1-\nu)}\frac{y(x^2-y^2)}{(x^2+y^2)^2}$$

$$\tau_{xy} = \frac{Gb}{2\pi(1-\nu)}\frac{x(x^2-y^2)}{(x^2+y^2)^2}$$

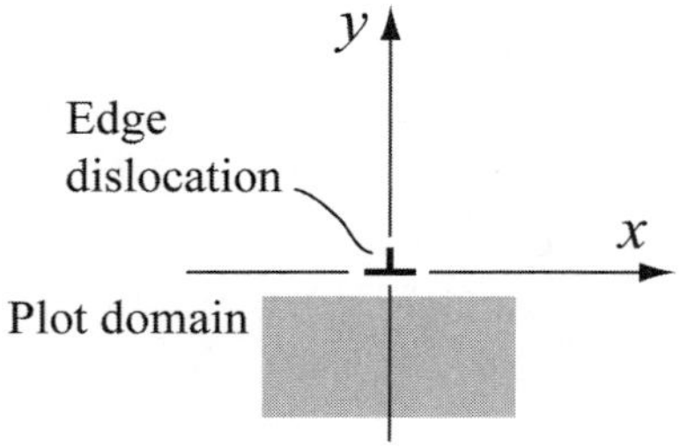

where G is the shear modulus, b is the Burgers vector, and ν is Poisson's ratio. Plot the stress components (each in a separate figure) due to an edge dislocation in aluminum, for which $G = 27.7 \times 10^9$ Pa, $b = 0.286 \times 10^{-9}$m, and $\nu = 0.334$. Plot the stresses in the domain $-5 \times 10^{-9} \le x \le 5 \times 10^{-9}$ m and $-5 \times 10^{-9} \le y \le -1 \times 10^{-9}$ m. Plot the coordinates x and y in the horizontal plane, and the stresses in the vertical direction.

15. The current I flowing through a semiconductor diode is given by

$$I = I_S\left(e^{\frac{qv_D}{kT}} - 1\right)$$

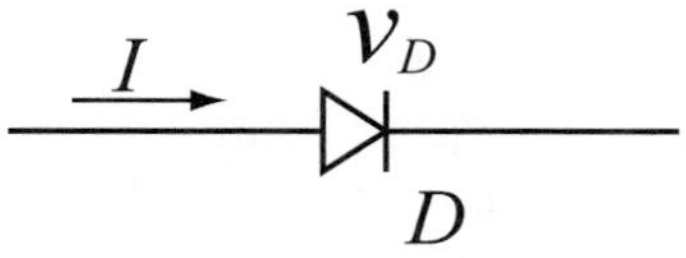

where $I_S = 10^{-12}$ A is the saturation current, $q = 1.6 \times 10^{-19}$ C is the elementary charge value, $k = 1.38 \times 10^{-23}$ J/K is Boltzmann's constant, v_D is the voltage drop across the diode, and T is the temperature in kelvins. Make a 3-D plot of I (z axis) versus v_D (x axis) for $0 \le v_D \le 0.4$, and versus T (y axis) for $290 \le T \le 320$K.

16. The equation for the streamlines for uniform flow over a cylinder is

$$\psi(x, y) = y - \frac{y}{x^2+y^2}$$

where ψ is the stream function. For example, if $\psi = 0$, then $y = 0$. Since the equation is satisfied for all x, the x axis is the zero ($\psi = 0$) streamline. Observe that the collection of points where $x^2 + y^2 = 1$ is also a streamline. Thus, the stream function above is for a cylinder of radius 1. Make a 2-D contour plot of the streamlines around a cylinder with 1 in. radius. Set up the domain for x and y to range between –3 and 3. Use 100 for the number of contour levels. Add to the figure a plot of a circle with a radius of 1. Note that MATLAB also plots streamlines inside the cylinder. This is a mathematical artifact.

17. The deflection w of a clamped circular membrane of radius r_d subjected to pressure P is given by (small deformation theory)

$$w(r) = \frac{Pr_d^4}{64K}\left[1 - \left(\frac{r}{r_d}\right)^2\right]^2$$

where r is the radial coordinate, and $K = \dfrac{Et^3}{12(1-\upsilon^2)}$, where E, t, and υ are the elastic modulus, thickness, and Poisson's ratio of the membrane, respectively. Consider a membrane with $P = 15$ psi, $r_d = 15$ in., $E = 18 \times 10^6$ psi, $t = 0.08$ in., and $\upsilon = 0.3$. Make a surface plot of the membrane.

18. The Verhulst model, given in the following equation, describes the growth of a population that is limited by various factors such as overcrowding and lack of resources:

$$N(t) = \frac{N_\infty}{1 + \left(\dfrac{N_\infty}{N_0} - 1\right)e^{-rt}}$$

where $N(t)$ is the number of individuals in the population, N_0 is the initial population size, N_∞ is the maximum population size possible due to the various limiting factors, and r is a rate constant. Make a surface plot of $N(t)$ versus t and N_∞ assuming $r = 0.1\ \text{s}^{-1}$, and $N_0 = 10$. Let t vary between 0 and 100 and N_∞ between 100 and 1,000.

19. The geometry of a ship hull (Wigley hull) can be modeled by the equation

$$y = \mp\frac{B}{2}\left[1 - \left(\frac{2x}{L}\right)^2\right]\left[1 - \left(\frac{2z}{T}\right)^2\right]$$

where x, y, and z are the length, width, and height, respectively. Use MATLAB to make a 3-D figure of the hull as shown. Use $B = 1.2$, $L = 4$, $T = 0.5$, $-2 \le x \le 2$, and $-0.5 \le z \le 0$.

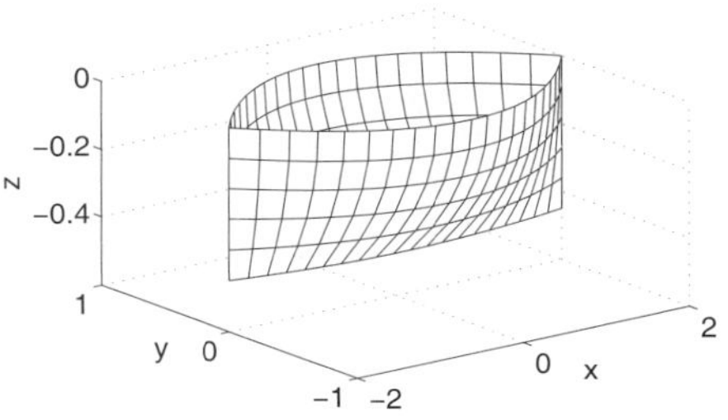

20. The stresses fields near a crack tip of a linear elastic isotropic material for mode I loading are given by:

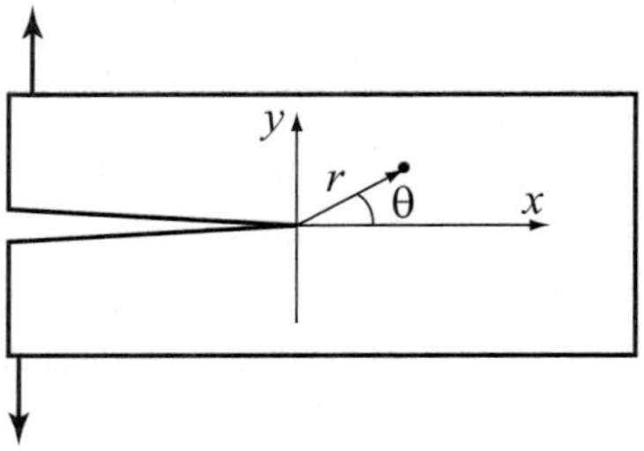

$$\sigma_{xx} = \frac{K_I}{\sqrt{2\pi r}}\cos\left(\frac{\theta}{2}\right)\left[1 - \sin\left(\frac{\theta}{2}\right)\sin\left(\frac{3\theta}{2}\right)\right]$$

$$\sigma_{yy} = \frac{K_I}{\sqrt{2\pi r}}\cos\left(\frac{\theta}{2}\right)\left[1 + \sin\left(\frac{\theta}{2}\right)\sin\left(\frac{3\theta}{2}\right)\right]$$

$$\tau_{xy} = \frac{K_I}{\sqrt{2\pi r}}\cos\left(\frac{\theta}{2}\right)\sin\left(\frac{\theta}{2}\right)\cos\left(\frac{3\theta}{2}\right)$$

For $K_I = 300$ ksi$\sqrt{\text{in}}$ plot the stresses (each in a separate figure) in the domain $0 \le \theta \le 90°$ and $0.02 \le r \le 0.2$ in. Plot the coordinates x and y in the horizontal plane, and the stresses in the vertical direction.

21. A ball thrown up falls back to the floor and bounces many times. For a ball thrown up in the direction shown in the figure, the position of the ball as a function of time is given by:

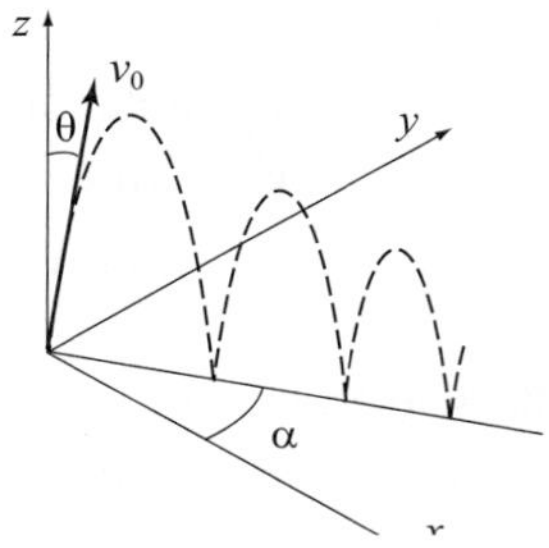

$$x = v_x t \qquad y = v_y t \qquad z = v_z t - \frac{1}{2}gt^2$$

The velocities in the x and y directions are constants throughout the motion and are given by $v_x = v_0\sin(\theta)\cos(\alpha)$ and $v_y = v_0\sin(\theta)\sin(\alpha)$. In the vertical z direction the initial velocity is $v_z = v_0\cos(\theta)$, and when the ball impacts the floor its rebound velocity is 0.8 of the vertical velocity at the start of the previous bounce. The time between bounces is given by $t_b = (2v_z)/g$, where v_z is the vertical component of the velocity at the start of the bounce. Make a 3-D plot (shown in the figure) that shows the trajectory of the ball during the first five bounces. Take $v_0 = 20$ m/s, $\theta = 30°$, $\alpha = 25°$, and $g = 20\,\text{m/s}^2$.

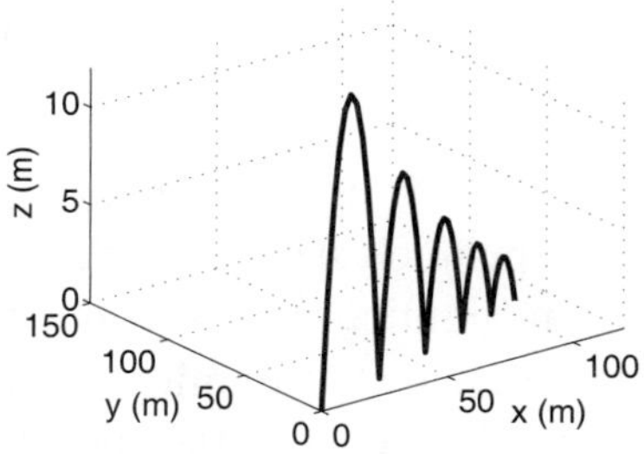

Chapter 11
Symbolic Math

All of the mathematical operations done with MATLAB in the first 10 chapters were numerical. The operations were carried out by writing numerical expressions that could contain numbers and variables with preassigned numerical values. When a numerical expression is executed by MATLAB, the outcome is also numerical (a single number or an array with numbers). The number, or numbers, are either exact or a floating point–approximated value. For example, typing 1/4 gives 0.2500—an exact value, and typing 1/3 gives 0.3333—an approximated value.

Many applications in math, science, and engineering require symbolic operations, which are mathematical operations with expressions that contain symbolic variables (variables that don't have specific numerical values when the operation is executed). The result of such operations is also a mathematical expression in terms of the symbolic variables. One simple example involves solving an algebraic equation that contains several variables and solving for one variable in terms of the others. If a, b, and x are symbolic variables, and $ax - b = 0$, x can be solved in terms of a and b to give $x = b/a$. Other examples of symbolic operations are analytical differentiation or integration of mathematical expressions. For instance, the derivative of $2t^3 + 5t - 8$ with respect to t is $6t^2 + 5$.

MATLAB has the capability of carrying out many types of symbolic operations. The numerical part of the symbolic operation is carried out by MATLAB exactly, with no approximation of numerical values. For example, the result of adding $\frac{x}{4}$ and $\frac{x}{3}$ is $\frac{7}{12}x$ and not $0.5833x$.

Symbolic operations can be performed by MATLAB once the Symbolic Math Toolbox is installed. The Symbolic Math Toolbox is a collection of MATLAB functions that are used for execution of symbolic operations. The commands and functions for the symbolic operations have the same style and syntax as those for the numerical operations. The symbolic operations themselves are executed primarily by MuPad®, which is mathematical software designed for this purpose. The MuPad software is embedded within MATLAB and is automatically activated when a symbolic MATLAB function is executed. MuPad can also be used as separate independent software. That software uses the MuPAD language, which has a completely different structure and commands than MATLAB. The Symbolic Math Toolbox is included in the student version of MATLAB. In the standard version, the toolbox is purchased separately. To check if the Symbolic Math Toolbox is installed on a computer, the user can type the command `ver` in the Command Window. In response, MATLAB displays information about the version that is used as well as a list of the toolboxes that are installed.

The starting point for symbolic operations is symbolic objects. Symbolic objects are made of variables and numbers that, when used in mathematical expressions, tell MATLAB to execute the expression symbolically. Typically, the user first defines (creates) the symbolic variables (objects) that are needed, and then uses them to create symbolic expressions that are subsequently used in symbolic operations. If needed, symbolic expressions can be used in numerical operations

The first section in this chapter describes how to define symbolic objects and how to use them to create symbolic expressions. The second section shows how to change the form of existing expressions. Once a symbolic expression has been created, it can be used in mathematical operations. MATLAB has a large selection of functions for this purpose. The next four sections (11.3–11.6) describe how to use MATLAB to solve algebraic equations, to carry out differentiation and integration, and to solve differential equations. Section 11.7 covers plotting symbolic expressions. How to use symbolic expressions in subsequent numerical calculations is explained in the following section.

11.1 Symbolic Objects and Symbolic Expressions

A symbolic object can be a variable (without a preassigned numerical value), a number, or an expression made of symbolic variables and numbers. A symbolic expression is a mathematical expression containing one or more symbolic objects. When typed, a symbolic expression may look like a standard numerical expression. However, because the expression contains symbolic objects, it is executed by MATLAB symbolically.

11.1.1 Creating Symbolic Objects

Symbolic objects can be variables or numbers. They can be created with the `sym` and/or `syms` commands. A single symbolic object can be created with the `sym` command:

```
object_name = sym('string')
```

where the string, which is the symbolic object, is assigned to a name. The string can be:

- A single letter or a combination of several letters (no spaces). Examples: `'a'`, `'x'`, `'yad'`.

- A combination of letters and digits starting with a letter and with no spaces Examples: `'xh12'`,`'r2d2'`.
- A number. Examples: `'15'`, `'4'`.

In the first two cases (where the string is a single letter, a combination of several letters, or a combination of letters and digits), the symbolic object is a symbolic variable. In this case it is convenient (but not necessary) to give the object the same name as the string. For example, *a*, *bb*, and *x*, can be defined as symbolic variables as follows:

```
>> a=sym('a')          Create a symbolic object a and assign it to a.
a =
a
>> bb=sym('bb')        The display of a symbolic
bb =                   object is not indented.
bb
>> x=sym('x');         The symbolic variable x is created but not displayed,
>>                     since a semicolon is typed at the end of the command.
```

The name of the symbolic object can be different from the name of the variable. For example:

```
>> g=sym('gamma')      The symbolic object is gamma, and
g =                    the name of the object is g.
gamma
```

As mentioned, symbolic objects can also be numbers. The numbers don't have to be typed as strings. For example, the `sym` command is used next to create symbolic objects from the numbers 5 and 7 and assign them to the variables *c* and *d*, respectively.

```
>> c=sym(5)            Create a symbolic object from the number 5 and assign it to c.
c =
5
>> d=sym(7)            The display of a symbolic
d =                    object is not indented.
7
```

As shown, when a symbolic object is created and a semicolon is not typed at the end of the command, MATLAB displays the name of the object and the object itself in the next two lines. The display of symbolic objects starts at the beginning of the line and is not indented as is the display of numerical variables. The difference is illustrated below, where a numerical variable is created.

```
>> e=13
e =
    13
```

13 is assigned to e (numerical variable).

The display of the value of a numerical variable is indented.

Several symbolic variables can be created in one command by using the syms command, which has the form:

```
syms variable_name variable_name variable_name
```

The command creates symbolic objects that have the same names as the symbolic variables. For example, the variables *y*, *z*, and *d* can all be created as symbolic variables in one command by typing:

```
>> syms y z d
>> y
y =
y
```

The variables created by the syms command are not displayed automatically. Typing the name of the variable shows that the variable was created.

When the syms command is executed, the variables it creates are not displayed automatically—even if a semicolon is not typed at the end of the command.

11.1.2 Creating Symbolic Expressions

Symbolic expressions are mathematical expressions written in terms of symbolic variables. Once symbolic variables are created, they can be used for creating symbolic expressions. The symbolic expression is a symbolic object (the display is not indented). The form for creating a symbolic expression is:

```
Expression_name = Mathematical expression
```

A few examples are:

```
>> syms a b c x y
>> f=a*x^2+b*x + c
f =
a*x^2 + b*x + c
```

Define a, b, c, x, and y as symbolic variables.

Create the symbolic expression $ax^2 + bx + c$ and assign it to f.

The display of the symbolic expression is not indented.

When a symbolic expression, which includes mathematical operations that can be executed (addition, subtraction, multiplication, and division), is entered, MATLAB executes the operations as the expression is created. For example:

```
>> g=2*a/3+4*a/7-6.5*x+x/3+4*5/3-1.5
```

$\frac{2a}{3} + \frac{4a}{7} - 6.5x + \frac{x}{3} + 4 \cdot \frac{5}{3} - 1.5$ is entered.

```
g =
(26*a)/21 - (37*x)/6 + 31/6
```
$\frac{26a}{21} - \frac{37x}{6} + \frac{31}{6}$ is displayed.

Notice that all the calculations are carried out exactly, with no numerical approximation. In the last example, $\frac{2a}{3}$ and $\frac{4a}{7}$ were added by MATLAB to give $\frac{26a}{21}$, and $-6.5x + \frac{x}{3}$ was added to $\frac{37x}{6}$. The operations with the terms that contain only numbers in the symbolic expression are carried out exactly. In the last example, $4 \cdot \frac{5}{3} + 1.5$ is replaced by $\frac{31}{6}$.

The difference between exact and approximate calculations is demonstrated in the following example, where the same mathematical operations are carried out—once with symbolic variables and once with numerical variables.

```
>> a=sym(3); b=sym(5);      Define a and b as symbolic 3 and 5, respectively.
>> e=b/a+sqrt(2)            Create an expression that includes a and b.
e =
2^(1/2) + 5/3               An exact value of e is displayed as a symbolic
                            object (the display is not indented).
>> c=3; d=5;                Define c and d as numerical 3 and 5, respectively.
>> f=d/c+sqrt(2)            Create an expression that includes c and d.
f =
    3.0809                  An approximated value of f is displayed
                            as a number (the display is indented).
```

An expression that is created can include both symbolic objects and numerical variables. However, if an expression includes a symbolic object (or several), all the mathematical operations will be carried out exactly. For example, if *c* is replaced by *a* in the last expression, the result is exact, as it was in the first example.

```
>> g=d/a+sqrt(2)
g =
2^(1/2) + 5/3
```

Additional facts about symbolic expressions and symbolic objects:

- Symbolic expressions can include numerical variables that have been obtained from the execution of numerical expressions. When these variables are inserted in symbolic expressions their exact value is used, even if the variable was displayed before with an approximated value. For example:

```
>> h=10/3                   h is defined to be 10/3 (a numerical variable).
```

```
h =
    3.3333          An approximated value of h (numerical variable) is displayed.
>> k=sym(5); m=sym(7);   Define k and m as symbolic 5 and 7, respectively.
>> p=k/m+h               h, k, and m are used in an expression.
p =                      The exact value of h is used in the determination of p.
85/21                    An exact value of p (symbolic object) is displayed.
```

- The double(S) command can be used to convert a symbolic expression (object) S that is written in an exact form to numerical form. (The name "double" comes from the fact that the command returns a double-precision floating-point number representing the value of S.) Two examples are shown. In the first, the p from the last example is converted into numerical form. In the second, a symbolic object is created and then converted into numerical form.

```
>> pN=double(p)          p is converted to numerical form (assigned to pN).
pN =
    4.0476
>> y=sym(10)*cos(5*pi/6)         Create a symbolic expression y.
y =
-5*3^(1/2)                       Exact value of y is displayed.
>> yN=double(y)
yN =                     y is converted to numerical form (assigned to yN).
   -8.6603
```

- A symbolic object that is created can also be a symbolic expression written in terms of variables that were not first created as symbolic objects. For example, the quadratic expression $ax^2 + bx + c$ can be created as a symbolic object named f by using the sym command:

```
>> f=sym('a*x^2+b*x+c')
f =
a*x^2 + b*x +c
```

It is important to understand that in this case, the variables a, b, c, and x included in the object do not exist individually as independent symbolic objects (the whole expression is one object). This means that it is impossible to perform symbolic math operations associated with the individual variables in the object. For example, it will not be possible to differentiate f with respect to x. This is different from the way in which the quadratic expression was created in the first example in this section, where the individual variables are first created as symbolic objects and then used in the quadratic expression.

- Existing symbolic expressions can be used to create new symbolic expressions. This is done by simply using the name of the existing expression in the new expression. For example:

```
>> syms x y                      Define x and y as symbolic variables.
>> SA=x+y, SB=x-y                Create two symbolic expressions SA and SB.
SA =
x+y
SB =
x-y
>> F=SA^2/SB^3+x^2               Create a new symbolic expression F using SA and SB.
F =
(x+y)^2/(x-y)^3+x^2
```

$SA = x + y$

$SB = x - y$

$$F = (SA^2)/(SB^3) + x^2 = \frac{(x+y)^2}{(x-y)^3} + x^2$$

11.1.3 The `findsym` Command and the Default Symbolic Variable

The `findsym` command can be used to find which symbolic variables are present in an existing symbolic expression. The format of the command is:

`findsym(S)` or `findsym(S,n)`

The `findsym(S)` command displays the names of all the symbolic variables (separated by commas) that are in the expression `S` in alphabetical order. The `findsym(S,n)` command displays `n` symbolic variables that are in expression `S` in the default order. For one-letter symbolic variables, the default order starts with *x*, and followed by letters, according to their closeness to *x*. If there are two letters equally close to *x*, the letter that is after *x* in alphabetical order is first (*y* before *w*, and *z* before *v*). The default symbolic variable in a symbolic expression is the first variable in the default order. The default symbolic variable in an expression `S` can be identified by typing `findsym(S,1)`. Examples:

```
>> syms x h w y d t          Define x, h, w, y, d, and t as symbolic variables.
>> S=h*x^2+d*y^2+t*w^2       Create a symbolic expression S.
S =
t*w^2 + h*x^2 + d*y^2
>> findsym(S)                Use the findsym(S) command.
ans =                        The symbolic variables are displayed in alphabetical order.
d, h, t, w, x, y
>> findsym(S,5)              Use the findsym(S,n) command (n = 5).
ans =                        Five symbolic variables are displayed in the default order.
x,y,w,t,h
```

```
>> findsym(S,1)          Use the findsym(S,n) command with n = 1.
ans =                    The default symbolic variable is displayed.
x
```

11.2 CHANGING THE FORM OF AN EXISTING SYMBOLIC EXPRESSION

Symbolic expressions are either created by the user or by MATLAB as the result of symbolic operations. The expressions created by MATLAB might not be in the simplest form or in a form that the user prefers. The form of an existing symbolic expression can be changed by collecting terms with the same power, by expanding products, by factoring out common multipliers, by using mathematical and trigonometric identities, and by many other operations. The following subsections describe several of the commands that can be used to change the form of an existing symbolic expression.

11.2.1 The `collect`, `expand`, and `factor` Commands

The `collect`, `expand`, and `factor` commands can be used to perform the mathematical operations that are implied by their names.

The `collect` command:

The `collect` command collects the terms in the expression that have the variable with the same power. In the new expression, the terms will be ordered in decreasing order of power. The command has the forms

`collect(S)`   `collect(S, variable_name)`

where `S` is the expression. The `collect(S)` form works best when an expression has only one symbolic variable. If an expression has more than one variable, MATLAB will collect the terms of one variable first, then those of a second variable, and so on. The order of the variables is determined by MATLAB. The user can specify the first variable by using the `collect(S, variable_name)` form of the command. Examples:

```
>> syms x y                        Define x and y as symbolic variables.
>> S=(x^2+x-exp(x))*(x+3)          Create the symbolic expression
S =                                (x+3)(x-e^x+x^2) and assign it to S.
(x + 3)*(x - exp(x) + x^2)
>> F = collect(S)                  Use the collect command.
F =                                MATLAB returns the expression:
x^3+4*x^2+(3-exp(x))*x-3*exp(x)    x^3+4x^2+(3--e^x)x-3e^x .
>> T=(2*x^2+y^2)*(x+y^2+3)         Create the symbolic expression T
T =                                (2x^2+y^2)(y^2+x+3).
(2*x^2+y^2)*(y^2+x+3)
```

```
>> G=collect(T)
G =
2*x^3+(2*y^2+6)*x^2+y^2*x+y^2*(y^2+3)
>> H=collect(T,y)
H =
y^4+(2*x^2+x+3)*y^2+2*x^2*(x+3)
```

Use the `collect(T)` command.

MATLAB returns the expression $2x^3 + (2y^2 + 6)x^2 + y^2x + y^2(y^2 + 3)$.

Use the `collect(T,y)` command.

MATLAB returns the expression $y^4 + (2x^2 + x + 3)y^2 + 2x^2(x + 3)$.

Note that when `collect(T)` is used, the reformatted expression is written in order of decreasing powers of x, but when `collect(T,y)` is used, the reformatted expression is written in order of decreasing powers of y.

The `expand` command:

The `expand` command expands expressions in two ways. It carries out products of terms that include summation (used with at least one of the terms), and it uses trigonometric identities and exponential and logarithmic laws to expand corresponding terms that include summation. The form of the command is:

```
expand(S)
```

where `S` is the symbolic expression. Two examples are:

```
>> syms a x y
>> S=(x+5)*(x-a)*(x+4)
S =
-(a-x)*(x+4)*(x+5)
>> T=expand(S)
T =
20*x-20*a-9*a*x-a*x^2+9*x^2+x^3
>> expand(sin(x-y))
ans =
cos(y)*sin(x)-cos(x)*sin(y)
```

Define `a`, `x`, and `y` as symbolic variables.

Create the symbolic expression $-(a - x)(x + 4)(x + 5)$ and assign it to `S`.

Use the `expand` command.

MATLAB returns the expression $20x - 20a - 9ax - ax^2 + 9x^2 + x^3$.

Use the `expand` command to expand $\sin(x - y)$.

MATLAB uses trig identity for the expansion.

The `factor` command:

The `factor` command changes an expression that is a polynomial to a product of polynomials of a lower degree. The form of the command is:

```
factor(S)
```

where `S` is the symbolic expression. An example is:

```
>> syms x
```

Define `x` as a symbolic variable.

```
>> S=x^3+4*x^2-11*x-30
S =
x^3+4*x^2-11*x-30
>> factor(S)
ans =
(x+5)*(x-3)*(x+2)
```

Create the symbolic expression $x^3 + 4x^2 - 11x - 30$ and assign it to `S`.

Use the `factor` command.

MATLAB returns the expression $(x+5)(x-3)(x+2)$.

11.2.2 The `simplify` *and* `simple` *Commands*

The `simplify` and `simple` commands are both general tools for simplifying the form of an expression. The `simplify` command uses built-in simplification rules to generate a simpler form of the expression than the original. The `simple` command is programmed to generate a form of the expression with the least number of characters. Although there is no guarantee that the form with the least number of characters is the simplest, in actuality this is often the case.

The `simplify` command:

The `simplify` command uses mathematical operations (addition, multiplication, rules of fractions, powers, logarithms, etc.) and functional and trigonometric identities to generate a simpler form of the expression. The format of the `simplify` command is:

`simplify(S)`

where either `S` is the name of the existing expression to be simplified, or an expression to be simplified can be typed in for `S`.

Two examples are:

```
>> syms x y
>> S=(x^2+5*x+6)/(x+2)
S =
(x^2+5*x+6)/(x+2)
>> SA = simplify(S)
SA =
x+3
>> simplify((x+y)/(1/x+1/y))

ans =
x*y
```

Define `x` and `y` as symbolic variables.

Create the symbolic expression $(x^2 + 5x + 6)/(x + 2)$, and assign it to `S`.

Use the `simplify` command to simplify `S`.

MATLAB simplifies the expression to $x + 3$.

Simplify $(x+y)/\left(\frac{1}{x}+\frac{1}{y}\right)$.

MATLAB simplifies the expression to xy).

The `simple` command:

The `simple` command finds the form of the expression with the fewest number of characters. In many cases this form is also the simplest. When the command is executed, MATLAB creates several forms of the expression by applying the `collect`, `expand`, `factor`, and `simplify` commands, and other simplification functions that are not covered here. Then MATLAB returns the expression with the shortest form. The `simple` command has the following three forms:

`F = simple(S)`	`simple(S)`	`[F how] = simple(S)`
The shortest form of `S` is assigned to `F`.	All the simplification trails are displayed. The shortest is assigned to `ans`.	The shortest form of `S` is assigned to `F`. The name (string) of the simplification method is assigned to `how`.

The difference between the forms is in the output. The use of two of the forms is shown next.

```
>> syms x
>> S=(x^3-4*x^2+16*x)/(x^3+64)
S =
(x^3-4*x^2+16*x)/(x^3+64)
>> F = simple(S)
F =
x/(x+4)
>> [G how] = simple(S)
G =
x/(x+4)
how =
simplify
```

Define `x` as a symbolic variable.

Create the symbolic expression $\frac{x^3 - 4x^2 + 16x}{x^3 + 64}$, and assign it to `S`.

Use the `F = simple(S)` command to simplify `S`.

The simplest form of `S`, $x/(x+4)$, is assigned to `F`.

Use the `[G how] = simple(S)` command.

The simplest form of `S`, $x/(x+4)$, is assigned to `G`.

The word "simplify" is assigned to `G`, which means that the shortest form was obtained using the `simplify` command.

The use of the `simple(S)` form of the command is not demonstrated because the display of the output is lengthy. MATLAB displays 10 different tries and assigns the shortest form to `ans`. The reader should try to execute the command and examine the output display.

11.2.3 The `pretty` Command

The `pretty` command displays a symbolic expression in a format resembling the mathematical format in which expressions are generally typed. The command has the form

`pretty(S)`

Example:

```
>> syms a b c x
>> S=sqrt(a*x^2 + b*x + c)
S =
(a*x^2+b*x+c)^(1/2)
>> pretty(S)
                 2           1/2
           (a x   + b x + c)
```

Define a, b, c, and x as symbolic variables.

Create the symbolic expression $\sqrt{ax^2 + bx + c}$, and assign it to S.

The pretty command displays the expression in a math format.

11.3 Solving Algebraic Equations

A single algebraic equation can be solved for one variable, and a system of equations can be solved for several variables with the solve function.

Solving a single equation:

An algebraic equation can have one or several symbolic variables. If the equation has one variable, the solution is numerical. If the equation has several symbolic variables, a solution can be obtained for any of the variables in terms of the others. The solution is obtained by using the solve command, which has the form

`h = solve(eq)` or `h = solve(eq,var)`

- The argument eq can be the name of a previously created symbolic expression, or an expression that is typed in. When a previously created symbolic expression S is entered for eq, or when an expression that does not contain the = sign is typed in for eq, MATLAB solves the equation eq = 0.
- An equation of the form $f(x) = g(x)$ can be solved by typing the equation (including the = sign) as a string for eq.
- If the equation to be solved has more than one variable, the solve(eq) command solves for the default symbolic variable (see Section 11.1.3). A solution for any of the variables can be obtained with the solve(eq,var) command by typing the variable name for var.
- If the user types solve(eq), the solution is assigned to the variable ans.
- If the equation has more than one solution, the output h is a symbolic column vector with a solution at each element. The elements of the vector are symbolic objects. When an array of symbolic objects is displayed, each row is enclosed with square brackets (see the following examples).

The following examples illustrate the use of the `solve` command.

```
>> syms a b x y z
>> h=solve(exp(2*z)-5)
h =
log(5)/2
>> S=x^2-x-6
S =
x^2-x-6
>> k=solve(S)
k =
 -2
  3
>> solve('cos(2*y)+3*sin(y)=2')
ans =
     pi/2
     pi/6
 (5*pi)/6
>> T= a*x^2+5*b*x+20
T =
a*x^2+5*b*x+20
>> solve(T)
ans =
 -(5*b+5^(1/2)*(5*b^2-16*a)^(1/2))/(2*a)
 -(5*b-5^(1/2)*(5*b^2-16*a)^(1/2))/(2*a)
>> M = solve(T,a)
M =
-(5*b*x+20)/x^2
```

- `syms a b x y z`: Define `a`, `b`, `x`, `y`, and `z` as symbolic variables.
- `h=solve(exp(2*z)-5)`: Use the `solve` command to solve $e^{2z} - 5 = 0$.
- `h =`: The solution is assigned to `h`.
- `S=x^2-x-6`: Create the symbolic expression $x^2 - x - 6$, and assign it to `S`.
- `k=solve(S)`: Use the `solve(S)` command to solve $x^2 - x - 6 = 0$.
- `k =`: The equation has two solutions. They are assigned to `k`, which is a column vector with symbolic objects.
- `solve('cos(2*y)+3*sin(y)=2')`: Use the `solve` command to solve $\cos(2y) + 3\sin(y) = 2$. (The equation is typed as a string in the command.)
- `ans =`: The solution is assigned to `ans`.
- `T= a*x^2+5*b*x+20`: Create the symbolic expression $ax^2 + 5bx + 20$, and assign it to `T`.
- `solve(T)`: Use the `solve(S)` command to solve $T = 0$.
- `ans =`: The equation $T = 0$ is solved for the variable x, which is the default variable.
- `M = solve(T,a)`: Use the `solve(eq,var)` command to solve $T = 0$.
- `M =`: The equation $T = 0$ is solved for the variable a.

- It is also possible to use the `solve` command by typing the equation to be solved as a string, without having the variables in the equation first created as symbolic objects. However, if the solution contains variables (when the equation has more than one variable), the variables do not exist as independent symbolic objects. For example:

```
>> ts=solve('4*t*h^2+20*t-5*g')
ts =
(5*g)/(4*h^2+20)
```

- `ts=solve('4*t*h^2+20*t-5*g')`: The expression $4th^2 + 20t - 5g$ is typed in the `solve` command. The variables t, h, and g were not created as symbolic variables before the expression was typed in the `solve` command.
- `ts =`: MATLAB solves the equation $4th^2 + 20t - 5g = 0$ for t.

The equation can also be solved for a different variable. For example, a solution for *g* is obtained by:

```
>> gs=solve('4*t*h^2+20*t-5*g','g')
gs =
(4*t*h^2)/5 + 4*t
```

Solving a system of equations:

The `solve` command can also be used for solving a system of equations. If the number of equations and the number of variables are the same, the solution is numerical. If the number of variables is greater than the number of equations, the solution is symbolic for the desired variables in terms of the other variables. A system of equations (depending on the type of equations) can have one or several solutions. If the system has one solution, each of the variables for which the system is solved has one numerical value (or expression). If the system has more than one solution, each of the variables can have several values.

The format of the `solve` command for solving a system of *n* equations is:

```
output = solve(eq1,eq2,....,eqn)
```

or

```
output = solve(eq1,eq2,...,eqn,var1,var2,...,varn)
```

- The arguments `eq1,eq2,...,eqn` are the equations to be solved. Each argument can be a name of a previously created symbolic expression, or an expression that is typed in as a string. When a previously created symbolic expression `S` is entered, the equation is `S = 0`. When a string that does not contain the = sign is typed in, the equation is `expression = 0`. An equation that contains the = sign must be typed as a string.
- In the first format, if the number of equations *n* is equal to the number of variables in the equations, MATLAB gives a numerical solution for all the variables. If the number of variables is greater than the number of equations *n*, MATLAB gives a solution for *n* variables in terms of the rest of the variables. The variables for which solutions are obtained are chosen by MATLAB according to the default order (Section 11.1.3).
- When the number of variables is greater than the number of equations *n*, the user can select the variables for which the system is solved. This is done by using the second format of the `solve` command and entering the names of the variables `var1,var2,...,varn`.

The `output` from the `solve` command, which is the solution of the system, can have two different forms. One is a cell array and the other is a structure. A cell array is an array in which each of the elements can be an array. A structure is an array in which the elements (called fields) are addressed by textual field designators. The fields of a structure can be arrays of different sizes and types. Cell arrays and structures are not presented in detail in this book, but a short explanation is given below so that the reader will be able to use them with the `solve` command.

When a cell array is used in the output of the `solve` command, the command has the following form (in the case of a system of three equations):

```
[varA, varB, varC] = solve(eq1,eq2,eq3)
```

- Once the command is executed, the solution is assigned to the variables `varA`, `varB`, and `varC`, and the variables are displayed with their assigned solution. Each of the variables will have one or several values (in a column vector) depending on whether the system of equations has one or several solutions.
- The user can select any names for `varA`, `varB`, and `varC`. MATLAB assigns the solution for the variables in the equations in alphabetical order. For example, if the variables for which the equations are solved are x, u, and t, the solution for t is assigned to `varA`, the solution for u is assigned to `varB`, and the solution for x is assigned to `varC`.

The following examples show how the `solve` command is used for the case where a cell array is used in the output:

```
>> syms x y t                        Define x, y, and t as symbolic variables.
>> S=10*x+12*y+16*t;                 Assign to S the expression 10x + 12y + 16t.
>> [xt yt]=solve(S, '5*x-y=13*t')    Use the solve command to solve
xt =                                 the system: 10x + 12y + 16t = 0
2*t                                              5x − y = 13t
yt =
-3*t                                 Output in a cell array with two cells named xt and yt.
            The solutions for x and y are assigned to xt and yt, respectively.
```

In the example above, notice that the system of two equations is solved by MATLAB for x and y in terms of t, since x and y are the first two variables in the default order. The system, however, can be solved for different variables. As an example, the system is solved next for y and t in terms of x (using the second form of the `solve` command:

```
>> [tx yx]=solve(S,'5*x-y=13*t',y,t)
                                    The variables for which the system
                                    is solved (y and t) are entered.
tx =
x/2             The solutions for the variables for which the system is
yx =            solved are assigned in alphabetical order. The first cell has
-(3*x)/2        the solution for t, and the second cell has the solution for y.
```

When a structure is used in the output of the `solve` command, the command has the form (in the case of a system of three equations)

`AN = solve(eq1,eq2,eq3)`

- `AN` is the name of the structure.
- Once the command is executed the solution is assigned to `AN`. MATLAB displays the name of the structure and the names of the fields of the structure, which are the names of the variables for which the equations are solved. The size and the type of each field is displayed next to the field name. The content of each field, which is the solution for the variable, is not displayed.
- To display the content of a field (the solution for the variable), the user has to type the address of the field. The form for typing the address is: `structure_name.field_name` (see example below).

As an illustration the system of equations solved in the last example is solved again using a structure for the output.

```
>> syms x y t
>> S=10*x+12*y+16*t;
>> AN=solve(S,'5*x-y=13*t')
AN =
   x: [1x1 sym]
   y: [1x1 sym]
>> AN.x
ans =
2*t
>> AN.y
ans =
-3*t
```

Use the `solve` command to solve the system: $10x + 12y + 16t = 0$
$5x - y = 13t$

MATLAB displays the name of the structure `AN` and the names of its fields `x` and `y` (size and type), which are the names of the variables for which the equations are solved.

Type the address of the field `x`.

The content of the field (the solution for x) is displayed.

Type the address of the field `y`.

The content of the field (the solution for y) is displayed.

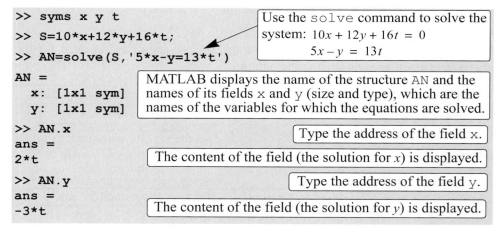

Sample Problem 11-1 shows the solution of a system of equations that has two solutions.

Sample Problem 11-1: Intersection of a circle and a line

The equation of a circle in the $x\,y$ plane with radius R and its center at point (2, 4) is given by $(x-2)^2 + (y-4)^2 = R^2$. The equation of a line in the plane is given by $y = \frac{x}{2} + 1$. Determine the coordinates of the points (as a function of R) where the line intersects the circle.

Solution

The solution is obtained by solving the system of the two equations for x and y in terms of R. To show the difference in the output between using cell array and structure output forms of the `solve` command, the system is solved twice. The first solution has the output in a cell array:

```
>> syms x y R
>> [xc,yc]=solve('(x-2)^2+(y-4)^2=R^2','y=x/2+1')
xc =
((4*R^2)/5 - 64/25)^(1/2) + 14/5
 14/5 - ((4*R^2)/5 - 64/25)^(1/2)
yc =
((4*R^2)/5 - 64/25)^(1/2)/2 + 12/5
 12/5 - ((4*R^2)/5 - 64/25)^(1/2)/2
```

The two equations are typed in the `solve` command.

Output in a cell array.

Output in a cell array with two cells named `xc` and `yc`. Each cell contains two solutions in a symbolic column vector.

The second solution has the output in a structure:

```
>> COORD=solve('(x-2)^2+(y-4)^2=R^2','y = x/2+1')
COORD =
    x: [2x1 sym]
    y: [2x1 sym]
>> COORD.x
ans =
((4*R^2)/5 - 64/25)^(1/2) + 14/5
14/5 - ((4*R^2)/5 - 64/25)^(1/2)
>> COORD.y
ans =
((4*R^2)/5 - 64/25)^(1/2)/2 + 12/5
12/5 - ((4*R^2)/5 - 64/25)^(1/2)/2
```

Output in a structure.

Output in a structure named `COORD` that has two fields, `x` and `y`. Each field is a 2 by 1 symbolic vector.

Type the address of the field `x`.

The content of the field (the solution for x) is displayed.

Type the address of the field `y`.

The content of the field (the solution for y) is displayed.

11.4 DIFFERENTIATION

Symbolic differentiation can be carried out by using the `diff` command. The form of the command is:

`diff(S)` or `diff(S,var)`

- Either `S` can be the name of a previously created symbolic expression, or an expression can be typed in for `S`.
- In the `diff(S)` command, if the expression contains one symbolic variable, the differentiation is carried out with respect to that variable. If the expression

contains more than one variable, the differentiation is carried out with respect to the default symbolic variable (Section 11.1.3).

- In the `diff(S,var)` command (which is used for differentiation of expressions with several symbolic variables) the differentiation is carried out with respect to the variable `var`.
- The second or higher (nth) derivative can be determined with the `diff(S,n)` or `diff(S,var,n)` command, where `n` is a positive number. $n = 2$ for the second derivative, $n = 3$ for the third, and so on.

Some examples are:

```
>> syms x y t                     Define x, y, and t as symbolic variables.
>> S=exp(x^4);                    Assign to S the expression e^(x^4).
>> diff(S)                        Use the diff(S) command to differentiate S.
ans =
4*x^3*exp(x^4)                    The answer 4x^3 e^(x^4) is displayed.
>> diff((1-4*x)^3)                Use the diff(S) command to differentiate (1 – 4x)^3.
ans =
-12*(1-4*x)^2                     The answer –12(1 – 4x)^2 is displayed.
>> R=5*y^2*cos(3*t);              Assign to R the expression 5y^2 cos(3t).
>> diff(R)                        Use the diff(R) command to differentiate R.
ans =                             MATLAB differentiates R with respect to y (default
10*y*cos(3*t)                     symbolic variable); the answer 10y cos(3t) is displayed.
>> diff(R,t)                      Use the diff(R,t) command to differentiate R w.r.t. t.
ans =
-15*y^2*sin(3*t)                  The answer –15y^2 sin(3t) is displayed.
>> diff(S,2)                      Use diff(S,2) command to obtain the second derivative of S.
ans =
12*x^2*exp(x^4)+16*x^6*exp(x^4)   The answer 12x^2 e^(x^4) + 16x^6 e^(x^4)
                                  is displayed.
```

- It is also possible to use the `diff` command by typing the expression to be differentiated as a string directly in the command without having the variables in the expression first created as symbolic objects. However, the variables in the differentiated expression do not exist as independent symbolic objects.

11.5 INTEGRATION

Symbolic integration can be carried out by using the `int` command. The command can be used for determining indefinite integrals (antiderivatives) and definite integrals. For indefinite integration the form of the command is:

`int(S)` or `int(S,var)`

- Either `S` can be the name of a previously created symbolic expression, or an expression can be typed in for `S`.
- In the `int(S)` command, if the expression contains one symbolic variable, the integration is carried out with respect to that variable. If the expression contains more than one variable, the integration is carried out with respect to the default symbolic variable (Section 11.1.3).
- In the `int(S,var)` command, which is used for integration of expressions with several symbolic variables, the integration is carried out with respect to the variable `var`.

Some examples are:

```
>> syms x y t
>> S=2*cos(x)-6*x;
>> int(S)
ans =
2*sin(x)-3*x^2
>> int(x*sin(x))
ans =
sin(x)-x*cos(x)
>>R=5*y^2*cos(4*t);
>> int(R)
ans =
(5*y^3*cos(4*t))/3
>> int(R,t)
ans =
(5*y^2*sin(4*t))/4
```

Define `x`, `y`, and `t` as symbolic variables.

Assign to `S` the expression $2\cos(x) - 6x$.

Use the `int(S)` command to integrate `S`.

The answer $2\sin(x) - 3x^2$ is displayed.

Use the `int(S)` command to integrate $x\sin(x)$.

The answer $\sin(x) - x\cos(x)$ is displayed.

Assign to `R` the expression $5y^2\cos(4t)$.

Use the `int(R)` command to integrate `R`.

MATLAB integrates `R` with respect to y (default symbolic variable); the answer $5y^3\cos(4t)/3$ is displayed.

Use the `int(R,t)` command to integrate `R` w.r.t. t.

The answer $5y^2\sin(4t)/4$ is displayed.

For definite integration the form of the command is:

`int(S,a,b)` or `int(S,var,a,b)`

- `a` and `b` are the limits of integration. The limits can be numbers or symbolic variables.

For example, determination of the definite integral $\int_0^{\pi}(\sin y - 5y^2)dy$ with MATLAB is:

```
>> syms y
>> int(sin(y)-5*y^2,0,pi)
ans =
2 - (5*pi^3)/3
```

- It is possible also to use the `int` command by typing the expression to be integrated as a string without having the variables in the expression first created as symbolic objects. However, the variables in the integrated expression do not exist as independent symbolic objects.
- Integration can sometimes be a difficult task. A closed-form answer may not exist, or if it exists, MATLAB might not be able to find it. When that happens MATLAB returns `int(S)` and the message `Explicit integral could not be found.`

11.6 Solving an Ordinary Differential Equation

An ordinary differential equation (ODE) can be solved symbolically with the `dsolve` command. The command can be used to solve a single equation or a system of equations. Only single equations are addressed here. Chapter 10 discusses using MATLAB to solve first-order ODEs numerically. The reader's familiarity with the subject of differential equations is assumed. The purpose of this section is to show how to use MATLAB for solving such equations.

A first-order ODE is an equation that contains the derivative of the dependent variable. If t is the independent variable and y is the dependent variable, the equation can be written in the form

$$\frac{dy}{dt} = f(t, y)$$

A second-order ODE contains the second derivative of the dependent variable (it can also contain the first derivative). Its general form is:

$$\frac{d^2y}{dt^2} = f\left(t, y, \frac{dy}{dt}\right)$$

A solution is a function $y = f(t)$ that satisfies the equation. The solution can be general or particular. A general solution contains constants. In a particular solution the constants are determined to have specific numerical values such that the solution satisfies specific initial or boundary conditions.

The command `dsolve` can be used for obtaining a general solution or, when the initial or boundary conditions are specified, for obtaining a particular solution.

General solution:

For obtaining a general solution, the `dsolve` command has the form:

`dsolve('eq')` or `dsolve('eq','var')`

- `eq` is the equation to be solved. It has to be typed as a string (even if the variables are symbolic objects).
- The variables in the equation don't have to first be created as symbolic objects. (If they have not been created, then, in the solution the variables will not be symbolic objects.)
- Any letter (lowercase or uppercase), except `D` can be used for the dependent variable.
- In the `dsolve('eq')` command the independent variable is assumed by MATLAB to be `t` (default).
- In the `dsolve('eq','var')` command the user defines the independent variable by typing it for `var` (as a string).
- In specifying the equation the letter `D` denotes differentiation. If y is the dependent variable and t is the independent variable, `Dy` stands for $\frac{dy}{dt}$. For example, the equation $\frac{dy}{dt} + 3y = 100$ is typed in as `'Dy + 3*y = 100'`.
- A second derivative is typed as `D2`, third derivative as `D3`, and so on. For example, the equation $\frac{d^2y}{dt^2} + 3\frac{dy}{dt} + 5y = \sin(t)$ is typed in as: `'D2y + 3*Dy + 5*y = sin(t)'`.
- The variables in the ODE equation that is typed in the `dsolve` command do not have to be previously created symbolic variables.
- In the solution MATLAB uses `C1`, `C2`, `C3`, and so on, for the constants of integration.

For example, a general solution of the first-order ODE $\frac{dy}{dt} = 4t + 2y$ is obtained by:

```
>> dsolve('Dy=4*t+2*y')
ans =
C1*exp(2*t) - 2*t - 1
```

The answer $y = C_1e^{2t} - 2t - 1$ is displayed.

A general solution of the second-order ODE $\frac{d^2x}{dt^2} + 2\frac{dx}{dt} + x = 0$ is obtained by:

```
>> dsolve('D2x+2*Dx+x=0')
```

```
ans =
C1/exp(t)+(C2*t)/exp(t)
```

The answer $x = C_1e^{-t} + C_2te^{-t}$ is displayed.

The following examples illustrate the solution of differential equations that contain symbolic variables in addition to the independent and dependent variables.

```
>> dsolve('Ds=a*x^2')

ans =
a*t*x^2 + C1
```

The independent variable is t (default). MATLAB solves the equation $\frac{ds}{dt} = ax^2$.

The solution $s = ax^2t + C_1$ is displayed.

```
>> dsolve('Ds=a*x^2','x')

ans =
(a*x^3)/3 + C1
```

The independent variable is defined to be x. MATLAB solves the equation $\frac{ds}{dx} = ax^2$.

The solution $s = \frac{1}{3}ax^3 + C_1$ is displayed.

```
>> dsolve('Ds=a*x^2','a')

ans =
(a^2*x^2)/2 + C2
```

The independent variable is defined to be a. MATLAB solves the equation $\frac{ds}{da} = ax^2$.

The solution $s = \frac{1}{2}a^2x^2 + C_1$ is displayed.

Particular solution:

A particular solution of an ODE can be obtained if boundary (or initial) conditions are specified. A first-order equation requires one condition, a second-order equation requires two conditions, and so on. For obtaining a particular solution, the `dsolve` command has the form

First-order ODE: `dsolve('eq','cond1','var')`

Higher-order ODE: `dsolve('eq','cond1','cond2',....,'var')`

- For solving equations of higher order, additional boundary conditions have to be entered in the command. If the number of conditions is less than the order of the equation, MATLAB returns a solution that includes constants of integration (`C1`, `C2`, `C3`, and so on).
- The boundary conditions are typed in as strings in the following:

Math form	**MATLAB form**
$y(a) = A$	`'y(a)=A'`
$y'(a) = A$	`'Dy(a)=A'`
$y''(a) = A$	`'D2y(a)=A'`

- The argument `'var'` is optional and is used to define the independent variable in the equation. If none is entered, the default is t.

For example, the first-order ODE $\frac{dy}{dt} + 4y = 60$, with the initial condition $y(0) = 5$ is solved with MATLAB by:

```
>> dsolve('Dy+4*y=60','y(0)=5')
ans =
15 - 10/exp(4*t)
```

The answer $y = 15 - (10/e^{4t})$ is displayed.

The second-order ODE $\frac{d^2y}{dt^2} - 2\frac{dy}{dt} + 2y = 0$, $y(0) = 1$, $\left.\frac{dy}{dt}\right|_{t=0} = 0$, can be solved with MATLAB by:

```
>> dsolve('D2y-2*Dy+2*y=0','y(0)=1','Dy(0)=0')

ans =
exp(t)*cos(t)-exp(t)*sin(t)
```

The answer $y = e^t\cos(t) - e^t\sin(t)$ is displayed.

```
>> factor(ans)
```

The answer can be simplified with the `factor` command.

```
ans =
exp(t)*(cos(t)-sin(t))
```

The simplified answer $y = e^t(\cos(t) - \sin(t))$ is displayed.

Additional examples of solving differential equations are shown in Sample Problem 11-5.

If MATLAB cannot find a solution, it returns an empty symbolic object and the message `Warning: explicit solution could not be found.`

11.7 Plotting Symbolic Expressions

In many cases, there is a need to plot a symbolic expression. This can easily be done with the `ezplot` command. For a symbolic expression `S` that contains one variable `var`, MATLAB considers the expression to be a function $S(var)$, and the command creates a plot of $S(var)$ versus *var*. For a symbolic expression that contains two symbolic variables `var1` and `var2`, MATLAB considers the expression to be a function in the form $S(var1,var2) = 0$, and the command creates a plot of one variable versus the other.

To plot a symbolic expression `S` that contains one or two variables, the `ezplot` command is:

```
ezplot(S)
```

or

```
ezplot(S,[min,max])
```

or

```
ezplot(S,[xmin,xmax,ymin,ymax])
```

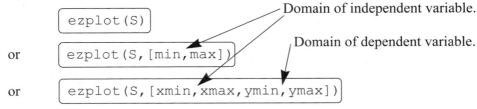

Domain of independent variable. (points to `[min,max]` and `[xmin,xmax`)

Domain of dependent variable. (points to `ymin,ymax]`)

- `S` is the symbolic expression to be plotted. It can be the name of a previously created symbolic expression, or an expression can be typed in for `S`.
- It is also possible to type the expression to be plotted as a string without having the variables in the expression first created as symbolic objects.
- If `S` has one symbolic variable, a plot of $S(var)$ versus (var) is created, with the values of *var* (the independent variable) on the abscissa (horizontal axis), and the values of $S(var)$ on the ordinate (vertical axis).
- If the symbolic expression `S` has two symbolic variables, `var1` and `var2`, the expression is assumed to be a function with the form $S(var1{,}var2) = 0$. MATLAB creates a plot of one variable versus the other variable. The variable that is first in alphabetic order is taken to be the independent variable. For example, if the variables in *S* are *x* and *y*, then *x* is the independent variable and is plotted on the abscissa and *y* is the dependent variable plotted on the ordinate. If the variables in *S* are *u* and *v*, then *u* is the independent variable and *v* is the dependent variable.
- In the `ezplot(S)` command, if `S` has one variable ($S(var)$), the plot is over the domain $-2\pi < var < 2\pi$ (default domain) and the range is selected by MATLAB. If `S` has two variables ($S(var1{,}var2)$), the plot is over $-2\pi < var1 < 2\pi$ and $-2\pi < var2 < 2\pi$.
- In the `ezplot(S,[min,max])` command the domain for the independent variable is defined by `min` and `max`:—$min < var < max$—and the range is selected by MATLAB.
- In the `ezplot(S,[xmin,xmax,ymin,ymax])` command the domain for the independent variable is defined by `xmin` and `xmax`, and the domain of the dependent variable is defined by `ymin` and `ymax`.

The `ezplot` command can also be used to plot a function that is given in a parametric form. In this case two symbolic expressions, `S1` and `S2`, are involved, where each expression is written in terms of the same symbolic variable (independent parameter). For example, for a plot of y versus x where $x = x(t)$ and $y = y(t)$, the form of the `ezplot` command is:

```
ezplot(S1,S2)
```

or

```
ezplot(S1,S2,[min,max])
```

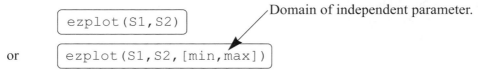

Domain of independent parameter. (points to `[min,max]`)

- `S1` and `S2` are symbolic expressions containing the same single symbolic variable, which is the independent parameter. `S1` and `S2` can be the names of previously created symbolic expressions, or expressions can be typed in.

- The command creates a plot of $S2(var)$ versus $S1(var)$. The symbolic expression that is typed first in the command (`S1` in the definition above) is used for the horizontal axis, and the expression that is typed second (`S2` in the definition above) is used for the vertical axis.
- In the `ezplot(S1,S2)` command the domain of the independent variable is $0 < var < 2\pi$ (default domain).
- In the `ezplot(S1,S2,[min,max])` command the domain for the independent variable is defined by `min` and `max`: $min < var < max$.

Additional comments:

Once a plot is created, it can be formatted in the same way as plots created with the `plot` or `fplot` format. This can be done in two ways: by using commands or by using the Plot Editor (see Section 5.4). When the plot is created, the expression that is plotted is displayed automatically at the top of the plot. MATLAB has additional plot functions for plotting two-dimensional polar plots and for plotting three-dimensional plots. For more information, the reader is referred to the Help menu of the Symbolic Math Toolbox.

Several examples of using the `ezplot` command are shown in Table 11-1.

Table 11-1: Plots with the `ezplot` command

Command	Plot

`>> syms x`
`>> S=(3*x+2)/(4*x-1)`
`S =`
`(3*x+2)/(4*x-1)`
`>> ezplot(S)`
	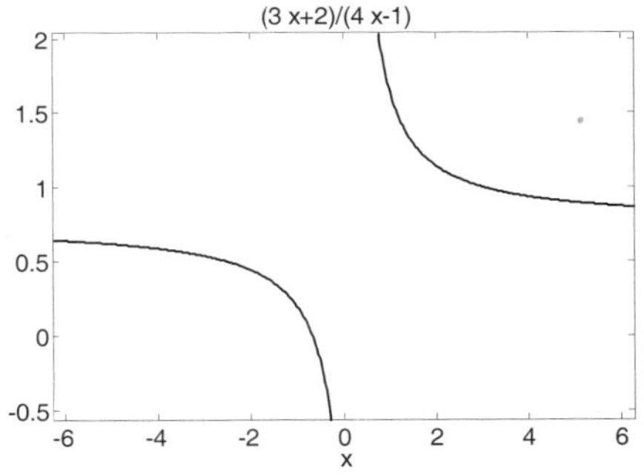

`>> syms x y`
`>> S=4*x^2-18*x+4*y^2+12*y-11`
`S =`
`4*x^2-18*x+4*y^2+12*y-11`
`>> ezplot(S)`
	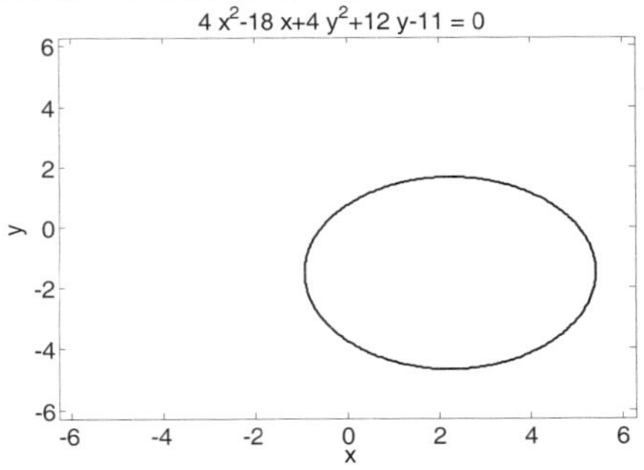

Table 11-1: Plots with the `ezplot` command (Continued)

Command	Plot

`>> syms t`
`>> x=cos(2*t)`
`x =`
`cos(2*t)`
`>> y=sin(4*t)`
`y =`
`sin(4*t)`
`>> ezplot(x,y)`
	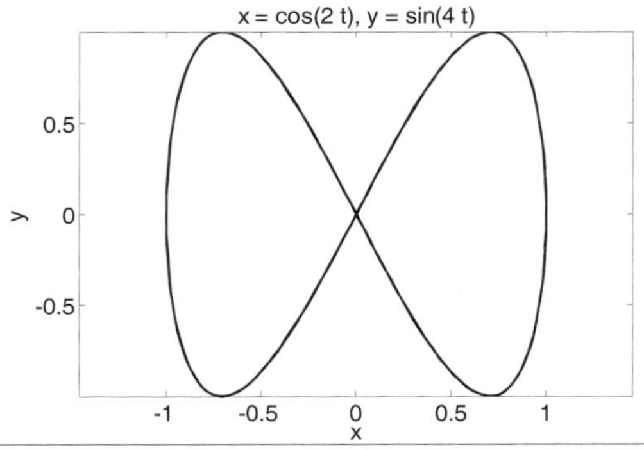

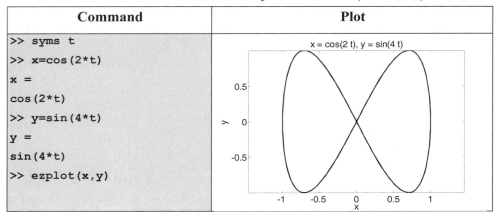

11.8 NUMERICAL CALCULATIONS WITH SYMBOLIC EXPRESSIONS

Once a symbolic expression is created by the user or by the output from any of MATLAB's symbolic operations, there may be a need to substitute numbers for the symbolic variables and calculate the numerical value of the expression. This can be done by using the `subs` command. The `subs` command has several forms and can be used in different ways. The following describes several forms that are easy to use and are suitable for most applications. In one form, the variable (or variables) for which a numerical value is substituted and the numerical value itself are typed inside the `subs` command. In another form, each variable is assigned a numerical value in a separate command and then the variable is substituted in the expression.

The `subs` command in which the variable and its value are typed inside the command is shown first. Two cases are presented—one for substituting a numerical value (or values) for one symbolic variable, and the other for substituting numerical values for two or more symbolic variables.

Substituting a numerical value for one symbolic variable:

A numerical value (or values) can be substituted for one symbolic variable when a symbolic expression has one or more symbolic variables. In this case the `subs` command has the form:

```
R = subs(S,var,number)
```

The name of the symbolic expression. (`S`)

The variable for which a numerical value is substituted. (`var`)

The numerical value (or values) assigned to `var`. (`number`)

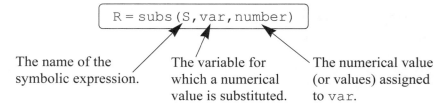

- `number` can be one number (a scalar), or an array with many elements (a vector or a matrix).

- The value of S is calculated for each value of number and the result is assigned to R, which will have the same size as number (scalar, vector, or matrix).
- If S has one variable, the output R is numerical. If S has several variables and a numerical value is substituted for only one of them, the output R is a symbolic expression.

An example with an expression that includes one symbolic variable is:

```
>> syms x
>> S=0.8*x^3+4*exp(0.5*x)
S =
4*exp(x/2) + (4*x^3)/5
>> SD=diff(S)
SD =
2*exp(x/2)+(12*x^2)/5)
>> subs(SD, x, 2)
ans =
   15.0366
>> SDU=subs(SD, x, [2:0.5:4])

SDU =
   15.0366   21.9807   30.5634   40.9092   53.1781
```

Define x as a symbolic variable.

Assign to S the expression $0.8x^3 + 4e^{(0.5x)}$.

Use the diff(S) command to differentiate S.

The answer $2e^{x/2} + 12x^2/5$ is assigned to SD.

Use the subs command to substitute $x = 2$ in SD.

The value of SD is displayed.

Use the subs command to substitute $x = [2, 2.5, 3, 3.5, 4]$ (vector) in SD.

The values of SD (assigned to SDU) for each value of x are displayed in a vector.

In the last example, notice that when the numerical value of the symbolic expression is calculated, the answer is numerical (the display is indented). An example of substituting numerical values for one symbolic variable in an expression that has several symbolic variables is:

```
>> syms a g t v
>> Y=v^2*exp(a*t)/g
Y =
v^2*exp(a*t)/g
>> subs(Y,t,2)
ans =
v^2*exp(2*a)/g
>> Yt=subs(Y,t,[2:4])
```

Define a, g, t, and v as symbolic variables.

Create the symbolic expression $v^2e^{(at)}/g$ and assign it to Y.

Use the subs command to substitute $t = 2$ in SD.

The answer $v^2e^{(2a)}/g$ is displayed.

Use the subs command to substitute $t = [2, 3, 4]$ (vector) in Y.

```
Yt =
[ v^2*exp(2*a)/g, v^2*exp(3*a)/g, v^2*exp(4*a)/g]
```

The answer is a vector with elements of symbolic expressions for each value of t.

Substituting a numerical value for two or more symbolic variables:

A numerical value (or values) can be substituted for two or more symbolic variables when a symbolic expression has several symbolic variables. In this case the `subs` command has the following form (it is shown for two variables, but it can be used in the same form for more):

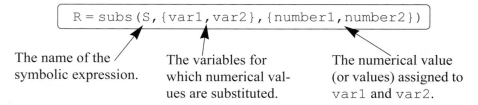

```
R = subs(S,{var1,var2},{number1,number2})
```

The name of the symbolic expression. (S)

The variables for which numerical values are substituted. ({var1,var2})

The numerical value (or values) assigned to `var1` and `var2`. ({number1,number2})

- The variables `var1` and `var2` are the variables in the expression `S` for which the numerical values are substituted. The variables are typed as a cell array (inside curly braces `{ }`). A cell array is an array of cells where each cell can be an array of numbers or text.
- The numbers `number1,number2` substituted for the variables are also typed as a cell array (inside curly braces `{ }`). The numbers can be scalars, vectors, or matrices. The first cell in the numbers cell array (`number1`) is substituted for the variable that is in the first cell of the variable cell array (`var1`), and so on.
- If all the numbers that are substituted for variables are scalars, the outcome will be one number or one expression (if some of the variables are still symbolic).
- If, for at least one variable, the substituted numbers are an array, the mathematical operations are executed element-by-element and the outcome is an array of numbers or expressions. It should be emphasized that the calculations are performed element-by-element even though the expression `S` is not typed in the element-by-element notation. This also means that all the arrays substituted for different variables must be of the same size.
- It is possible to substitute arrays (of the same size) for some of the variables and scalars for other variables. In this case, in order to carry out element-by-element operations, MATLAB expands the scalars (array of 1s times the scalar) to produce an array result.

The substitution of numerical values for two or more variables is demonstrated in the next examples.

```
>> syms a b c e x                        Define a, b, c, e, and x as symbolic variables.
>> S=a*x^e+b*x+c
                                         Create the symbolic expression
S =                                      ax^e + bx + c and assigned it to S.
a*x^e+b*x+c
>> subs(S,{a,b,c,e,x},{5,4,-20,2,3})     Substitute in S scalars for all
          Cell array.  Cell array.       the symbolic variables.

ans =
   37                                    The value of S is displayed.
>> T=subs(S,{a,b,c},{6,5,7})             Substitute in S scalars for the
                                         symbolic variables a, b, and c.
T =
5*x+ 6*x^e+7        The result is an expression with the variables x and e.
>> R=subs(S,{b,c,e},{[2 4 6],9,[1 3 5]}) Substitute in S a scalar for c,
                                         and vectors for b and e.
R =                                      The result is a vector of
[  2*x+a*x+9, a*x^3+4*x+9, a*x^5+6*x+9]  symbolic expressions.
>> W=subs(S,{a,b,c,e,x},{[4 2 0],[2 4 6],[2 2 2],[1 3 5],[3 2 1]})
                                Substitute in S vectors for all the variables.
W =
    20    26     8                  The result is a vector of numerical values.
```

A second method for substituting numerical values for symbolic variables in a symbolic expression is to first assign numerical values to the variables and then use the `subs` command. In this method, once the symbolic expression exists (at which point the variables in the expression are symbolic) the variables are assigned numerical values. Then the subs command is used in the form:

`R = subs(S)` ← The name of the symbolic expression.

Once the symbolic variables are redefined as numerical variables they can no longer be used as symbolic. The method is demonstrated in the following examples.

```
>> syms A c m x y                        Define A, c, m, x, and y as symbolic variables.
>> S=A*cos(m*x)+c*y
                                         Create the symbolic expression
S =                                      A cos(mx) + cy and assign it to S.
c*y+A*cos(m*x)
>> A=10; m=0.5; c=3;            Assign numerical values to variables A, m, and c.
>> subs(S)                      Use the subs command with the expression S.
ans =                                    The numerical values of variables
3*y + 10*cos(x/2)                        A, m, and c are substituted in S.
```

```
>> x=linspace(0,2*pi,4);
>> T = subs(S)
T =
[ 3*y+10, 3*y+5, 3*y-5, 3*y-10]
```

Assign numerical values (vector) to variable `x`.

Use the `subs` command with the expression `S`.

The numerical values of variables `A`, `m`, `c`, and `x` are substituted. The result is a vector of symbolic expressions.

11.9 *EXAMPLES OF MATLAB APPLICATIONS*

Sample Problem 11-2: Firing angle of a projectile

A projectile is fired at a speed of 210 m/s and an angle θ. The projectile's intended target is 2,600 m away and 350 m above the firing point.

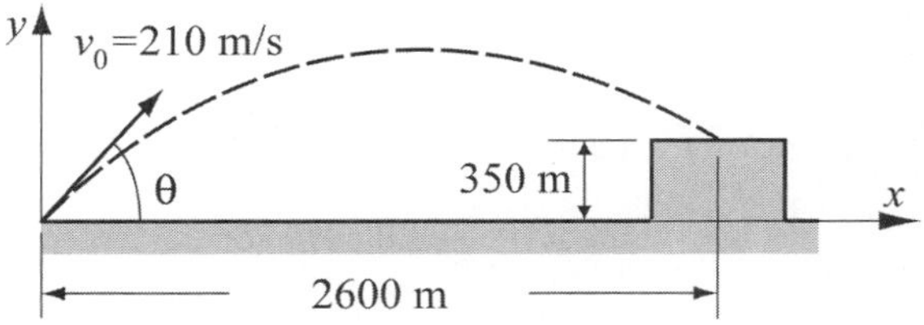

(*a*) Derive the equation that has to be solved in order to determine the angle θ such that the projectile will hit the target.

(*b*) Use MATLAB to solve the equation derived in part (*a*).

(*c*) For the angle determined in part (*b*), use the `ezplot` command to make a plot of the projectile's trajectory.

Solution

(*a*) The motion of the projectile can be analyzed by considering the horizontal and vertical components. The initial velocity v_0 can be resolved into horizontal and vertical components:

$$v_{0x} = v_0\cos(\theta) \quad \text{and} \quad v_{0y} = v_0\sin(\theta)$$

In the horizontal direction the velocity is constant, and the position of the projectile as a function of time is given by:

$$x = v_{0x}t$$

Substituting $x = 2600$ m for the horizontal distance that the projectile travels to reach the target and $210\cos(\theta)$ for v_{0x}, and solving for t gives:

$$t = \frac{2600}{210\cos(\theta)}$$

In the vertical direction the position of the projectile is given by:

$$y = v_{0y}t - \frac{1}{2}gt^2$$

Substituting $y = 350$ m for the vertical coordinate of the target, $210\sin(\theta)$ for v_{0x}, $g = 9.81$, and t gives:

$$350 = 210\sin(\theta)\frac{2600}{210\cos(\theta)} - \frac{1}{2}9.81\left(\frac{2600}{210\cos(\theta)}\right)^2$$

or:

$$350 = \frac{2600\sqrt{1-\cos^2(\theta)}}{\cos(\theta)} - \frac{1}{2}9.81\left(\frac{2600}{210\cos(\theta)}\right)^2$$

The solution of this equation gives the angle θ at which the projectile has to be fired.

(*b*) A solution of the equation derived in part (*a*) obtained by using the `solve` command (in the Command Window) is:

```
>> syms th
Angle = solve('2600*sqrt(1 - cos(th)^2)/cos(th) - 0.5*9.81*(2600/
(210*cos(th)))^2 = 350')
Angle =
    1.2453544972374161683138135806 56
   0.45925280703207121277786452037279
  -0.45925280703207121277786452037279
   -1.2453544972374161683138135806 56
```

MATLAB displays four solutions. The two positive ones are relevant to the problem.

```
>> Angle1 = Angle(1)*180/pi
```

Converting the solution in the first element of `Angle` from radians to degrees.

```
Angle1 =
224.16380950273491029648644451808/pi
```

MATLAB displays the answer as a symbolic object in terms of π.

```
>> Angle1=double(Angle1)
Angle1 =
   71.3536
```

Use the `double` command to obtain numerical values for `Angle1`.

```
>> Angle2=Angle(2)*180/pi
```

Converting the solution in the second element of `Angle` from radians to degrees.

```
Angle2 =
82.665505265772818300015613667102/pi
```

MATLAB displays the answer as a symbolic object in terms of π.

```
>> Angle2=double(Angle2)
Angle2 =
   26.3132
```

Use the `double` command to obtain numerical values for `Angle2`.

(*c*) The solution from part (*b*) shows that there are two possible angles and thus two trajectories. In order to make a plot of a trajectory, the x and y coordinates of the projectile are written in terms of t (parametric form):

$$x = v_0\cos(\theta)t \quad \text{and} \quad y = v_0\sin(\theta)t - \frac{1}{2}gt^2$$

The domain for t is $t = 0$ to $t = \dfrac{2600}{210\cos(\theta)}$.

These equations can be used in the `ezplot` command to make the plots shown in the following program written in a script file.

```
xmax=2600; v0=210; g=9.81;
theta1=1.24535; theta2=.45925;
t1=xmax/(v0*cos(theta1));
t2=xmax/(v0*cos(theta2));
syms t
X1=v0*cos(theta1)*t;
X2=v0*cos(theta2)*t;
Y1=v0*sin(theta1)*t-0.5*g*t^2;
Y2=v0*sin(theta2)*t-0.5*g*t^2;
ezplot(X1,Y1,[0,t1])
hold on
ezplot(X2,Y2,[0,t2])
hold off
```

Assign the two solutions from part (*b*) to `theta1` and `theta2`.

Plot one trajectory.

Plot a second trajectory.

When this program is executed, the following plot is generated in the Figure Window:

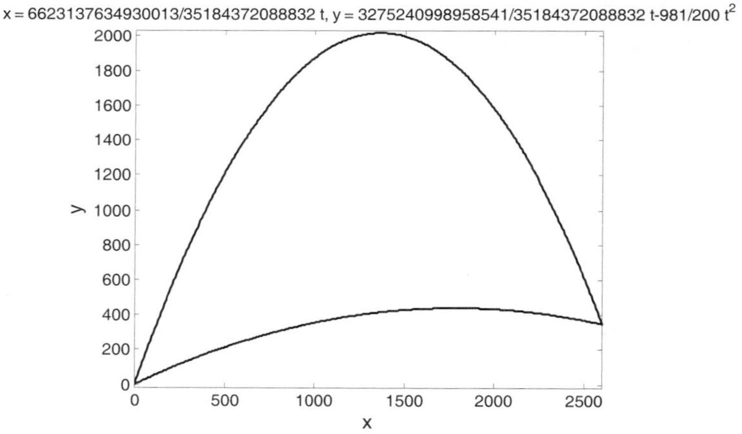

Sample Problem 11-3: Bending resistance of a beam

The bending resistance of a rectangular beam of width b and height h is proportional to the beam's moment of inertia I, defined by $I = \frac{1}{12}bh^3$. A rectangular beam is cut out of a cylindrical log of radius R. Determine b and h (as a function of R) such that the beam will have maximum I.

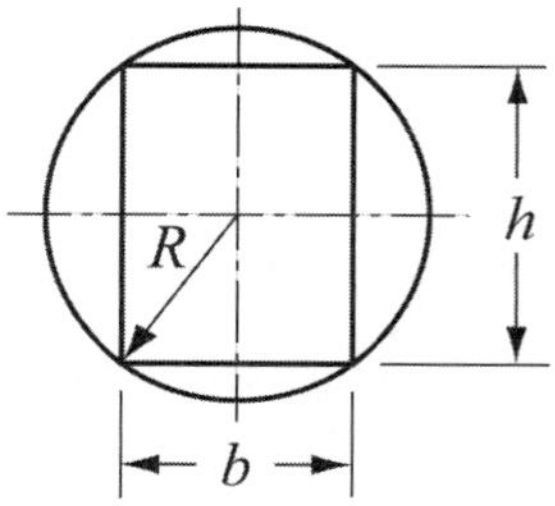

Solution

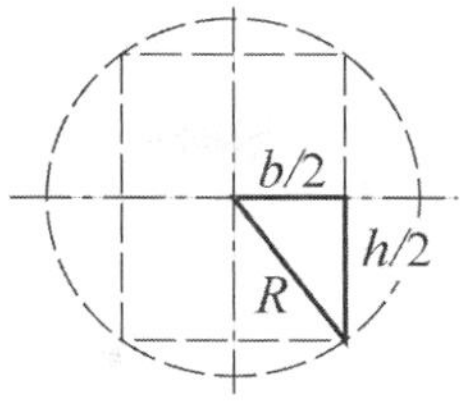

The problem is solved by following these steps:

1. Write an equation that relates *R*, *h*, and *b*.
2. Derive an expression for *I* in terms of *h*.
3. Take the derivative of *I* with respect to *h*.
4. Set the derivative equal to zero and solve for *h*.
5. Determine the corresponding *b*.

The first step is carried out by looking at the triangle in the figure. The relationship between *R*, *h*, and *b* is given by the Pythagorean theorem as $\left(\frac{b}{2}\right)^2 + \left(\frac{h}{2}\right)^2 = R^2$. Solving this equation for *b* gives $b = \sqrt{4R^2 - h^2}$.

The rest of the steps are done using MATLAB:

```
>> syms b h R
>> b=sqrt(4*R^2-h^2);
```
Create a symbolic expression for *b*.

```
>> I=b*h^3/12
```
Step 2: Create a symbolic expression for *I*.

```
I =
(h^3*(4*R^2-h^2)^(1/2))/12
```
MATLAB substitutes *b* in *I*.

```
>> ID=diff(I,h)
```
Step 3: Use the `diff(R)` command to differentiate `I` with respect to *h*.

```
ID =
(h^2*(4*R^2-h^2)^(1/2))/4-h^4/(12*(4*R^2-h^2)^(1/2))
```
The derivative of *I* is displayed.

```
>> hs=solve(ID,h)
```
Step 4: Use the `solve` command to solve the equation *ID* = 0 for *h*. Assign the answer to *hs*.

```
hs =
           0
  3^(1/2)*R
 -3^(1/2)*R
```
MATLAB displays three solutions. The positive non zero solution $\sqrt{3}R$ is relevant to the problem.

```
>> bs=subs(b,hs(2))
```
Step 5: Use the `subs` command to determine *b* by substituting the solution for *h* in the expression for *b*.

```
bs =
(R^2)^(1/2)
```
The answer for *b* is displayed. (The answer is *R*, but MATLAB displays $(R^2)^{1/2}$.)

Sample Problem 11-4: Fuel level in a tank

The horizontal cylindrical tank shown is used to store fuel. The tank has a diameter of 6 m and is 8 m long. The amount of fuel in the tank can be estimated by looking at the level of the fuel through a narrow vertical glass window at the front of the tank. A scale that is marked next to the window shows the levels of the fuel corresponding to 40, 60, 80, 120, and 160 thousand liters. Determine the vertical positions (measured from the ground) of the lines of the scale.

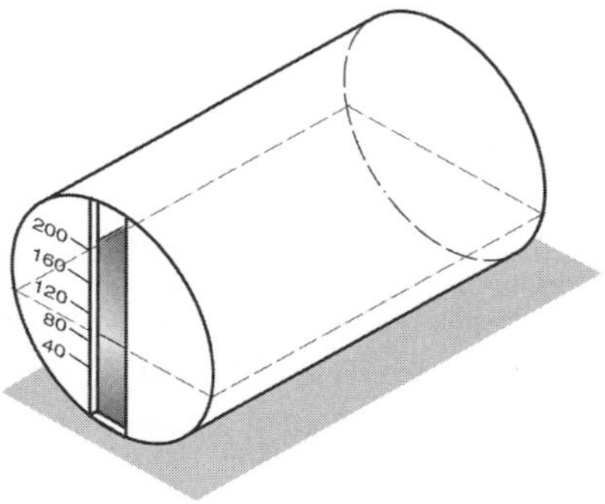

Solution

The relationship between the level of the fuel and its volume can be written in the form of a definite integral. Once the integration is carried out, an equation is obtained for the volume in terms of the fuel's height. The height corresponding to a specific volume can then be determined from solving the equation for the height.

The volume of the fuel V can be determined by multiplying the area of the cross section of the fuel A (the shaded area) by the length of the tank L. The cross-sectional area can be calculated by integration.

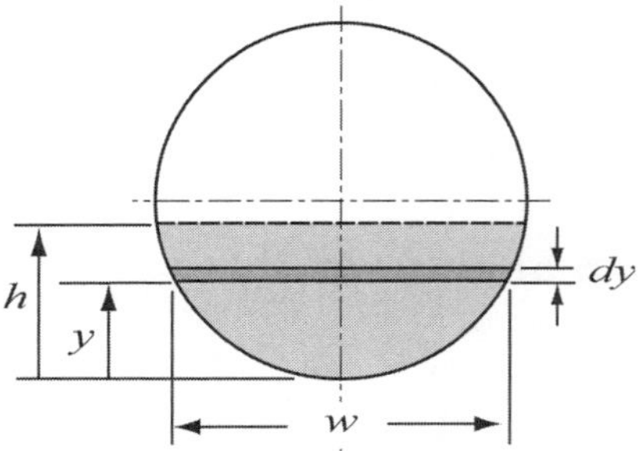

$$V = AL = L\int_0^h w\,dy$$

The width w of the top surface of the fuel can be written as a function of y. From the triangle in the figure on the right, the variables y, w, and R are related by:

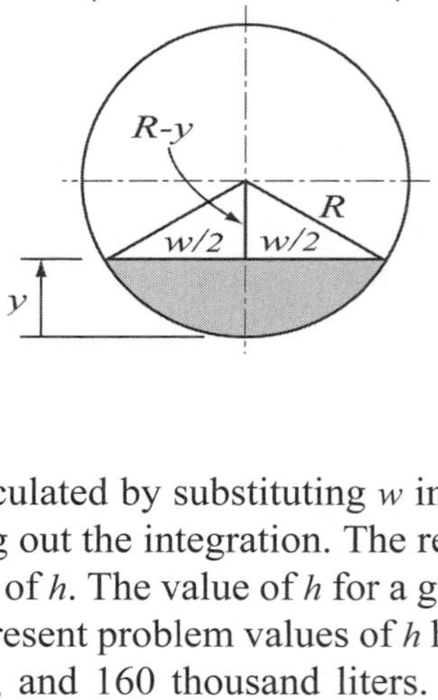

$$\left(\frac{w}{2}\right)^2 + (R-y)^2 = R^2$$

Solving this equation for w gives:

$$w = 2\sqrt{R^2 - (R-y)^2}$$

The volume of the fuel at height h can now be calculated by substituting w in the integral in the equation for the volume and carrying out the integration. The result is an equation that gives the volume V as a function of h. The value of h for a given V is obtained by solving the equation for h. In the present problem values of h have to be determined for volumes of 40, 60, 80, 120, and 160 thousand liters. The solution is given in the following MATLAB program (script file):

```
R=3; L=8;
syms w y h
w=2*sqrt(R^2-(R-y)^2)
S = L*w
V = int(S,y,0,h)
Vscale=[40:40:200]
for i=1:5
    Veq=V-Vscale(i);
    h_ans(i)=solve(Veq);
end
h_scale=double(h_ans)
```

- `w=2*sqrt(R^2-(R-y)^2)`: Create a symbolic expression for *w*.
- `S = L*w`: Create the expression that will be integrated.
- `V = int(S,y,0,h)`: Use the `int` command to integrate `S` from 0 to *h*. The result gives *V* as a function of *h*.
- `Vscale=[40:40:200]`: Create a vector with the values of *V* in the scale.
- `for i=1:5`: Each pass in the loop solves *h* for one value of *V*.
- `Veq=V-Vscale(i);`: Create the equation for *h* that has to be solved.
- `h_ans(i)=solve(Veq);`: Use the `solve` command to solve for *h*.
- `end`: `h_ans` is a vector (symbolic with numbers) with the values of *h* that correspond to the values of *V* in the vector `Vscale`.
- `h_scale=double(h_ans)`: Use the `double` command to obtain numerical values for the elements of vector `h_ans`.

When the script file is executed, the outcomes from commands that don't have a semicolon at the end are displayed. The display in the Command Window is:

```
>> w =
2*(9-(y-3)^2)^(1/2)
S =
16*(9-(y-3)^2)^(1/2)
V =
36*pi+72*asin(h/3-1)+8*(9-(h-3)^2)^(1/2)*(h-3)
Vscale =
  40   80   120   160   200
h_scale =
  1.3972   2.3042   3.1439   3.9957   4.9608
```

- `w =`: The symbolic expression for *w* is displayed.
- `S =`: `S` is the expression that will be integrated.
- `V =`: The result from the integration; *V* as a function of *h*.
- `Vscale =`: The values of *V* in the scale are displayed.
- `h_scale =`: The positions of the lines in the scale are displayed.

Units: The unit for length in the solution is meters, which correspond to m^3 for the volume (1 m^3 = 1,000 L).

Sample Problem 11-5: Amount of medication in the body

The amount M of medication present in the body depends on the rate at which the medication is consumed by the body and on the rate at which the medication enters the body, where the rate at which the medication is consumed is proportional to the amount present in the body. A differential equation for M is

$$\frac{dM}{dt} = -kM + p$$

where k is the proportionality constant and p is the rate at which the medication is injected into the body.

(*a*) Determine k if the half-life of the medication is 3 hours.

(*b*) A patient is admitted to a hospital and the medication is given at a rate of 50 mg per hour. (Initially there is no medication in the patient's body.) Derive an expression for M as a function of time.

(*c*) Plot M as a function of time for the first 24 hours.

Solution

(*a*) The proportionality constant can be determined from considering the case in which the medication is consumed by the body and no new medication is given. In this case the differential equation is:

$$\frac{dM}{dt} = -kM$$

The equation can be solved with the initial condition $M = M_0$ at $t = 0$:

```
>> syms M M0 k t
>> Mt=dsolve('DM=-k*M','M(0)=M0')
Mt =
M0/exp(k*t)
```

Use the `dsolve` command to solve $\frac{dM}{dt} = -kM$.

The solution gives M as a function of time:

$$M(t) = \frac{M_0}{e^{kt}}$$

A half-life of 3 hours means that at $t = 3$ hours $M(t) = \frac{1}{2}M_0$. Substituting this information in the solution gives $0.5 = \frac{1}{e^{3k}}$, and the constant k is determined from solving this equation:

```
ks=solve('0.5=1/exp(k*3)')
ks =
.231049060186648436472410707715273
```

Use the `solve` command to solve $0.5 = e^{-3k}$.

(*b*) For this part the differential equation for M is:

$$\frac{dM}{dt} = -kM + p$$

The constant k is known from part (*a*), and $p = 50$ mg/h is given. The initial condition is that in the beginning there is no medication in the patient's body, or $M = 0$ at $t = 0$. The solution of this equation with MATLAB is:

```
>> syms p
>> Mtb=dsolve('DM=-k*M+p','M(0)=0')
Mtb =
(p-p/exp(k*t))/k)
```

Use the `dsolve` command to solve $\frac{dM}{dt} = -kM + p$.

(*c*) A plot of `Mtb` as a function of time for $0 \le t \le 24$ can be done by using the `ezplot` command:

```
>> pgiven=50;
>> Mtt=subs(Mtb,{p,k},{pgiven,ks})
Mtt =
216.404-216.404/exp(0.231049*t)
>> ezplot(Mtt,[0,24])
```

Substitute numerical values for p and k.

In the actual display of the last expression that was generated by MATLAB (`Mtt` = . . .) the numbers have many more decimal digits than shown above. The numbers were shortened so that they will fit on the page.
The plot that is generated is:

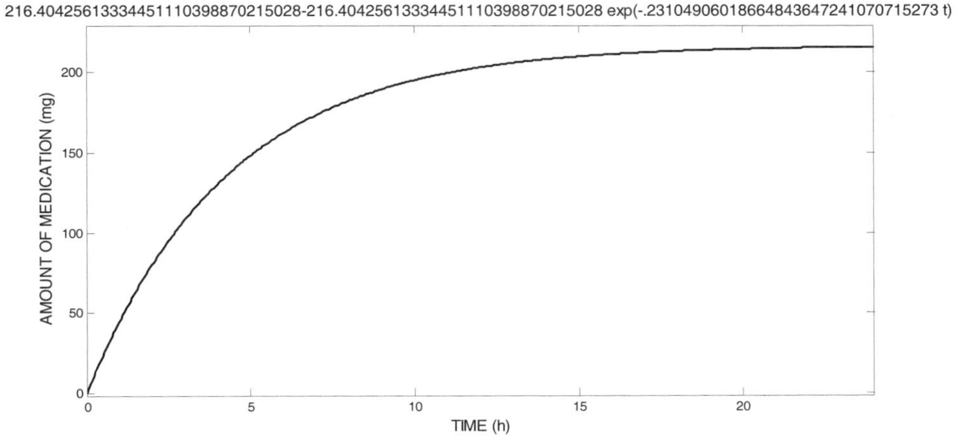

11.10 PROBLEMS

1. Define x as a symbolic variable and create the two symbolic expressions

$$S_1 = x^2(x-6)+4(3x-2) \text{ and } S_2 = (x+2)^2 - 8x$$

Use symbolic operations to determine the simplest form of each of following expressions:

(*a*) $S_1 \cdot S_2$ (*b*) $\frac{S_1}{S_2}$ (*c*) $S_1 + S_2$

(*d*) Use the `subs` command to evaluate the numerical value of the result from part (*c*) for $x = 5$.

2. Define y as a symbolic variable and create the two symbolic expressions

$$S_1 = x(x^2+6x+12)+8 \text{ and } S_2 = (x-3)^2 + 10x - 5$$

Use symbolic operations to determine the simplest form of each of following expressions:

(*a*) $S_1 \cdot S_2$ (*b*) $\frac{S_1}{S_2}$ (*c*) $S_1 + S_2$

(*d*) Use the `subs` command to evaluate the numerical value of the result from part (*c*) for $x = 3$.

3. Define x and y as symbolic variables and create the two symbolic expressions

$$S = x + \sqrt{x}y^2 + y^4 \text{ and } T = \sqrt{x} - y^2$$

Use symbolic operations to determine the simplest form of $S \cdot T$. Use the `subs` command to evaluate the numerical value of the result for $x = 9$ and $y = 2$.

4. Define x as a symbolic variable.

(*a*) Derive the equation of the polynomial that has the roots $x = -2$, $x = -0.5$, $x = 2$, and $x = 4.5$.

(*b*) Determine the roots of the polynomial

$$f(x) = x^6 - 6.5x^5 - 58x^4 + 167.5x^3 + 728x^2 - 890x - 1400$$

by using the `factor` command.

5. Use the commands from Section 11.2 to show that:

(*a*) $\sin(4x) = 4\sin x\cos x - 8\sin^3 x\cos x$

(*b*) $\cos x\cos y = \frac{1}{2}[\cos(x-y) + \cos(x+y)]$

6. Use the commands from Section 11.2 to show that:

(*a*) $\tan(3x) = \dfrac{3\tan x - \tan^3 x}{1 - 3\tan^2 x}$

(*b*) $\sin(x+y+z) = \sin x\cos y\cos z + \cos x\sin y\cos z + \cos x\cos y\sin z - \sin x\sin y\sin z$

7. The folium of Descartes is the graph shown in the figure. In parametric form its equation is given by:

$$x = \frac{3t}{1+t^3} \quad \text{and} \quad y = \frac{3t^2}{1+t^3} \quad \text{for } t \neq -1$$

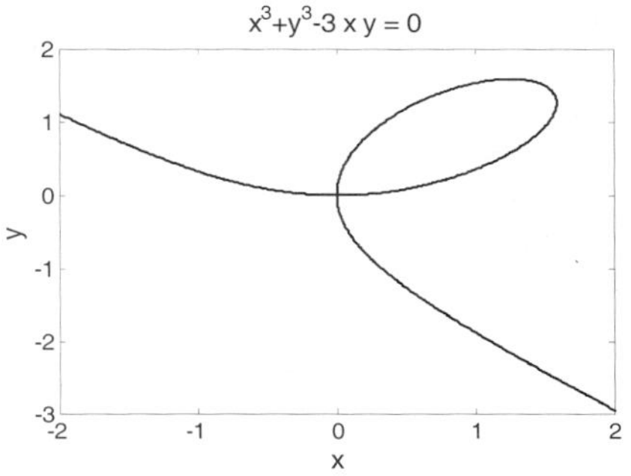

(*a*) Use MATLAB to show that the equation of the folium of Descartes can also be written as:

$$x^3 + y^3 = 3xy$$

(*b*) Make a plot of the folium for the domain shown in the figure by using the `ezplot` command.

8. A water tower has the geometry shown in the figure (the lower part is a cylinder with radius *R* and height *h*, and the upper part is a half sphere with radius *R*). Determine the radius *R* if $h = 10$ m and the volume is 1,050 m³. (Write an equation for the volume in terms of the radius and the height. Solve the equation for the radius, and use the `double` command to obtain a numerical value.

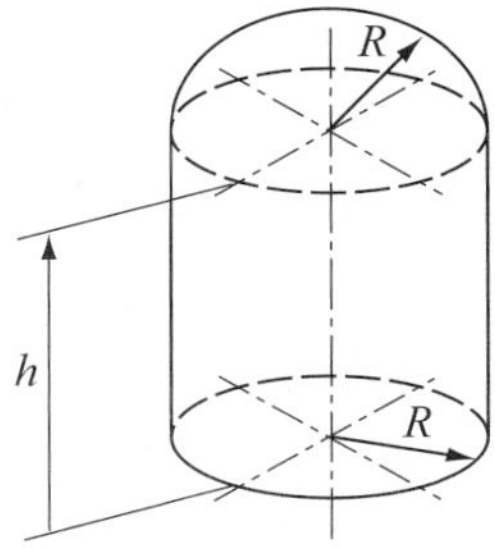

9. The relation between the tension *T* and the steady shortening velocity *v* in a muscle is given by the Hill equation:

$$(T+a)(v+b) = (T_0+a)b$$

where a and b are positive constants and T_0 is the isometric tension, i.e., the tension in the muscle when $v = 0$. The maximum shortening velocity occurs when $T = 0$.

(*a*) Using symbolic operations, create the Hill equation as a symbolic expression. Then use `subs` to substitute $T = 0$, and finally solve for v to show that $v_{max} = (bT_0)/a$.

(*b*) Use v_{max} from part (*a*) to eliminate the constant b from the Hill equation, and show that $v = \dfrac{a(T_0 - T)}{T_0(T+a)}v_{max}$.

10. Consider the two ellipses in the $x\,y$ plane given by the equations

$$\frac{(x-1)^2}{6^2}+\frac{y^2}{3^2} = 1 \quad \text{and} \quad \frac{(x+2)^2}{2^2}+\frac{(y-5)^2}{4^2} = 1$$

(*a*) Use the `ezplot` command to plot the two ellipses in the same figure.
(*b*) Determine the coordinates of the points where the ellipses intersect.

11. A 120 in.–long beam AB is attached to the wall with a pin at point A and to a 66 in.–long cable CD. A load $W = 200$ lb is attached to the beam at point B. The tension in the cable T and the x and y components of the force at A (F_{Ax} and F_{Ay}) can be calculated from the equations:

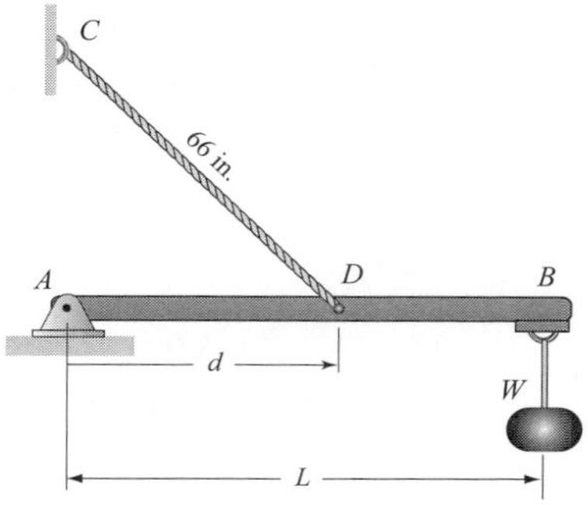

$$F_{Ax} - T\frac{d}{L_c} = 0$$

$$F_{Ay} + T\frac{\sqrt{L_c^2-d^2}}{L_c} - W = 0$$

$$T\frac{\sqrt{L_c^2-d^2}}{L_c}d - WL = 0$$

where L and L_c are the lengths of the beam and the cable, respectively, and d is the distance from point A to point D where the cable is attached.

(*a*) Use MATLAB to solve the equations for the forces T, F_{Ax}, and F_{Ay} in terms of d, L, L_c, and W. Determine F_A given by $F_A = \sqrt{F_{Ax}^2+F_{Ay}^2}$.
(*b*) Use the `subs` command to substitute $W = 200$ lb, $L = 120$ in., and $L_c = 66$ in. into the expressions derived in part (*a*). This will give the forces as a function of the distance d.
(*c*) Use the `ezplot` command to plot the forces T and F_A (both in the same figure as functions of d, for d starting at 20 and ending at 70 in.
(*d*) Determine the distance d where the tension in the cable is the smallest. Determine the value of this force.

12. A box of mass m is being pulled by a rope as shown. The force F in the rope as a function of x can be calculated from the equations:

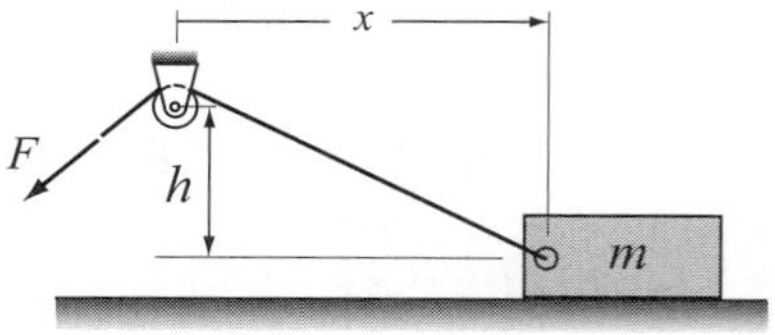

$$-F\frac{x}{\sqrt{x^2+h^2}} + \mu N = 0$$

$$-mg + N + F\frac{h}{\sqrt{x^2+h^2}} = 0$$

where N and μ are the normal force and friction coefficient between the box and surface, respectively. Consider the case where $m = 18$ kg, $h = 10$ m, $\mu = 0.55$, and $g = 9.81$ m/s².

(*a*) Use MATLAB to derive an expression for F, in terms of x, h, m, g, and μ.

(*b*) Use the `subs` command to substitute $m = 18$ kg, $h = 10$ m, $\mu = 0.55$, and $g = 9.81$ m/s² into the expressions that were derived in part (*a*). This will give the force as a function of the distance x.

(*c*) Use the `ezplot` command to plot the force F as a function of x, for x starting at 5 and ending at 30 m.

(*d*) Determine the distance x where the force that is required to pull the box is the smallest, and determine the magnitude of that force.

13. The mechanical power output P in a contracting muscle is given by:

$$P = Tv = \frac{kvT_0\left(1 - \frac{v}{v_{max}}\right)}{k + \frac{v}{v_{max}}}$$

where T is the muscle tension, v is the shortening velocity (max of v_{max}), T_0 is the isometric tension (i.e., tension at zero velocity), and k is a non-dimensional constant that ranges between 0.15 and 0.25 for most muscles. The equation can be written in non-dimensional form:

$$p = \frac{ku(1-u)}{k+u}$$

where $p = (Tv)/(T_0 v_{max})$, and $u = v/v_{max}$. Consider the case $k = 0.25$.

(*a*) Plot p versus u for $0 \le u \le 1$.

(*b*) Use differentiation to find the value of u where p is maximum.

(*c*) Find the maximum value of p.

14. The equation of a circle is $x^2 + y^2 = R^2$, where R is the radius of the circle. Write a program in a script file that first derives the equation (symbolically) of the tangent line to the circle at the point (x_0, y_0) on the upper part of the circle (i.e., for $-R < x_0 < R$ and $0 < y_0$). Then for specific values of R, x_0, and y_0 the program makes a plot, like the one shown on the right, of the circle and the tangent line. Execute the program with $R = 10$ and $x_0 = 7$.

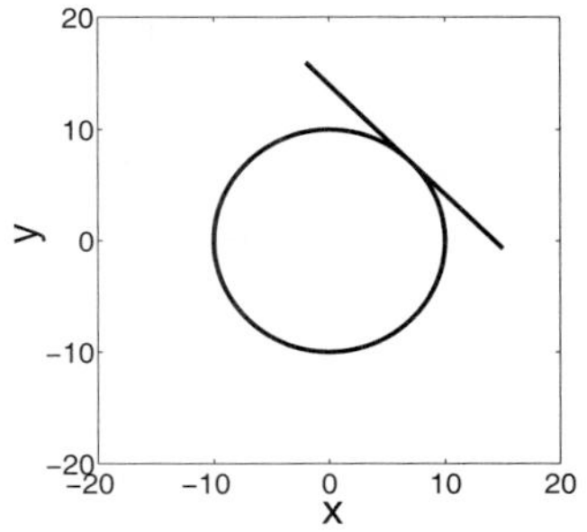

15. A tracking radar antenna is locked on an airplane flying at a constant altitude of 5 km, and a constant speed of 540 km/h. The airplane travels along a path that passes exactly above the radar station. The radar starts the tracking when the airplane is 100 km away.

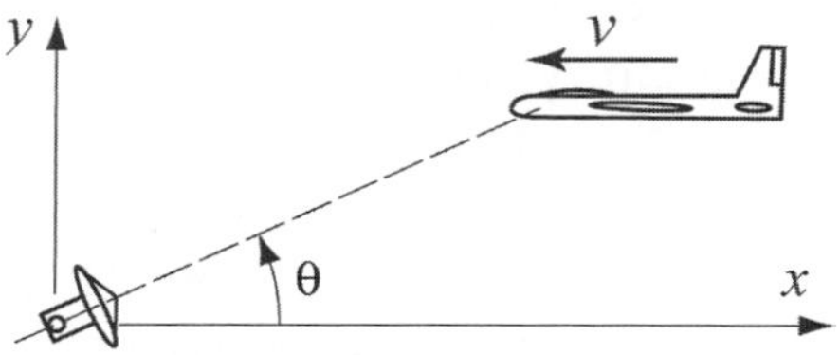

(*a*) Derive an expression for the angle θ of the radar antenna as a function of time.

(*b*) Derive an expression for the angular velocity of the antenna, $\frac{d\theta}{dt}$, as a function of time.

(c) Make two plots on the same page, one of θ versus time and the other of $\frac{d\theta}{dt}$ versus time, where the angle is in degrees and the time is in minutes for $0 \le t \le 20$ min.

16. Evaluate the following indefinite integrals:

(*a*) $I = \int \frac{x^3}{\sqrt{1-x^2}}dx$ (*b*) $I = \int x^2 \cos x \, dx$

17. Define *x* as a symbolic variable and create the symbolic expression

$$S = \frac{\cos^2 x}{1 + \sin^2 x}$$

Plot *S* in the domain $0 \le x \le \pi$ and calculate the integral $I = \int_0^{\pi} \frac{\cos^2 x}{1 + \sin^2 x} dx$.

18. The parametric equations of an ellipsoid are:

$x = a\cos u \sin v$, $y = b\cos u \sin v$, $z = c\cos v$

where $0 \le u \le 2\pi$ and $-\pi \le v \le 0$

Show that the differential volume element of the ellipsoid shown is given by:

$$dV = -\pi abc \sin^3 v dv$$

Use MATLAB to evaluate the integral of *dV* from $-\pi$ to 0 symbolically and show that the volume of the ellipsoid is $V = \frac{4}{3}\pi abc$.

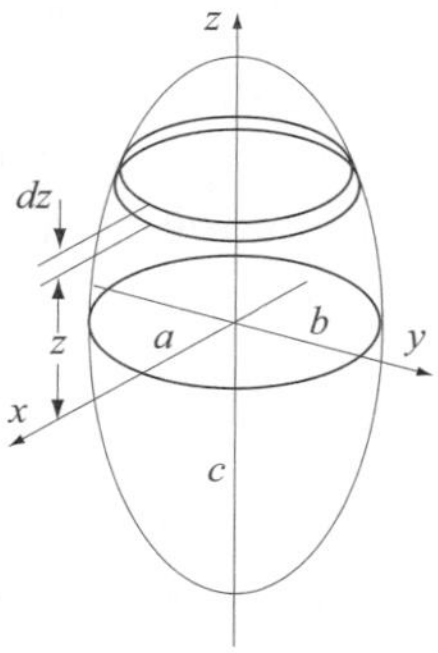

19. The one-dimensional diffusion equation is given by:

$$\frac{\partial u}{\partial t} = m\frac{\partial^2 u}{\partial x^2}$$

Show that the following are solutions to the diffusion equation.

(*a*) $u = A\frac{1}{\sqrt{t}}\exp\left(\frac{-x^2}{4mt}\right) + B$, where A and B are constants.

(*b*) $u = A\exp(-\alpha x)\cos(\alpha x - 2m\alpha^2 t + B) + C$, where A, B, C, and α are constants.

20. A ceramic tile has the design shown in the figure. The shaded area is painted red and the rest of the tile is white. The border line between the red and the white areas follows the equation

$$y = -kx^2 + 12kx$$

Determine k such that the areas of the white and the red colors will be the same.

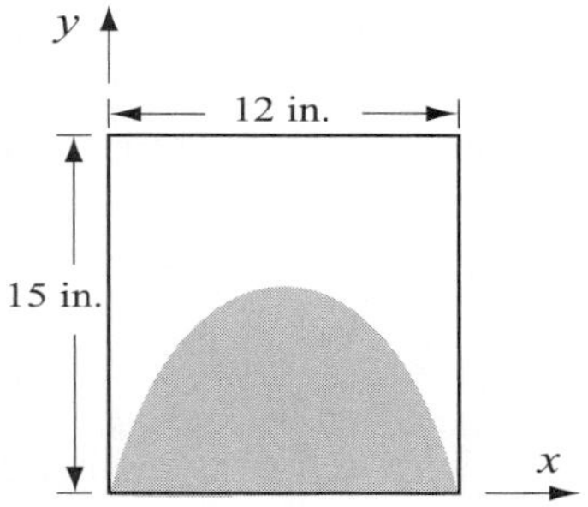

21. Show that the location of the centroid y_c of the half-circle area shown is given by $y_c = \frac{4R}{3\pi}$. The coordinate y_c can be calculated by:

$$y_c = \frac{\int_A \bar{y}\,dA}{\int_A dA}$$

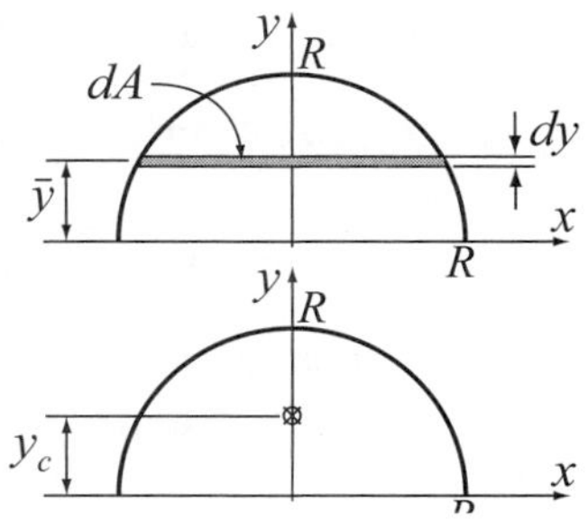

22. For the half-circle area shown in the previous problem, show that the moment of inertia about the x axis, I_x, is given by $I_x = \frac{1}{8}\pi R^4$. The moment of inertia I_x can be calculated by:

$$I_x = \int_A y^2\,dA$$

23. The *rms* value of an AC voltage is defined by

$$v_{rms} = \sqrt{\frac{1}{T}\int_0^T v^2(t')dt'}$$

where T is the period of the waveform.

(*a*) A voltage is given by $v(t) = V\cos(\omega t)$. Show that $v_{rms} = \frac{V}{\sqrt{2}}$ and is independent of ω. (The relationship between the period T and the radian frequency ω is $T = \frac{2\pi}{\omega}$.)

(*b*) A voltage is given by $v(t) = 2.5\cos(350t) + 3$ V. Determine v_{rms}.

24. The spread of an infection from a single individual to a population of N uninfected persons can be described by the equation

$$\frac{dx}{dt} = -Rx(N + 1 - x) \text{ with initial condition } x(0) = N$$

where x is the number of uninfected individuals and R is a positive rate constant. Solve this differential equation symbolically for $x(t)$. Also, determine symbolically the time t at which the infection rate dx/dt is maximum.

25. The Maxwell-Boltzmann probability density function $f(v)$ is given by

$$f(v) = \sqrt{\frac{2}{\pi}\left(\frac{m}{kT}\right)^3}\, v^2 \exp\left(\frac{-mv^2}{2kT}\right)$$

where m (kg) is the mass of each molecule, v (m/s) is the speed, T (K) is the temperature, and $k = 1.38 \times 10^{-23}$ J/K is Boltzmann's constant. The most probable speed v_p corresponds to the maximum value of $f(v)$ and can be determined from $\frac{df(v)}{dv} = 0$. Create a symbolic expression for $f(v)$, differentiate it with respect to v and show that $v_p = \sqrt{\frac{2kT}{m}}$. Calculate v_p for oxygen molecules ($m = 5.3 \times 10^{-26}$ kg) at $T = 300$ K ($k = 1.38 \times 10^{-23}$ J/K). Make a plot of $f(v)$ versus v for $0 \le v \le 2500$ m/s for oxygen molecules

26. The velocity of a skydiver whose parachute is still closed can be modeled by assuming that the air resistance is proportional to the velocity. From Newton's second law of motion the relationship between the mass m of the skydiver and his velocity v is given by (down is positive)

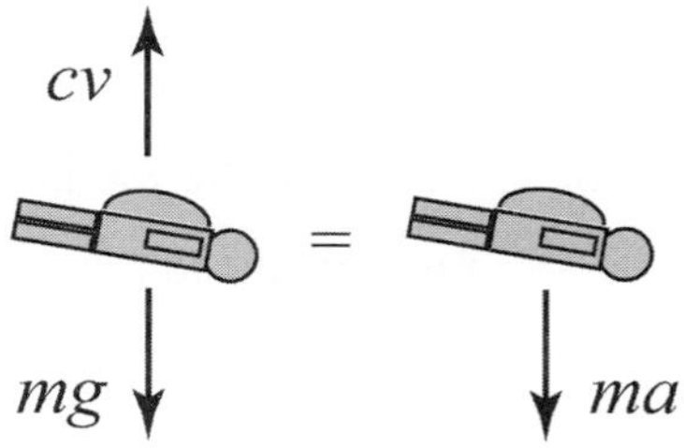

$$mg - cv = m\frac{dv}{dt}$$

where c is a drag constant and g is the gravitational constant ($g = 9.81$ m/s²).

(*a*) Solve the equation for v in terms of m, g, c, and t, assuming that the initial velocity of the skydiver is zero.

(*b*) It is observed that 4 s after a 90 kg skydiver jumps out of an airplane, his velocity is 28 m/s. Determine the constant c.

(*c*) Make a plot of the skydiver velocity as a function of time for $0 \le t \le 30$ s.

27. A resistor R ($R = 0.4\Omega$) and an inductor L ($L = 0.08\,\text{H}$) are connected as shown. Initially, the switch is connected to point A and there is no current in the circuit. At $t = 0$ the switch is moved from A to B, so that the resistor and the inductor are connected to v_S ($v_S = 6\,\text{V}$), and current starts flowing in the circuit. The switch remains connected to B until the voltage on the resistor reaches 5 V. At that time (t_{BA}) the switch is moved back to A.

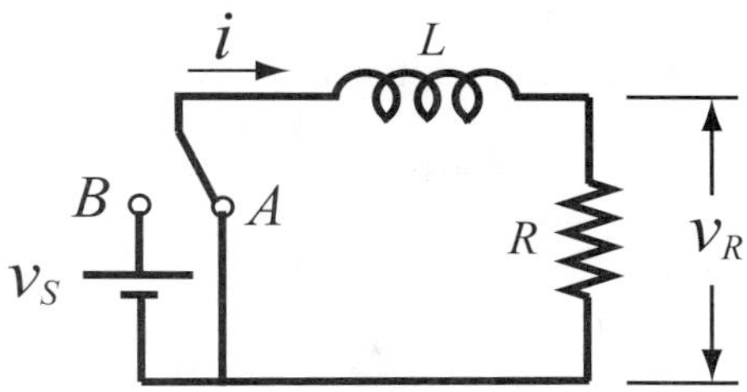

The current i in the circuit can be calculated from solving the differential equations:

$iR + L\frac{di}{dt} = v_S$ during the time from $t = 0$ and until the time when the switch is moved back to A.

$iR + L\frac{di}{dt} = 0$ from the time when the switch is moved back to A and on.

The voltage across the resistor, v_R, at any time is given by $v_R = iR$.

(*a*) Derive an expression for the current i in terms of R, L, v_S, and t for $0 \le t \le t_{BA}$ by solving the first differential equation.

(*b*) Substitute the values of R, L, and v_S in the solution for i, and determine the time t_{BA} when the voltage across the resistor reaches 5 V.

(*c*) Derive an expression for the current i in terms of R, L, and t, for $t_{BA} \le t$ by solving the second differential equation.

(*d*) Make two plots (on the same page), one for v_R versus t for $0 \le t \le t_{BA}$ and the other for v_R versus t for $t_{BA} \le t \le 2t_{BA}$.

28. Determine the general solution of the differential equation

$$\frac{dy}{dx} = \frac{x^4 - 2y}{2x}$$

Show that the solution is correct. (Derive the first derivative of the solution, and then substitute back into the equation.)

29. Determine the solution of the following differential equation that satisfies the given initial conditions. Plot the solution for $0 \le t \le 7$.

$$\frac{d^2y}{dt^2} - 0.08\frac{dy}{dy} + 0.6t = 0\,,\quad y(0) = 2\,,\quad \left.\frac{dy}{dx}\right|_{x=0} = 3$$

30. The current, i, in a series RLC circuit when the switch is closed at $t = 0$ can be determined from the solution of the 2nd-order ODE

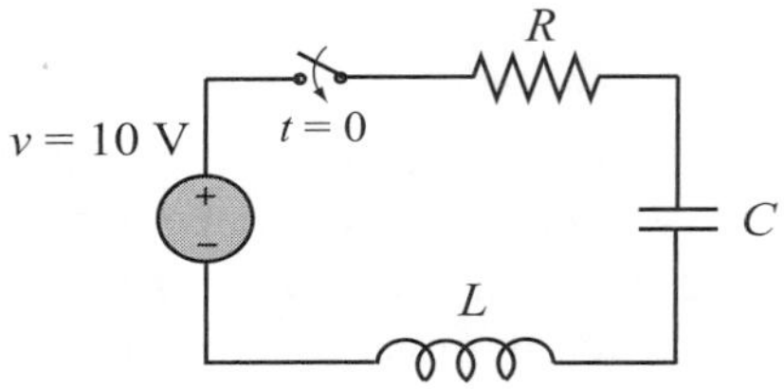

$$L\frac{d^2i}{dt^2} + R\frac{di}{dt} + \frac{1}{C}i = 0$$

where R, L, and C are the resistance of the resistor, the inductance of the inductor, and the capacitance of the capacitor, respectively.

(a) Solve the equation for i in terms of L, R, C, and t, assuming that at $t = 0$ $i = 0$ and $di/dt = 8$.

(b) Use the `subs` command to substitute $L = 3$ H, $R = 10\ \Omega$, and $C = 80\ \mu\text{F}$ into the expression that were derived in part (a). Make a plot of i versus t for $0 \le t \le 1$ s. (Underdamped response.)

(c) Use the `subs` command to substitute $L = 3$ H, $R = 200\ \Omega$, and $C = 1200\ \mu\text{F}$ into the expression that were derived in part (a). Make a plot of i versus t for $0 \le t \le 2$ s. (Overdamped response.)

(d) Use the `subs` command to substitute $L = 3$ H, $R = 201\ \Omega$, and $C = 300\ \mu\text{F}$ into the expression that were derived in part (a). Make a plot of i versus t for $0 \le t \le 2$ s. (Critically damped response.)

31. Damped free vibrations can be modeled by a block of mass m that is attached to a spring and a dashpot as shown. From Newton's second law of motion, the displacement x of the mass as a function of time can be determined by solving the differential equation

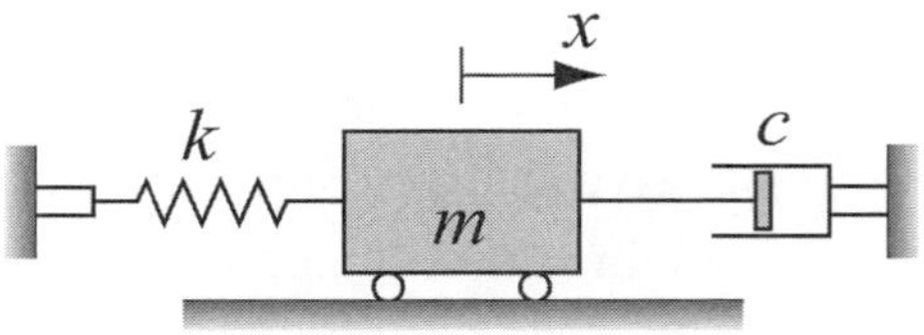

$$m\frac{d^2x}{dt^2} + c\frac{dx}{dt} + kx = 0$$

where k is the spring constant and c is the damping coefficient of the dashpot. If the mass is displaced from its equilibrium position and then released, it will start oscillating back and forth. The nature of the oscillations depends on the size of the mass and the values of k and c.

For the system shown in the figure, $m = 10$ kg and $k = 28$ N/m. At time $t = 0$ the mass is displaced to $x = 0.18$ m and then released from rest. Derive expressions for the displacement x and the velocity v of the mass, as a function of time. Consider the following two cases:

(a) $c = 3$ (N s)/m.

(b) $c = 50$ (N s)/m.

For each case, plot the position x and the velocity v versus time (two plots on one page). For case (a) take $0 \le t \le 20$ s, and for case (b) take $0 \le t \le 10$ s.

Appendix: Summary of Characters, Commands, and Functions

The following tables list MATLAB's characters, commands, and functions that are covered in the book. The items are grouped by subjects.

Characters and arithmetic operators

Character	Description	Page
+	Addition.	11, 64
–	Subtraction.	11, 64
*	Scalar and array multiplication.	11, 65
.*	Element-by-element multiplication of arrays.	72
/	Right division.	11, 71
\	Left division.	11, 70
./	Element-by-element right division.	72
.\	Element-by-element left division.	72
^	Exponentiation.	11
.^	Element-by-element exponentiation.	72
:	Colon; creates vectors with equally spaced elements, represents range of elements in arrays.	37, 44
=	Assignment operator.	16
( )	Parentheses; sets precedence, encloses input arguments in functions and subscripts of arrays.	11, 42, 44, 222
[ ]	Brackets; forms arrays. encloses output arguments in functions.	37, 38, 39, 222
,	Comma; separates array subscripts and function arguments, separates commands in the same line.	9, 17, 42-45, 222
;	Semicolon; suppresses display, ends row in array.	10, 39
'	Single quote; matrix transpose, creates string.	41, 53-55
...	Ellipsis; continuation of line.	9
%	Percent; denotes a comment, specifies output format.	10

Relational and logical operators

Character	Description	Page
<	Less than.	174
>	Greater than.	174
<=	Less than or equal.	174

Relational and logical operators (Continued)

Character	Description	Page
>=	Greater than or equal.	174
==	Equal.	174
~=	Not equal.	174
&	Logical AND.	177
|	Logical OR.	177
~	Logical NOT.	177

Managing commands

Command	Description	Page
`cd`	Changes current directory.	23
`clc`	Clears the Command Window.	10
`clear`	Removes all variables from the memory.	19
`clear x y z`	Removes variables `x`, `y`, and `z` from the memory.	19
`close`	Closes the active Figure Window.	158
`fclose`	Closes a file.	109
`figure`	Opens a Figure Window.	158
`fopen`	Opens a file.	108
`global`	Declares global variables.	225
`help`	Displays help for MATLAB functions.	224
`iskeyword`	Displays keywords.	19
`lookfor`	Search for specified word in all help entries.	224
`who`	Displays variables currently in the memory.	19, 96
`whos`	Displays information on variables in the memory.	19, 96

Predefined variables

Variable	Description	Page
`ans`	Value of last expression.	19
`eps`	The smallest difference between two numbers.	19
`i`	$\sqrt{-1}$	19
`inf`	Infinity.	19
`j`	Same as i.	19
`NaN`	Not a number.	19
`pi`	The number π.	19

Display formats in the Command Window

Command	Description	Page
`format bank`	Two decimal digits.	13
`format compact`	Eliminates empty lines.	13
`format long`	Fixed-point format with 14 decimal digits.	13
`format long e`	Scientific notation with 15 decimal digits.	13

Display formats in the Command Window (Continued)

Command	Description	Page
`format long g`	Best of 15-digit fixed or floating point.	13
`format loose`	Adds empty lines.	13
`format short`	Fixed-point format with 4 decimal digits.	13
`format short e`	Scientific notation with 4 decimal digits.	13
`format short g`	Best of 5-digit fixed or floating point.	13

Elementary math functions

Function	Description	Page
`abs`	Absolute value.	14
`exp`	Exponential.	14
`factorial`	The factorial function.	15
`log`	Natural logarithm.	14
`log10`	Base 10 logarithm.	14
`nthroot`	Real nth root or a real number.	14
`sqrt`	Square root.	14

Trigonometric math functions

Function	Description	Page	Function	Description	Page
`acos`	Inverse cosine.	15	`cos`	Cosine.	15
`acot`	Inverse cotangent.	15	`cot`	Cotangent.	15
`asin`	Inverse sine.	15	`sin`	Sine.	15
`atan`	Inverse tangent.	15	`tan`	Tangent.	15

Hyperbolic math functions

Function	Description	Page	Function	Description	Page
`cosh`	Hyperbolic cosine.	15	`sinh`	Hyperbolic sine.	15
`coth`	Hyperbolic cotangent.	15	`tanh`	Hyperbolic tangent.	15

Rounding

Function	Description	Page
`ceil`	Round towards infinity.	15
`fix`	Round towards zero.	15
`floor`	Round towards minus infinity.	15
`rem`	Returns the remainder after `x` is divided by `y`.	15
`round`	Round to the nearest integer.	15
`sign`	Signum function.	16

Creating arrays

Function	Description	Page
`diag`	Creates a diagonal matrix from a vector. Creates a vector from the diagonal of a matrix.	50
`eye`	Creates a unit matrix.	40, 68
`linspace`	Creates equally spaced vector.	38
`ones`	Creates an array with ones.	40
`rand`	Creates an array with random numbers.	77, 78
`randi`	Creates an array with random integers.	78
`randn`	Creates an array with normally distributed numbers.	79
`randperm`	Creates vector with permutation of integers.	78
`zeros`	Creates an array with zeros.	40

Handling arrays

Function	Description	Page
`length`	Number of elements in the vector.	49
`reshape`	Rearrange a matrix.	49
`size`	Size of an array.	49

Array functions

Function	Description	Page
`cross`	Calculates cross product of two vectors.	77
`det`	Calculates determinant.	70, 77
`dot`	Calculates scalar product of two vectors.	66, 77
`inv`	Calculates the inverse of a square matrix.	69, 77
`max`	Returns maximum value.	76
`mean`	Calculates mean value.	76
`median`	Calculates median value.	76
`min`	Returns minimum value.	76
`sort`	Arranges elements in ascending order.	76
`std`	Calculates standard deviation.	77
`sum`	Calculates sum of elements.	76

Input and output

Command	Description	Page
`disp`	Displays output.	101
`fprintf`	Displays or saves output.	103-110
`input`	Prompts for user input.	99
`load`	Retrieves variables to the workspace.	112
`save`	Saves the variables in the workspace.	111
`uiimport`	Starts the Import Wizard	116
`xlsread`	Imports data from Excel	114

Input and output

Command	Description	Page
`xlswrite`	Exports data to Excel	115

Two-dimensional plotting

Command	Description	Page
`bar`	Creates a vertical bar plot.	152
`barh`	Creates a horizontal bar plot.	152
`errorbar`	Creates a plot with error bars.	151
`fplot`	Plots a function.	140
`hist`	Creates a histogram.	154-156
`hold off`	Ends `hold on`.	142
`hold on`	Keeps current graph open.	142
`line`	Adds curves to existing plot.	143
`loglog`	Creates a plot with log scale on both axes.	149
`pie`	Creates a pie plot.	153
`plot`	Creates a plot.	134
`polar`	Creates a polar plot.	156
`semilogx`	Creates a plot with log scale on the *x* axis.	149
`semilogy`	Creates a plot with log scale on the *y* axis.	149
`stairs`	Creates a stairs plot.	153
`stem`	Creates a stem plot.	153

Three-dimensional plotting

Command	Description	Page
`bar3`	Creates a vertical 3-D bar plot.	331
`contour`	Creates a 2-D contour plot.	330
`contour3`	Creates a 3-D contour plot.	330
`cylinder`	Plots a cylinder.	331
`mesh`	Creates a mesh plot.	327, 328
`meshc`	Creates a mesh and a contour plot.	329
`meshgrid`	Creates a grid for a 3-D plot.	325
`meshz`	Creates a mesh plot with a curtain.	329
`pie3`	Creates a pie plot.	332
`plot3`	Creates a plot.	323
`pol2cart`	Convert the polar coordinates grid to a grid in Cartesian coordinates.	333
`scatter3`	Creates a scatter plot.	332
`sphere`	Plots a sphere.	331
`stem3`	Creates a stem plot	332
`surf`	Creates a surface plot.	327, 329
`surfc`	Creates a surface and a contour plot.	329

Three-dimensional plotting (Continued)

Command	Description	Page
surfl	Creates a surface plot with lighting.	330
waterfall	Creates a mesh plot with a waterfall effect.	330

Formatting plots

Command	Description	Page
axis	Sets limits to axes.	147
colormap	Sets color.	328
grid	Adds grid to a plot.	148, 328
gtext	Adds text to a plot.	145
legend	Adds legend to a plot.	145
subplot	Creates multiple plots on one page.	157
text	Adds text to a plot.	145
title	Adds title to a plot.	144
view	Controls the viewing direction of a 3-D plot.	333
xlabel	Adds label to x axis.	144
ylabel	Adds label to y axis.	144

Math functions (create, evaluate, solve)

Command	Description	Page
feval	Evaluates the value of a math function.	238
fminbnd	Determines the minimum of a function.	298
fzero	Solves an equation with one variable.	296
inline	Creates an inline function.	233

Numerical integration

Function	Description	Page
quad	Integrates a function.	300
quadl	Integrates a function.	301
trapz	Integrates a function.	302

Ordinary differential equation solvers

Command	Description	Page
ode113	Solves a first order ODE.	304
ode15s	Solves a first order ODE.	305
ode23	Solves a first order ODE.	304
ode23s	Solves a first order ODE.	305
ode23t	Solves a first order ODE.	305
ode23tb	Solves a first order ODE.	305
ode45	Solves a first order ODE.	304

Logical Functions

Function	Description	Page
all	Determines if all array elements are nonzero.	180
and	Logical AND.	179
any	Determines if any array elements are nonzero.	180
find	Finds indices of certain elements of a vector.	180
not	Logical NOT.	179
or	Logical OR.	179
xor	Logical exclusive OR.	180

Flow control commands

Command	Description	Page
break	Terminates execution of a loop.	200
case	Conditionally execute commands.	187
continue	Terminates a pass in a loop.	200
else	Conditionally execute commands.	184
elseif	Conditionally execute commands.	185
end	Terminates conditional statements and loops.	182, 187, 191, 195
for	Repeats execution of a group of commands.	191
if	Conditionally execute commands.	182
otherwise	Conditionally execute commands.	187
switch	Switches among several cases based on expression.	187
while	Repeats execution of a group of commands.	195

Polynomial functions

Function	Description	Page
conv	Multiplies polynomials.	265
deconv	Divides polynomials.	265
poly	Determines coefficients of a polynomial.	264
polyder	Determines the derivative of a polynomial.	266
polyval	Calculates the value of a polynomial.	262
roots	Determines the roots of a polynomial.	263

Curve fitting and interpolation

Function	Description	Page
interp1	One-dimensional interpolation.	267
polyfit	Curve fit polynomial to set of points.	269

Symbolic Math

Function	Description	Page
collect	Collects terms in an expression.	354

Symbolic Math (Continued)

Function	Description	Page
diff	Differentiates an equation.	363
double	Converts number from symbolic form to numerical form	352
dsolve	Solves an ordinary differential equation.	367
expand	Expands an expression.	355
ezplot	Plots an expression.	369
factor	Factors to product of lower order polynomials.	355
findsym	Displays the symbolic variables in an expression.	353
int	integrates an expression.	365
pretty	Displays expression in math format.	357
simple	Finds a form of an expression with fewest characters.	357
simplify	Simplifies an expression.	356
solve	Solves a single equation, or a system of equations.	358
subs	Substitutes numbers in an expression.	372
sym	Creates symbolic object.	348
syms	Creates symbolic object.	350

Answers to Selected Problems

Chapter 1

2. (*a*) 7.6412 (*b*) 6.8450

4. (*a*) 7.9842 (*b*) 80.0894

6. (*a*) 73.2258 (*b*) 26.0345

8. (*a*) 62.6899 (*b*) 2.1741

10. (*a*) 12.4378cm (*b*) 11.1663cm

16. (*a*) α=15.3245°
(*b*) β=31.909° γ=132.7665°

18. (*a*) γ=82.8192°
(*b*) and *c*) 66.1438 mm

20. 2.6042

22. 77

24. (*a*) \$1678.20 (*b*) \$1783.09
(*c*) \$1783.00

26. 2598960

28. 0.3815 A

32. 2.7778e-10 m

34. 193 days

36. (*a*) 92.0412 (*b*) 7.9057

38. 1.1838e6 watts

40. 30.1497 s, 1063.3 ft, 14635 ft

Chapter 2

2. 2.6163 32.0000 -12.1500 54.0000
40.4473 1.2962

4. 3.1250
0.3290
6.1000
6.7346
0.0055
11.3387
133.0000

6. 3.5000 12.2500 -0.5469 -22.4000
1.8708

8. 81.0000 72.3750 63.7500 55.1250
46.5000 37.8750 29.2500 20.6250
12.0000

10. -21.0000
-18.6429
-16.2857
....
9.6429
12.0000

Chapter 3

2. 7.0000 1.0000 -0.3333 -0.5000
-0.2000 0.3333 1.0000

4. 1.9933 10.9800 11.2161 10.8566
10.4286 10.0259 9.6652 9.3455
9.0616

6. 0 0.2410 0.3949 0.4669 0.4958
0.5066 0.5106 0.5120 m/s

8. (*a*) and (*b*) 29.6184

14. (*a*) 1.3333 9.3750 24.6154
47.0556 76.6957
(*b*) -2 8 76 250 578

16. 42

18. 106.9541°

20.

0	0	0
7.7863	214.42	1082.2
15.573	428.83	2164.5
...	...	...
70.077	1929.7	9740.1
77.863	2144.2	10822

22. 6.000000000000000
4.000000000000000
3.000000000000000
2.500000000000000
2.100000000000000
2.000999999999918
2.000010000000827

24. (*a*) 3.141593304503081
(*b*) 3.141592653595635
(*c*) 3.141592653589794

26. hm=575.3948m, xhm=309.6821m

30. (*c*) pmax05=0.095454545454545
pmax01=0.095491071428571
(*d*) E =0.038250669386692

32. u=-4, v=2.5, w=4, x=1, y=-2

34. $I_{R1} = 0.5185$, $I_{R2} = 1.8642$,
$I_{R3} = 1.7037$, $I_{R4} = 0.2716$,
$I_{R5} = 0.4074$

Chapter 4

2.

Years	Monthly Pay	Total Pay
10.00	1053.34	126400.61
11.00	979.04	129232.91
...	...	...
30.00	527.69	189969.06

4.

h(cm)	R1(cm)	R2(cm)	S(cm^2)
8.00	5.73	6.87	571.23
10.00	5.12	6.15	556.95
...	...	...	...
16.00	4.05	4.86	574.04

6.

Time(hrs)	No of Bacteria
0	1
1	2
2	4
...	...
23	8.3886e+006
24	1.6777e+007

8.

Time(s)	x (m)	v (m/s)
0	0	20.0000
0.0200	0.3693	17.0407
0.0400	0.6832	14.4510
...	...	...
0.5000	1.7337	-1.8957

10.

Interest Rate	Acc Value
2.00	12189.94
2.50	12800.85
...	...
6.00	17908.48

12. a=74.5 b=80.931

14. 153 ft

16.

t(s)	th (deg)	r(m)
0	90	500
4.488	66.401	559.35
8.976	51.029	707.62
...	...	...
62.832	51.34	3201.6

18.

T (C)	p(mmHg)
0	26.5741
2.0000	29.6487
4.0000	33.0268
...	...
42.0000	197.7684

20. Fractions of SO2, SO3, O2 and N2 are 0.1477, 0.4212, 0.1002, and 0.3308 respectively.

22. $F_1 = 11139\text{N}$, $F_2 = -8340.6\text{N}$, $F_3 = -7876.1\text{N}$, $F_4 = 7876.1\text{N}$, $F_5 = -9567.7\text{N}$, $F_6 = -1575.2\text{N}$, $F_7 = 6600\text{N}$, $F_8 = 1575.2\text{N}$, $F_9 = -2391.9\text{N}$.

24. $a = 0.5$, $b = -0.1$, $c = -10$, $d = -2$, $e = 10$

26. eagle 4, birdie 2, bogey -1, double -2

Chapter 5

2.

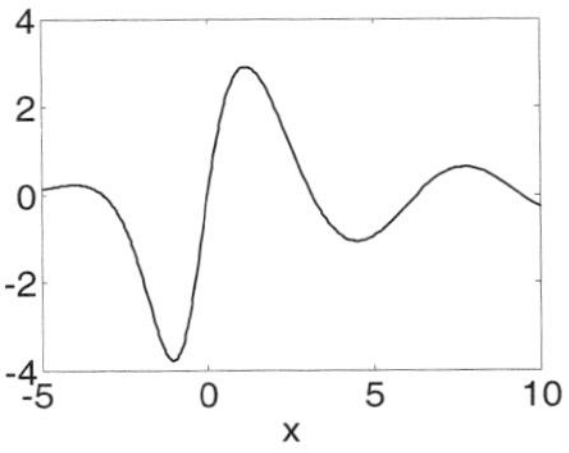

4.

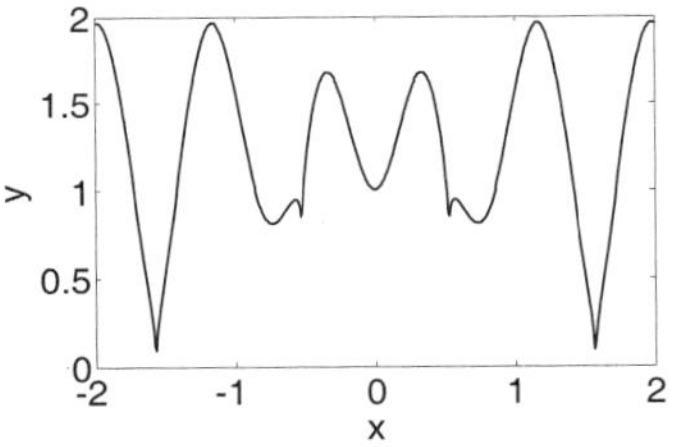

6.

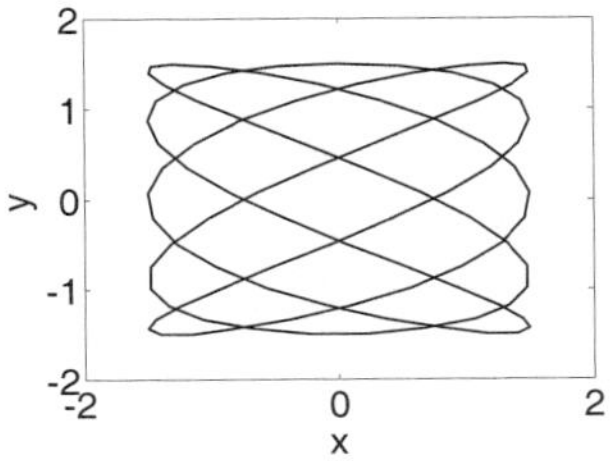

8.

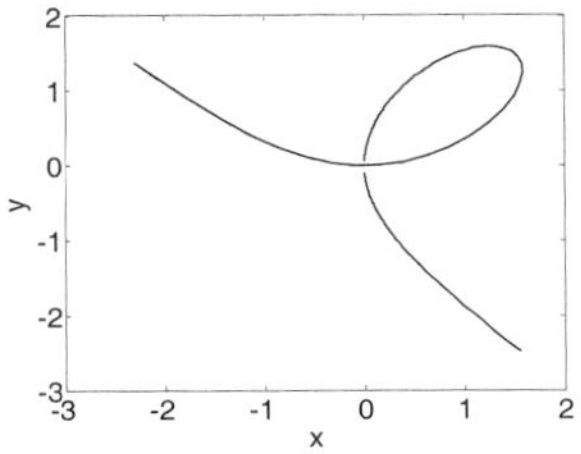

12.

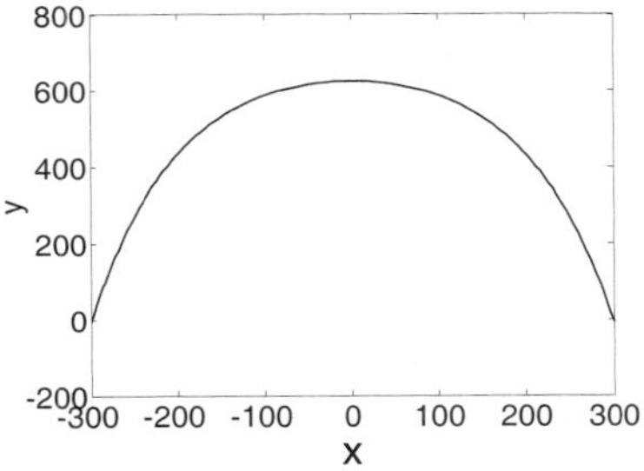

14.

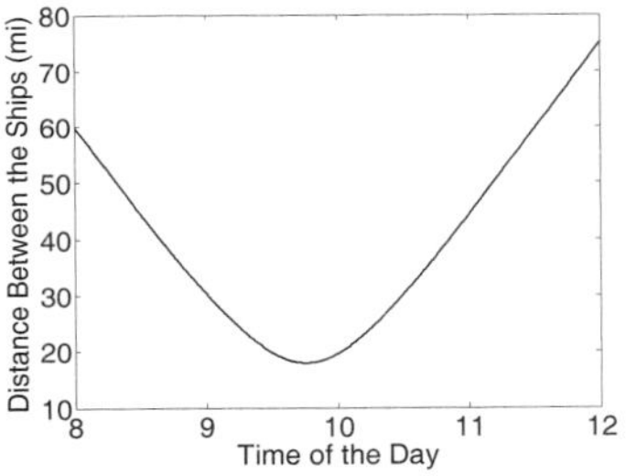

16.

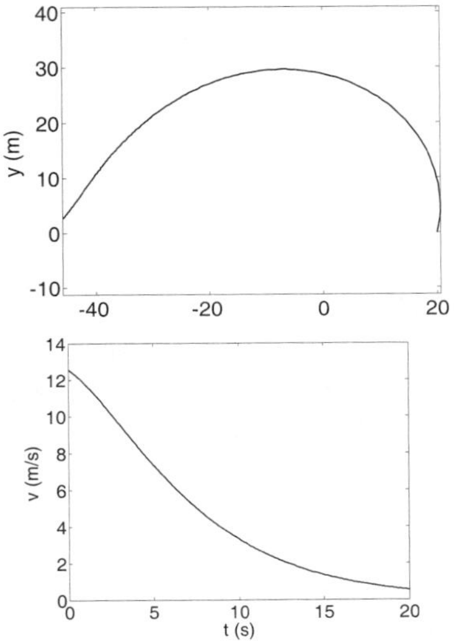

18.

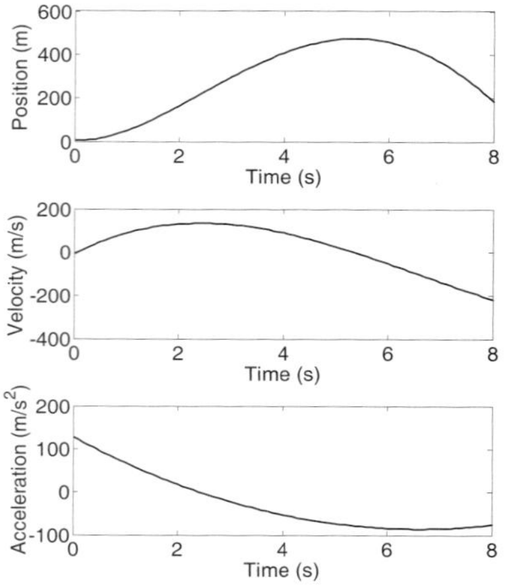

20.

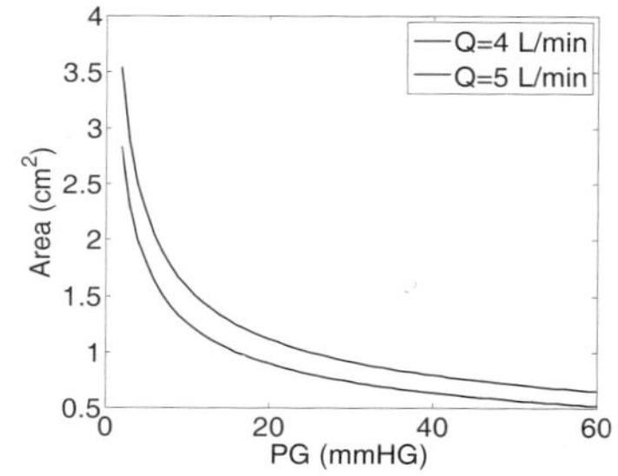

22.

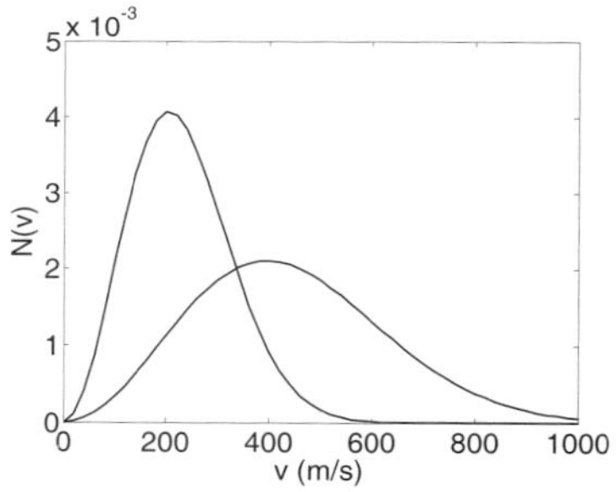

26.

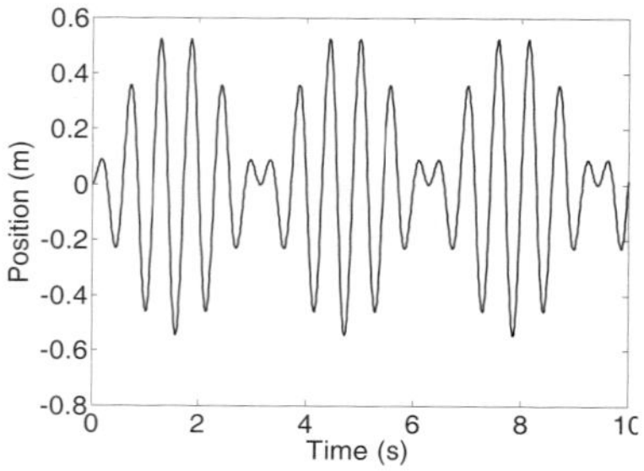

28.

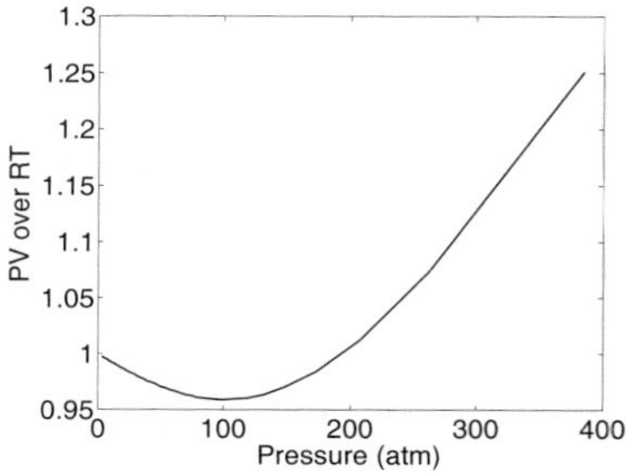

30.

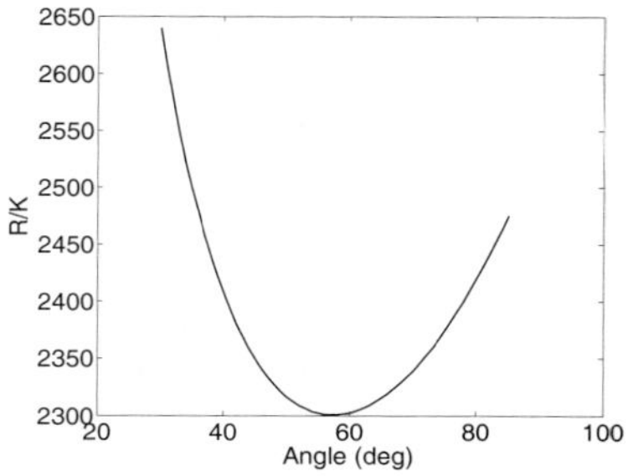

$s = 25.9763$ ft.

32.

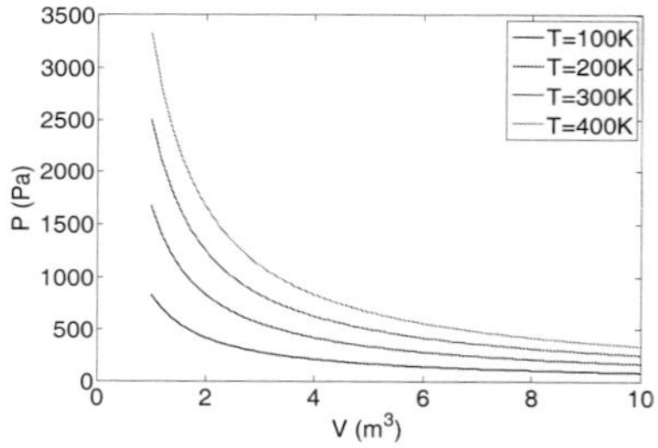

34.

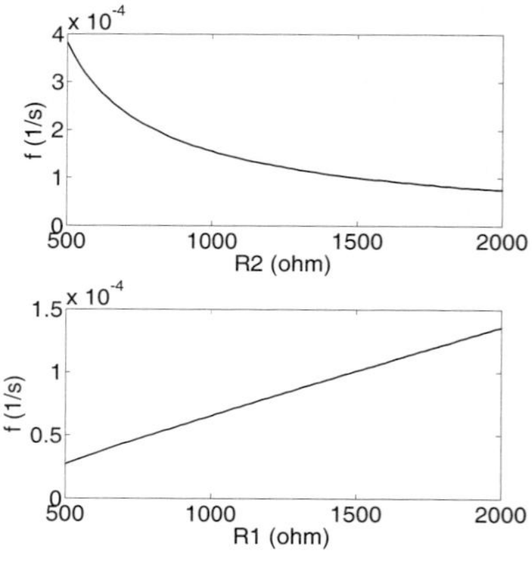

Chapter 6

6. (*a*) Chicago 79.1290°F
   Sun Francisco 74.5484 °F
   (*b*) Chicago 16
   Sun Franciscoe 113
   (*c*) 23 days, on days: 1 2 3 4 5 6 7 8 9 11 13 14 15 16 17 18 19 20 22 23 24 25 26
   (*d*) 1 day, on the 30th

8.

2.0000	0.7500	0.4444
3.0000	1.0000	0.5556
4.0000	1.2500	0.6667
5.0000	1.5000	0.7778

12. The required number is: 17435

14. For $m = 100$, 3.133787490628158

18. (*a*) 0.707106782936867
    (*b*) -0.258819047933546

20. (*a*) 137
    (*b*) 165

26. (*a*) 10.488088482190042
    (*b*) 3.056844778539776e+002
    (*c*) 4.821825380515788

28. (*a*) 924.602 USGalon
    (*b*) 7.06293 ft^3
    (*c*) 13.5921 m^3

Chapter 7

2. (*a*) -18.5991, 52.8245
   (*b*)

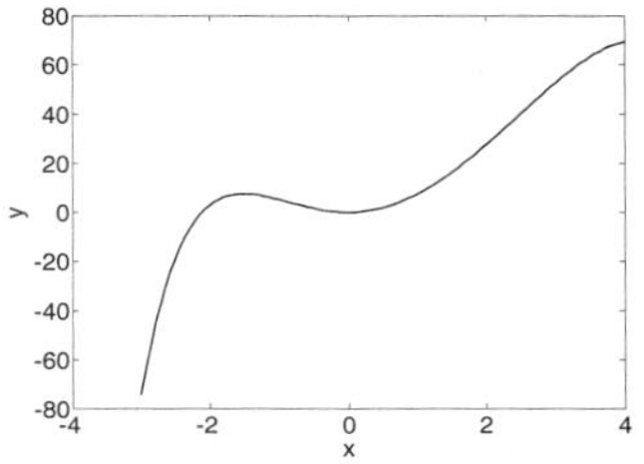

4. 24.5872 m/s

6. (*a*) 9.9216
   (*b*) 16.3459

8. 0.013518673497095 lb

10. (*a*) 134 °F
    (*b*) 195 °F

12. 2.4615

14. (*a*) [-3.5 14.2]
    (*b*) [13.4 -8.1 17.2]

16. (*a*) [0.68457 0.72894]
    (*b*) [-0.23337 0.77791 -0.58343]

18. (*a*) 38
    (*b*) 87.885

20.

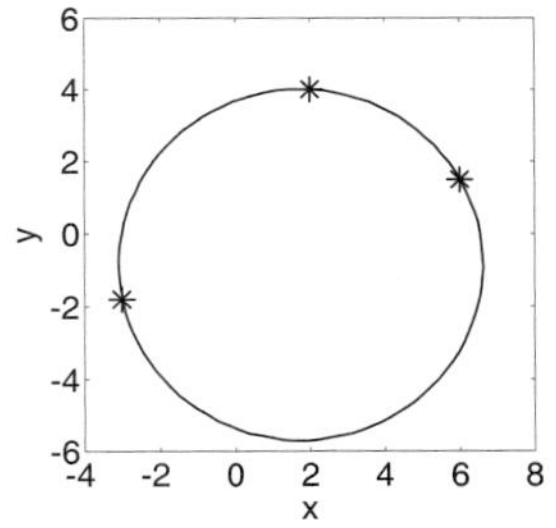

22. (*a*)

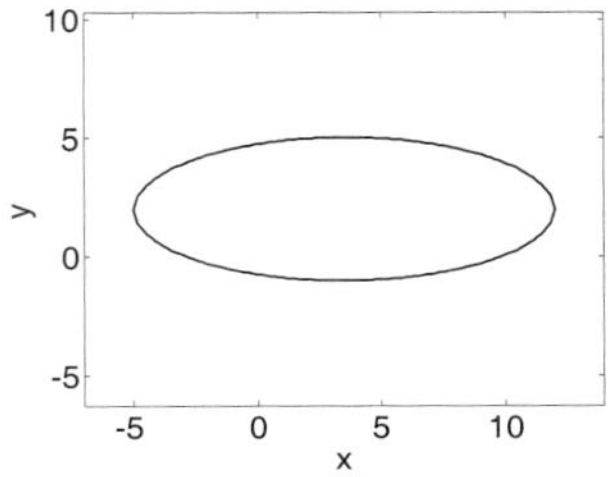

(*b*)

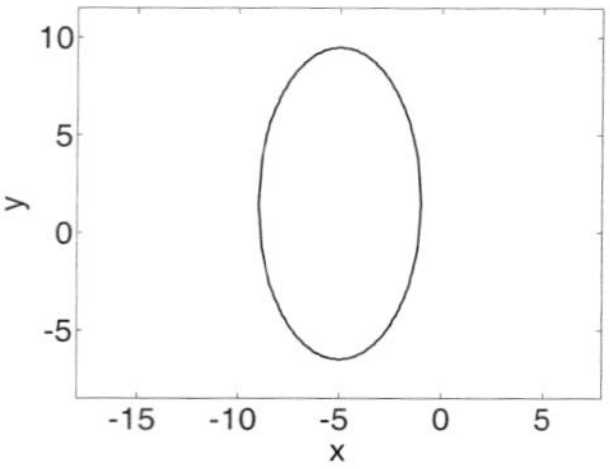

24. 1.0978

28. (*a*) -39
(*b*) -36.3

30. (*a*) 15.8 ° C, 56.7%
(*b*) 29.6 ° C, 69.7%

34. 258.2759 mm^4

36.

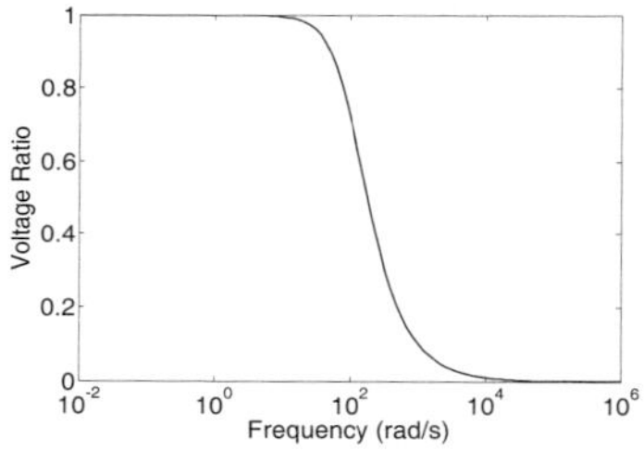

38. (*a*) 0.722263919605908 Numerical
0.722264296886855 Analytical
(*b*) 0.386396294708275 Numerical
0.386294361119891 Analytical

Chapter 8

2.

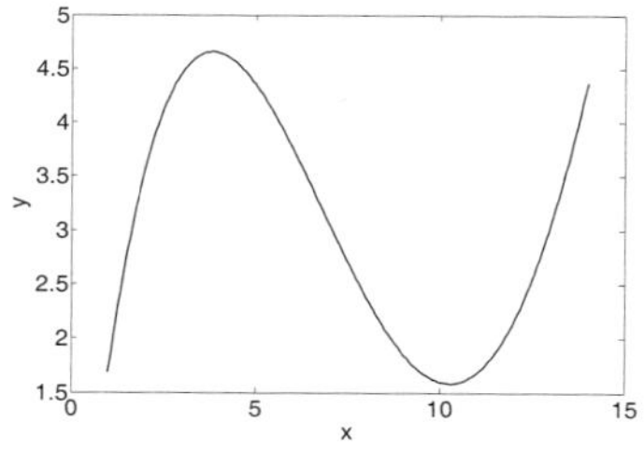

4. P = [1 0.2 -2.2 -0.392 0.4704]
$x^4 + 0.2x^3 - 2.2x^2 - 0.392x + 0.4704$

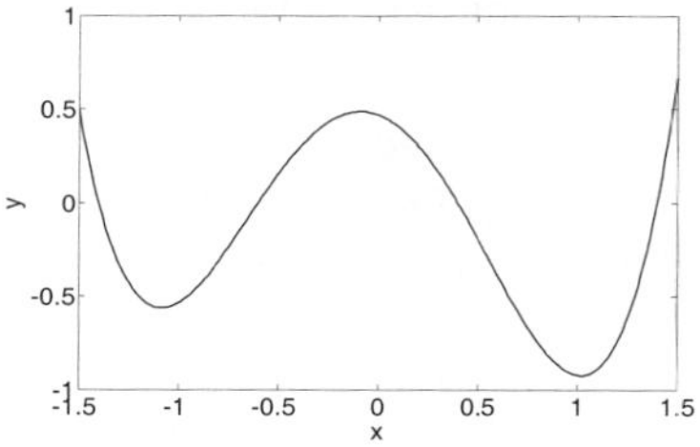

6. $x^2 - 3x + 2$

8. 8 10 12 14

10. 2.4829 cm

12. (*a*) p = [4 -124 880 0]
(*b*)

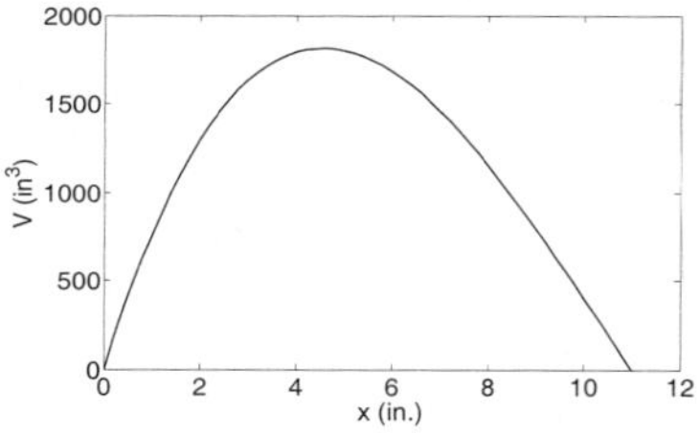

(*c*) 1.4001 in. or 8.4374 in.
(*d*) 4.5502 in. 1813.7 in^3.

14. (*a*) p = [-9.4248 94.248 0 0]
(*b*)

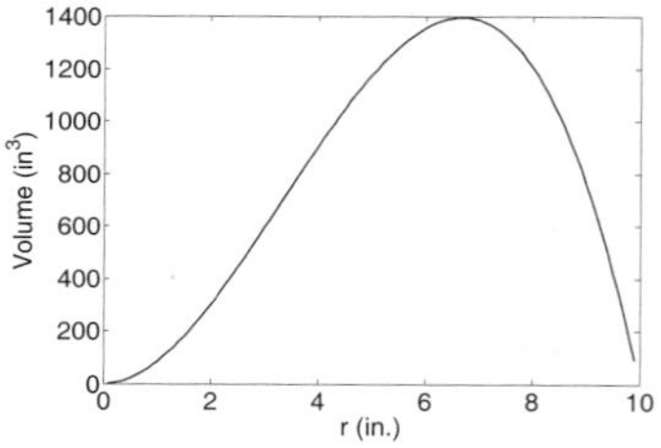

(*c*) 3.6586 in. or 8.9373 in.
(*d*) 6.6667 in. 1396.3 in^3.

18. $m = -0.0017042$, $b = 211.88$

$T_{B16000} = 184.61$

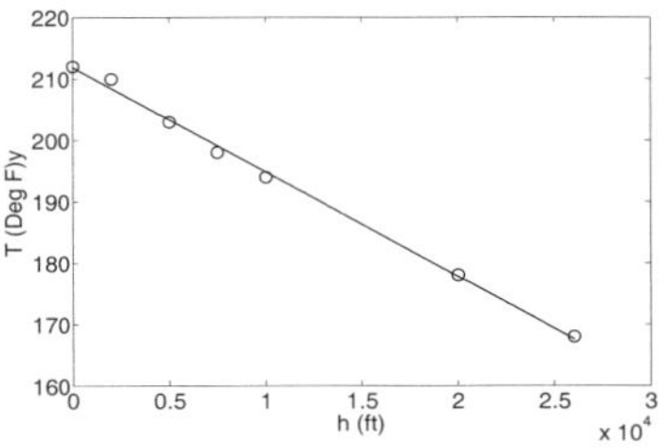

20. 1.1987 L

22.

(*a*)

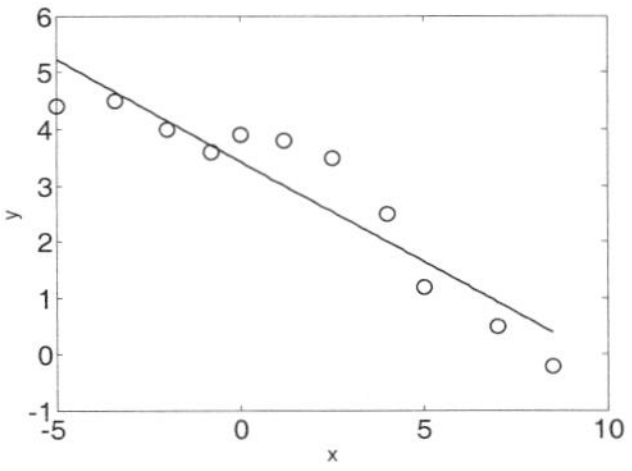

(*b*)

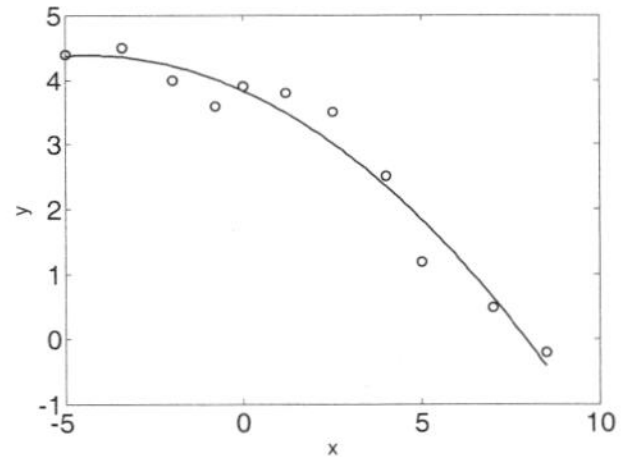

(*c*)

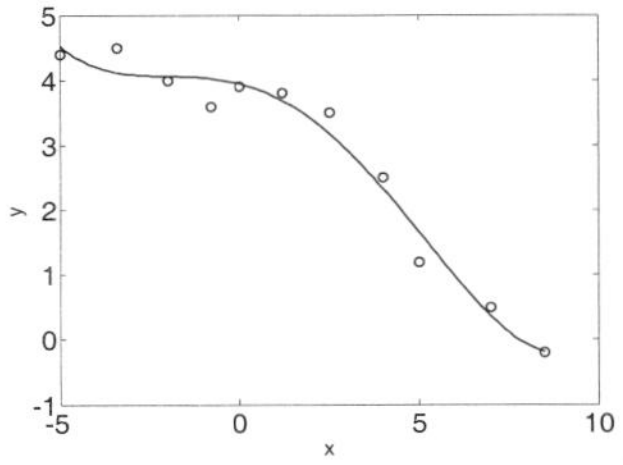

(*d*)

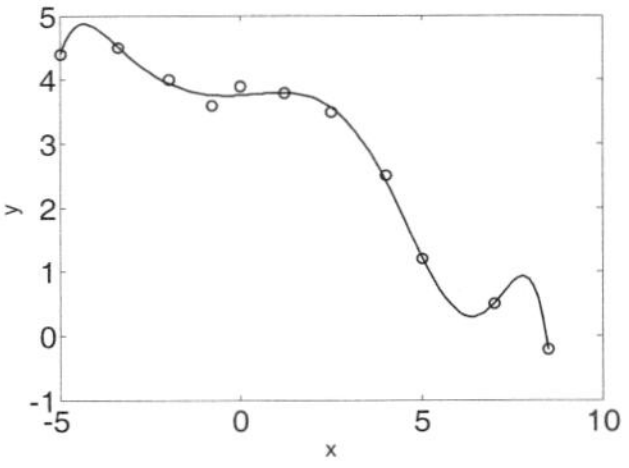

24.

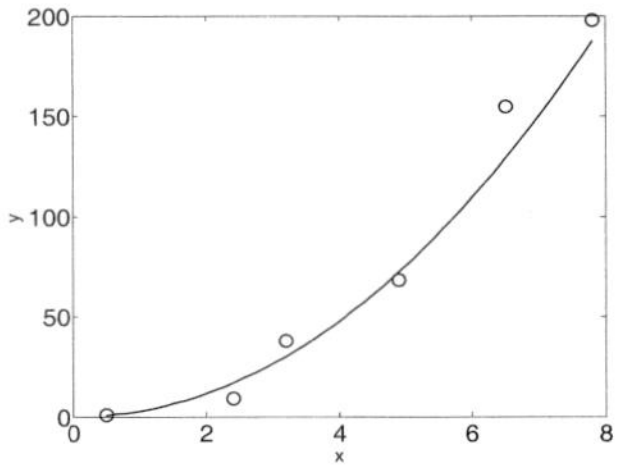

26. $C = 1.5682\text{e}{+}5$, $S = 148.16$

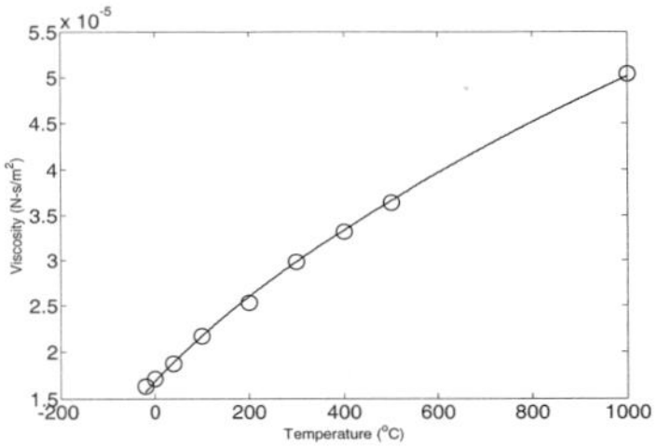

28. $m = 9.4157$, $b = 3.4418$

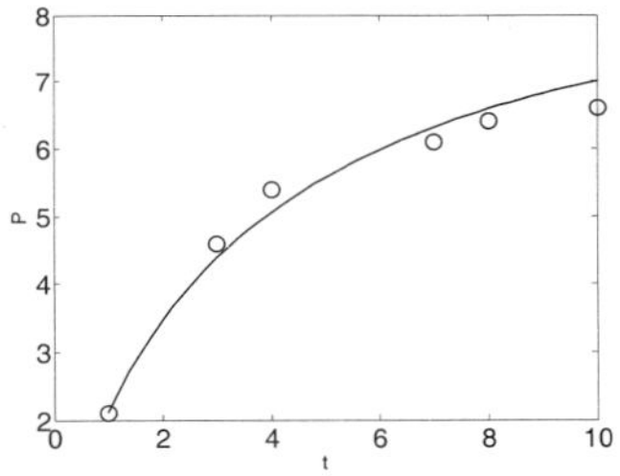

30. $m = -0.19897$, $b = 0.9062$

$k_{rd04} = 1.7194$

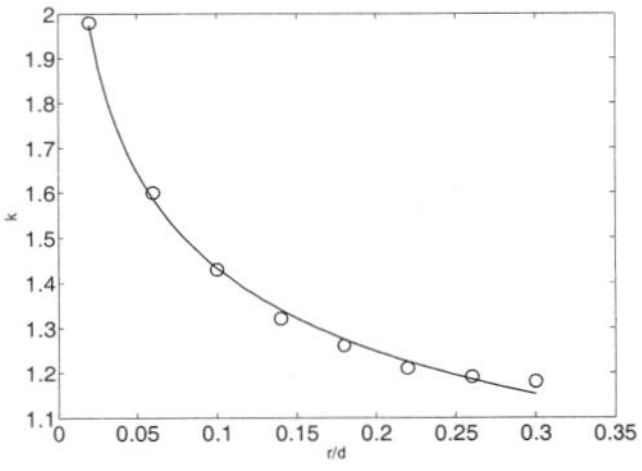

Chapter 9

2. 2.2112

4. 3.8011 3.4936 1.8387 1.3148

6. 0.17289 m

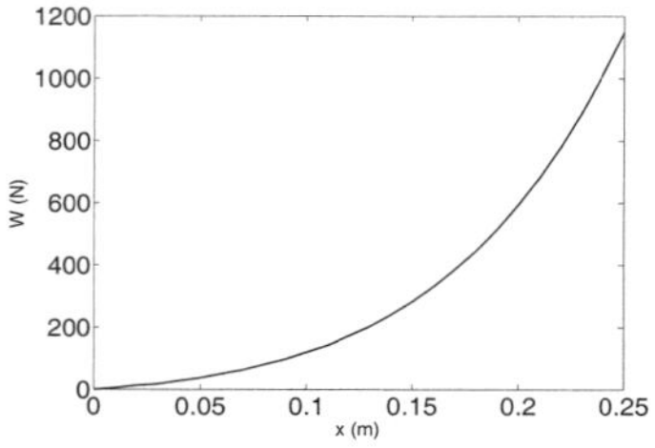

8. 0.5405 V

10. $r = 6.9632$ cm, $h = 4.9237$ cm.

12. $r = 11.431$ in., $h = 16.166$ in.

14. $\lambda_{max} = 1.9382 \times 10^{-6}$ m

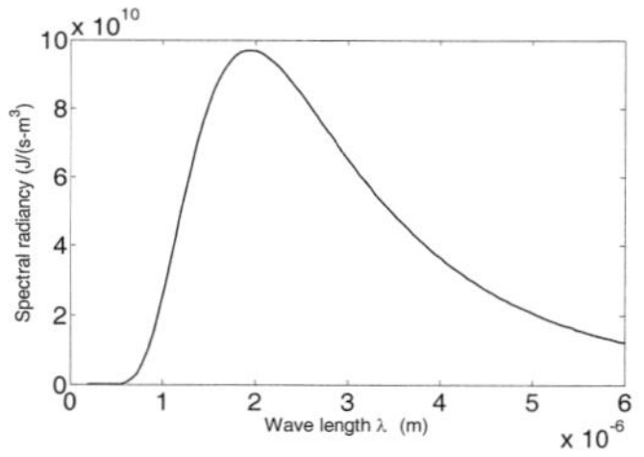

16. (*a*) 62.269
    (*b*) -0.5640

18. 776.6000 ft

20. 1.6035 in^3/s

22. 61.152 in.

24. 40800 ft^2

26. 5.839 psi, 5.306 psi, 5.012 psi

28.

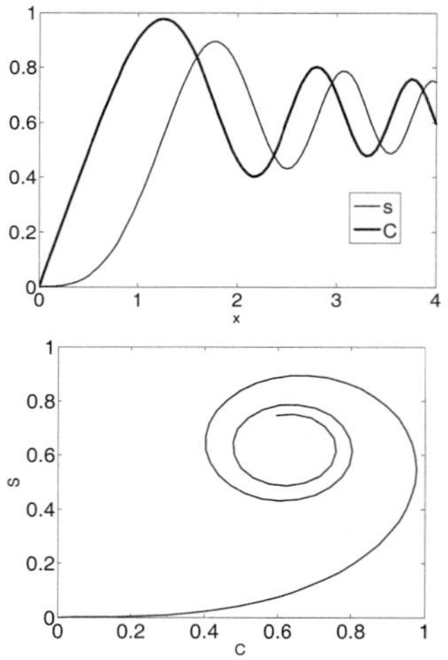

30.

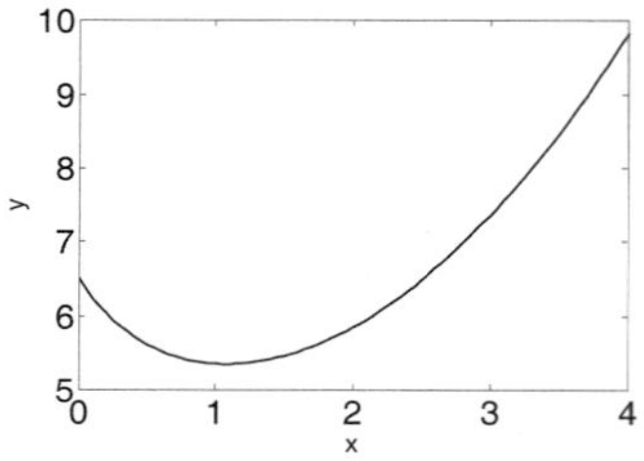

32. 5642.5 s

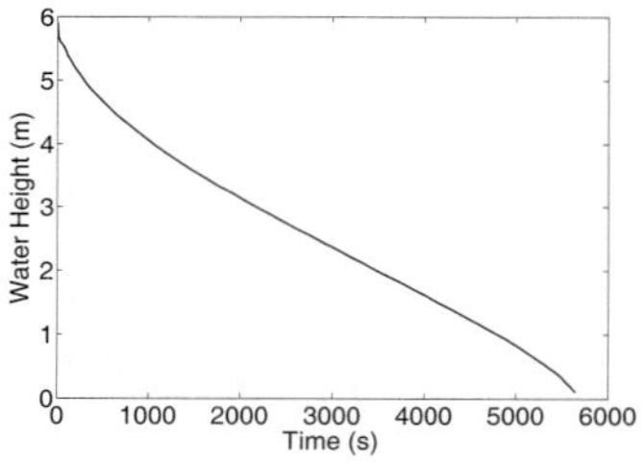

34.

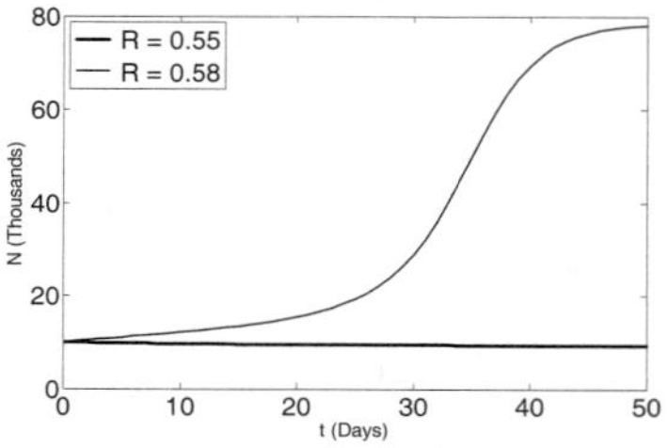

36. (*a*)

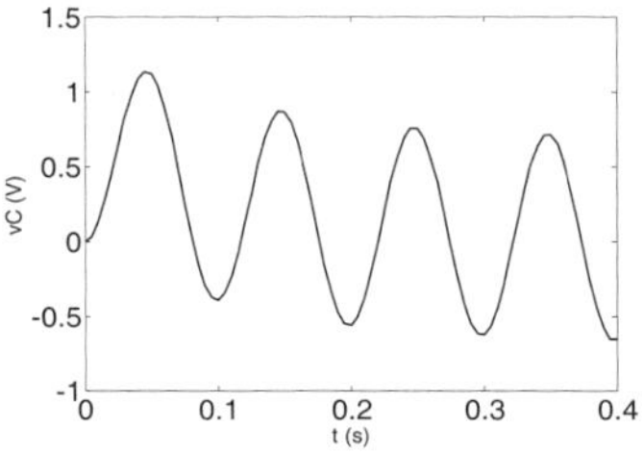

(*b*)

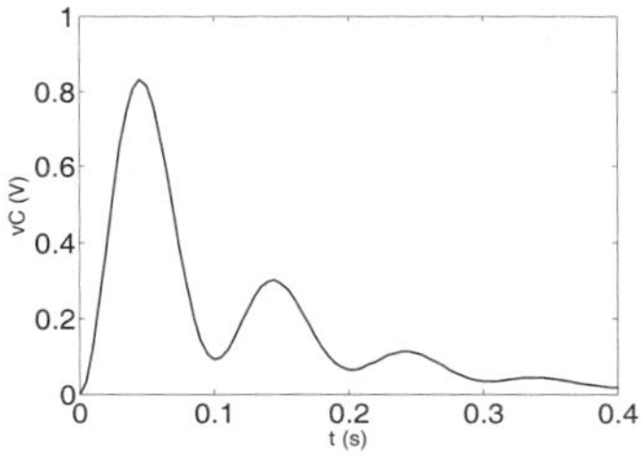

(*c*)

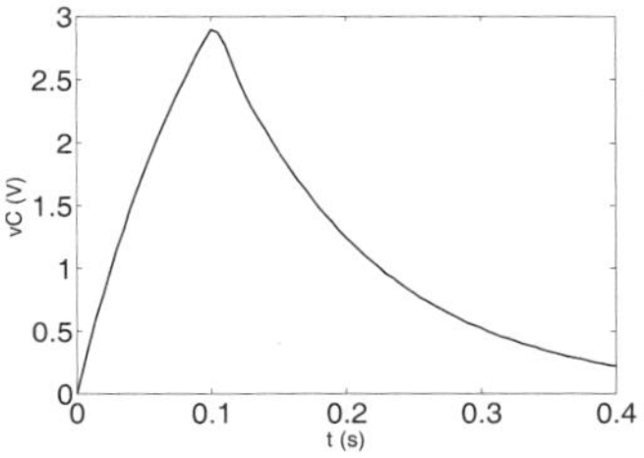

38.

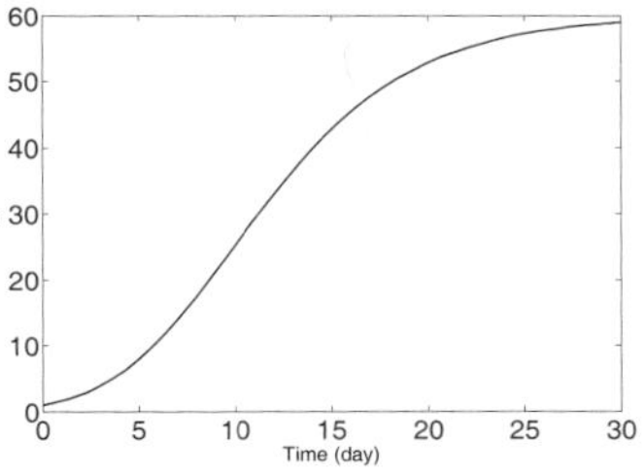

Chapter 10

4.

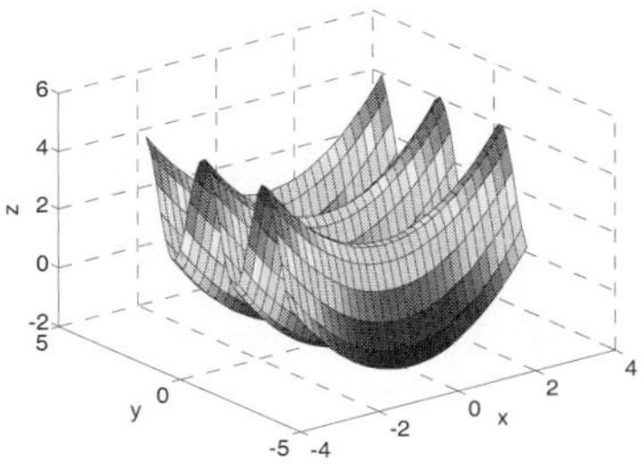

6.

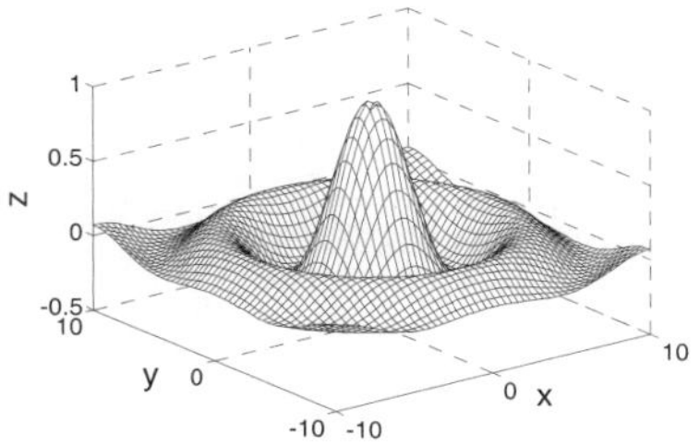

8.

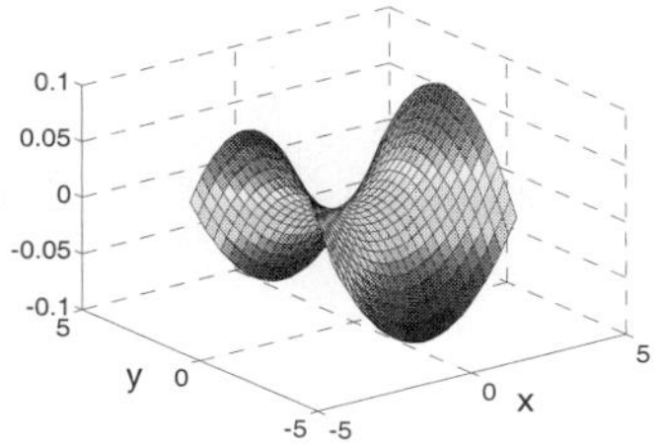

10.

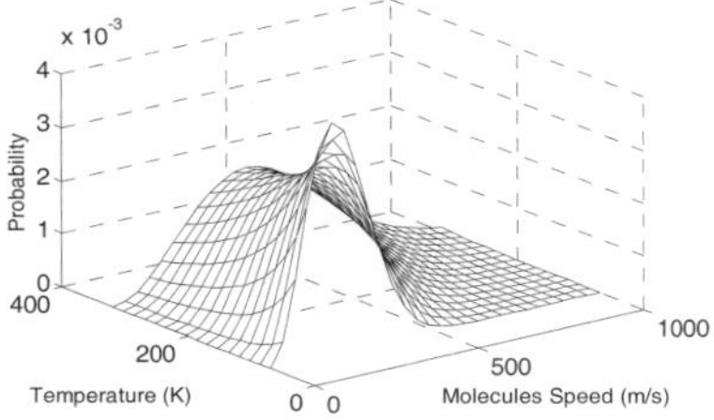

12.

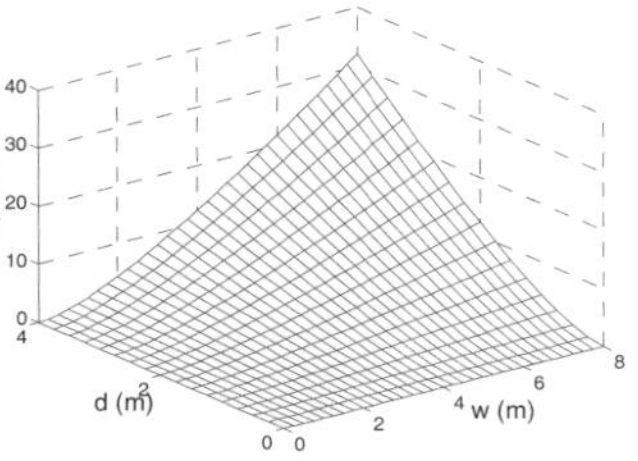

14.

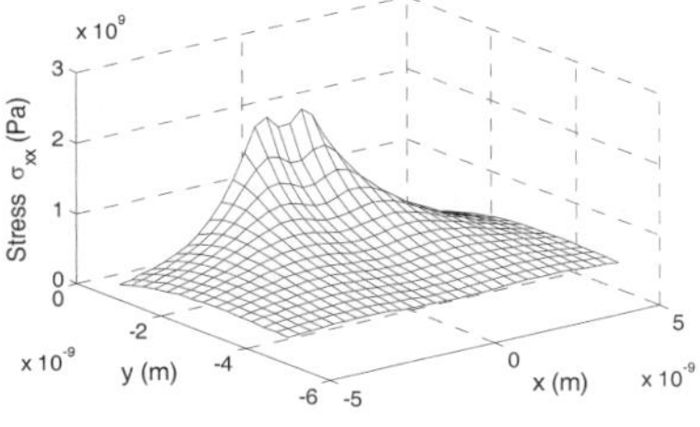

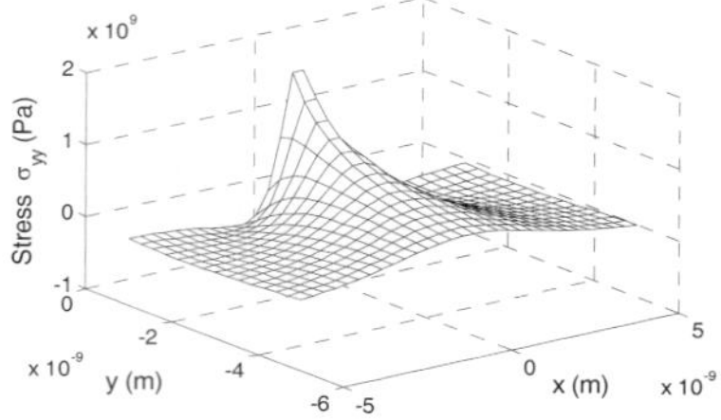

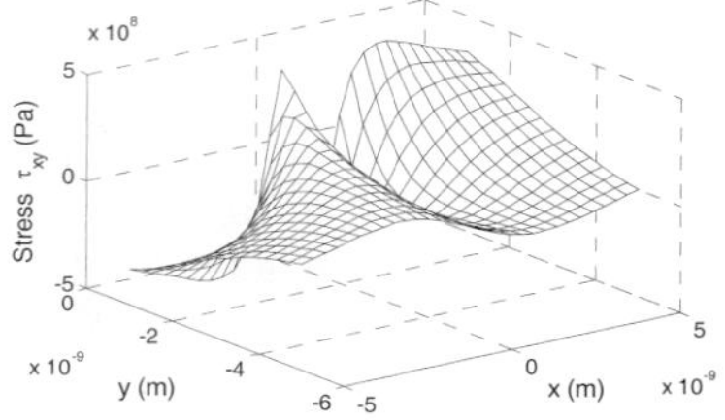

16.

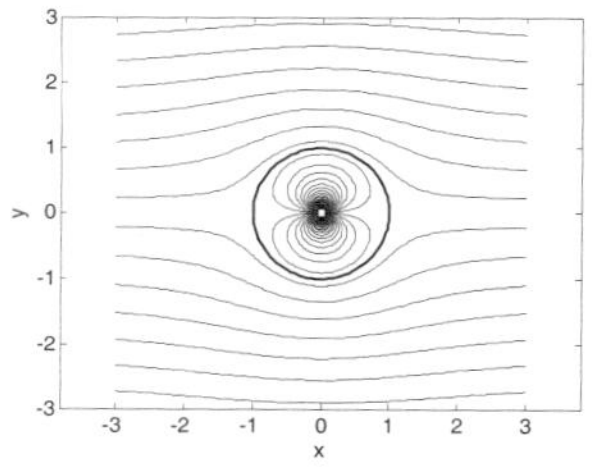

18.

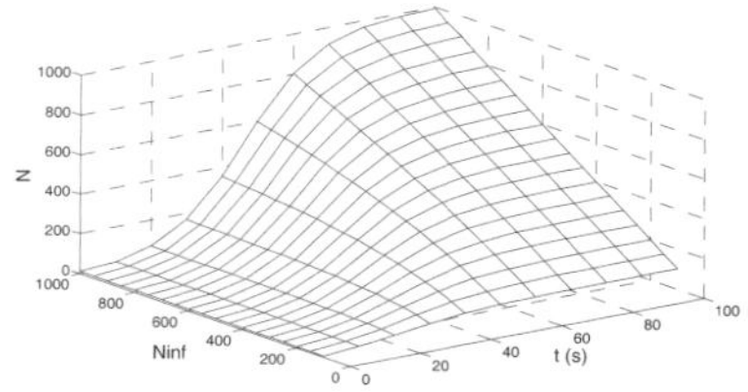

20.

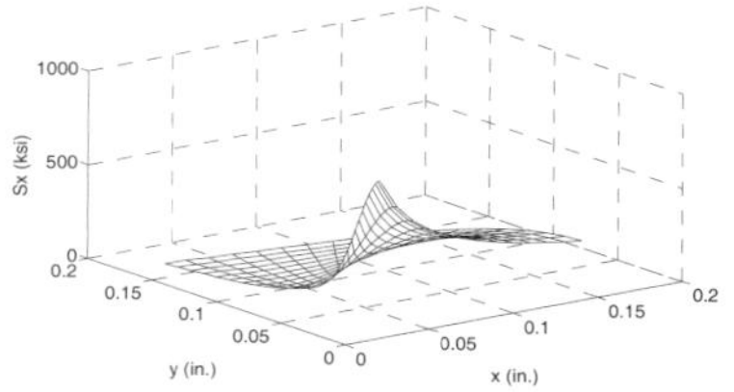

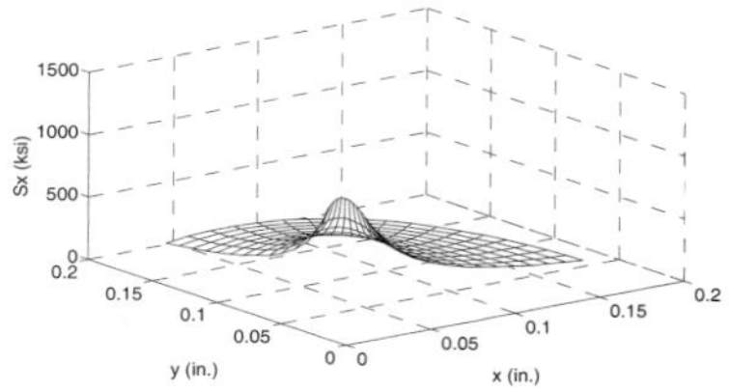

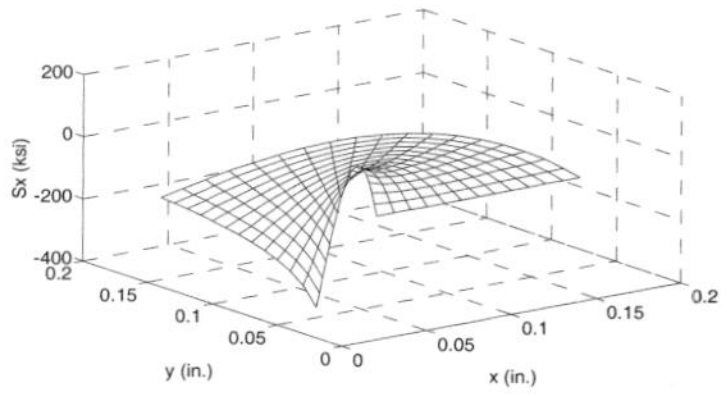

Chapter 11

2. (*a*) (x + 2)^5
   (*b*) x + 2
   (*c*) (x + 2)^2*(x + 3)
   (*d*) 150

4. (*a*) x^6 - (13*x^5)/2 - 58*x^4 + (335*x^3)/2 + 728*x^2 - 890*x - 1400
   (*b*) -5 -3.5 -1 2 4 10

8. 5.0059 m

10. (*a*)

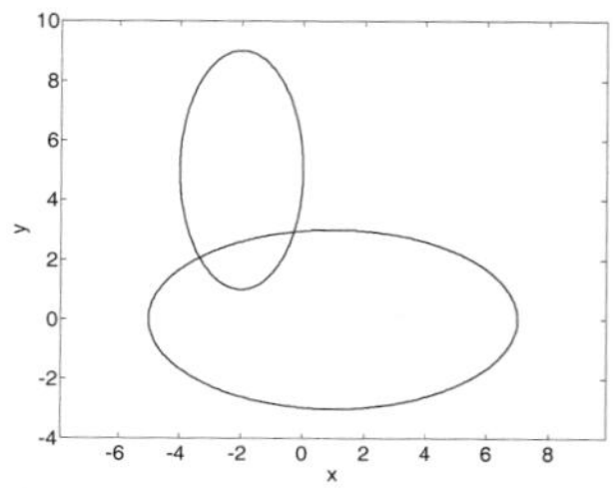

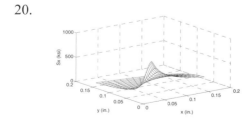

   (*b*) (-0.2886359424, 2.9299922102)
   (-3.3574030955, 2.0623432220)

12. (*a*) F =(g*m*mew*(h^2+x^2)^(12))/(x + h*mew)
    N =(g*m*x)/(x + h*mew)
    (*b*) (97119*(x^2 + 100)^(1/2))/(1000*(x + 11/2))

(*c*)

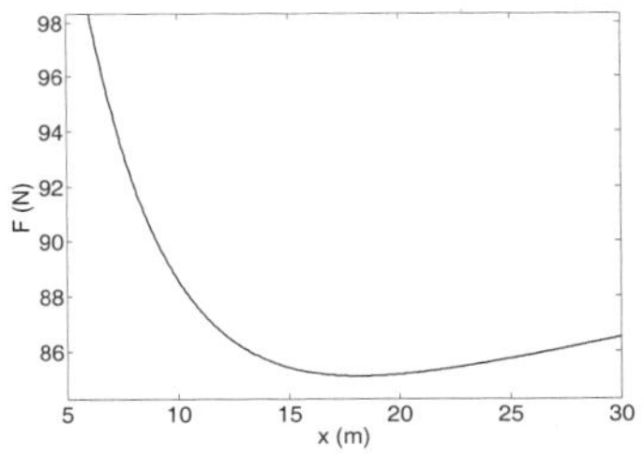

(*d*) 200/11 m, 85.0972 N

14. y =-(x*x0 - R^2)/((R + x0)^(1/2)*
(R - x0)^(1/2))

16. (*a*) -((1 - x^2)^(1/2)*(x^2 + 2))/3
(*b*) x^2*sin(x)-2*sin(x)+2*x*cos(x)

20 1/4

24. x = exp(-R*(N+1)*t)*N*(N+1)
/(1+exp(-R*(N+1)*t)*N)
t_max = log(N)/R/(N+1)

26. (*a*) g/c*m-exp(-c/m*t)*g/c*m
(*b*) 16.1489 kg/s
(*c*)

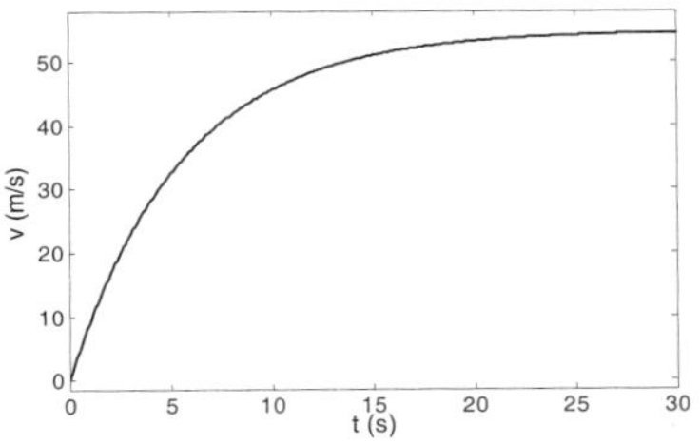

28. C2/x + x^4/10

30. (*a*) 10*C - (C*(8*L + 5*(C^2*R^2 -
4*C*L)^(1/2) - 5*C*R))/
(exp((t*((C^2*R^2 - 4*C*L)^(1/2) +
C*R))/(2*C*L))*(C^2*R^2 -
4*C*L)^(1/2)) -
(C*exp((t*((C^2*R^2 - 4*C*L)^(1/2)
- C*R))/(2*C*L))*(5*(C^2*R^2 -
4*C*L)^(1/2) - 8*L + 5*C*R))/
(C^2*R^2 - 4*C*L)^(1/2)

(*b*)

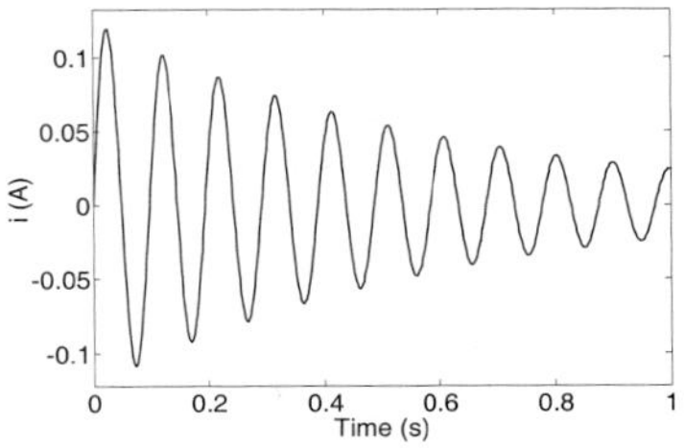

(*c*)

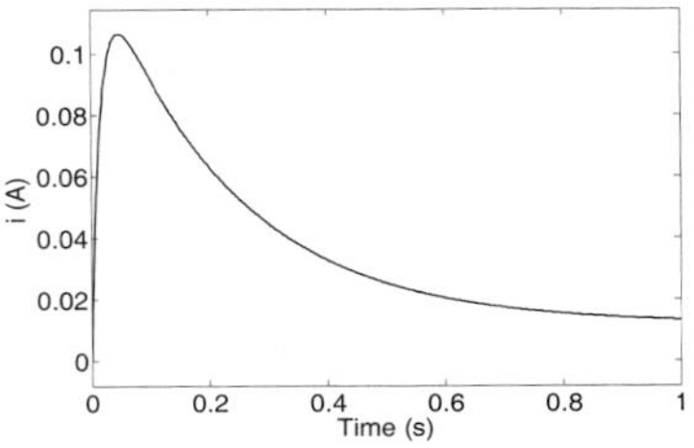

(*d*)

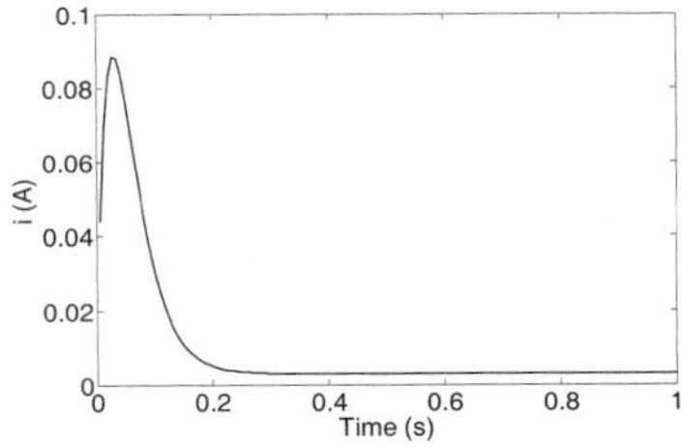

Index